BASIC ELECTRICITY:
A Text-Lab Manual

Other Books by Paul B. Zbar

Basic Electronics: A Text-Lab Manual, Third Edition
Basic Radio: Theory and Servicing, A Text-Lab Manual, Third Edition
*Basic Television: Theory and Servicing, A Text-Lab Manual, Second
    Edition (with P. Orne)*
Electricity—Electronics Fundamentals: A Text-Laboratory Manual
Electronic Instruments and Measurements: A Text-Lab Manual
Industrial Electronics: A Text-Lab Manual, Second Edition

BASIC ELECTRICITY:

A Text-Lab Manual

Fourth Edition

Paul B. Zbar

McGRAW-HILL BOOK COMPANY
New York   St. Louis   San Francisco   Düsseldorf   Johannesburg
Kuala Lumpur   London   Mexico   Montreal   New Delhi
Panama   Paris   São Paulo   Singapore   Sydney   Tokyo   Toronto

Basic Electricity

0-07-072787-2

1 2 3 4 5 6 7 8 9 0 B A B A 7 9 8 7 6 5 4

The editors for this book were Alan W. Lowe, Cynthia Newby, and Zivile Khoury, and its production was supervised by Laurence Charnow. It was set in Times Roman by Black Dot, Inc.
It was printed and bound by George Banta Company, Inc.

CONTENTS

Series Preface

Preface

Safety

NOTE ON EXPERIMENT CONTENT

Each of the experiments described below is set up in the following manner:

OBJECTIVES   The objectives are enumerated and clearly stated.

INTRODUCTORY INFORMATION   The theory and basic principles involved in the experiment are clearly stated.

SUMMARY   A summary of the salient points is given.

SELF-TEST   A self-test, based on the material included in Introductory Information, helps the student evaluate his understanding of the principles covered, prior to the experiment proper. The self-test should be taken before the experiment is undertaken. Answers to the self-test questions are given at the end of each experiment.

MATERIALS REQUIRED   All the materials required to do the experiment—including test equipment and components—are enumerated.

PROCEDURE   A detailed step-by-step procedure is given for performing the experiment.

QUESTIONS   The conclusions reached by the student are brought out by a series of pertinent questions.

EXTRA CREDIT   Design problems and questions are included for the more advanced student.

Series Preface

Faced with a need for increasing numbers of technicians for an expanding industry, as well as faced with a rapidly advancing technology, the Consumer Electronics Group Service Committee of the Electronic Industries Association (EIA), in association with various publishers, has long been active in developing and constantly revising educational materials to meet these challenges. In recent years, the pressing need for training programs to serve students of various backgrounds and abilities has also become evident.

These reasons, coupled with an increasingly critical shortage of trained technicians, especially for home-entertainment electronics servicing, have induced EIA to sponsor the preparation of a wide range of materials in three specific categories, tailored to meet a variety of requirements.

The following paragraphs explain these instructional materials and suggest their proper use to achieve the desired results.

The text-laboratory manuals in the Basic Electricity Electronics Series provide in-depth, detailed technical materials, including a closely coordinated program of experiments, each preceded by a comprehensive discussion of the objectives, theory, and underlying principles. *Electricity-Electronics Fundamentals* provides introductory electrical-electronic theory and experiments especially suitable for the basic preparation of television and

THE BASIC ELECTRICITY-ELECTRONICS SERIES

Title	Author	Publisher
Electricity-Electronics Fundamentals	Zbar	McGraw-Hill Book Company
Basic Electricity	Zbar	McGraw-Hill Book Company
Basic Electronics	Zbar	McGraw-Hill Book Company

audio service technicians or for other broad-based programs of instruction. *Basic Electricity* and *Basic Electronics* are planned as 270-hour courses, one to follow the other, providing a more thorough background for all levels of technician training. A related instructor's guide is available.

The Radio-Television Servicing Series includes materials in two categories: those designed to prepare apprentice technicians for performing in-home servicing and other apprenticeship functions, and those designed to prepare technicians for performing more sophisticated and complicated job skills such as bench-type servicing in the shop.

The apprenticeship servicing group of materials, entitled *Television Symptom Diagnosis*, includes a text, student response manual (workbook) and related film materials. This group is used for a semester course where trouble diagnosis skills are developed through audio and video symptom recog-

THE RADIO-TELEVISION SERVICING SERIES

Title	Author	Publisher
Television Symptom Diagnosis—An Entry into TV Servicing	Tinnell	Howard W. Sams & Co. Inc.
Television Symptom Diagnosis (film loops or slides)	Tinnell	Howard W. Sams & Co. Inc. & Hickok Teaching Systems, Inc.
Basic Radio: Theory and Servicing	Zbar	McGraw-Hill Book Company
Basic Television: Theory and Servicing	Zbar and Orne	McGraw-Hill Book Company

nition. The television receiver is presented on a functional block diagram basis, and related malfunctions are demonstrated. The student in turn is trained to recognize a symptom and relate it to a given functional block, for the purpose of effective servicing. A complete instructor's guide for this group of materials is provided.

The related film materials for *Television Symptom Diagnosis* come in two forms, a set of color-sound motion picture film loops or a set of color slides for use with any standard slide projector. The film medium is especially suitable for students who may have reading and, in turn, learning difficulties. The film materials can be used, along with the student response manual, as an integrated learning system to teach color-television adjustment and set-up procedures, trouble symptom diagnosis, and, in turn, the ability to isolate troubles to a given stage. Or the film materials can be used independently to supplement any color television servicing course.

Since only a minimum of theory is presented in *Television Symptom Diagnosis*, apprentices completing this program can be expected to progress to the following more comprehensive text-lab manuals for the purpose of deepening their understanding of electronics theory and procedures for the bench servicing of television and audio-electronics products.

The text-lab manuals for bench servicing are *Basic Radio: Theory and Servicing* and *Basic Television: Theory and Servicing*. These books provide a series of experiments, with preparatory theory, designed to teach the in-depth theory and procedures required for servicing all types of con-

sumer electronics products. An instructor's guide for these books is also available.

Materials for basic laboratory courses in industrial control and computer circuits and laboratory standard measuring equipment are provided by the Industrial Electronics Series and their related instructor's guides. *Industrial Electronics* is concerned with the fundamental building blocks in industrial electronics technology, giving the student an understanding of the basic control circuits and their applications. *Electronic Instruments and Measurements* fills the need for basic training in the complex field of industrial instrumentation. Prerequisites for both courses are *Basic Electricity* and *Basic Electronics*.

THE INDUSTRIAL ELECTRONICS SERIES

Title	Author	Publisher
Industral Electronics	Zbar	McGraw-Hill Book Company
Electronic Instruments and Measurements	Zbar	McGraw-Hill Book Company

The foreword to the first edition of the EIA co-sponsored basic series said: "The aim of this basic instructional series is to supply schools with a well-integrated, standardized training program, fashioned to produce a technician tailored to industry's needs." This statement is still the objective of the varied training program that has been developed through joint industry educator-publisher cooperation.

V. J. Adduci, President
Electronic Industries Association

PREFACE

The fourth edition of "Basic Electricity" is an updated and expanded text-laboratory manual which emphasizes self-learning. Employing the concepts of modern education, it features behavioral objectives and self-learning techniques. It is intended for use in vocational-technical schools, community colleges, and military and industrial training programs.

The manual provides a complete and tested laboratory program in electric circuits, dc and ac, which is closely integrated with a recommended outline for classroom instruction. As in the previous edition, emphasis is on scientific method, analysis, and logical deduction.

The 71 experiments in the fourth edition provide students with intensive experience in the use of test instruments that they will find in industry.

Significant changes mark the fourth edition. The first is related to the formulation of the experiment's objectives in measurable behavioral terms. At the conclusion of each experiment student and teacher can evaluate the results of that learning experience in terms of the stated objectives.

In a second major change, a detailed summary follows the introductory information in which the principles of the experiment are discussed. Immediately after the summary comes a self-test (questions with answers) to help students determine how well they have understood the principles of the experiment. The results of the self-test will indicate to students their readiness to undertake the experimental procedure which follows or possibly a need to review the basic principles. The presentation of objectives, introductory information, self-test, and answers to the self-test will therefore pace students progress and prepare them for a meaningful hands-on experience in the procedural steps of the experiment.

A third major change in the fourth edition is that it has been expanded from 49 to 71 experiments. This increase results both from the addition of new topics and from the organization of the experiments into smaller work units. The fact that there are more experiments and that these comprise smaller work units than those in the third edition, will make it possible for students to proceed at their own pace. Faster students will complete more; the slower will do fewer experiments.

All the topics in the 49 experiments of the preceding edition are treated in the fourth edition together with these *new* topics: voltage and current meter loading effects; magnetism and electromagnetism; design considerations at the technician's level; the operation and use of triggered oscilloscopes; the *j* operator (optional); low-pass and high-pass filters.

As in past editions, students are stimulated to discover for themselves the principles and laws of electricity. Scientific method, analysis, and logical deduction mark this aspect of their learning. As in previous editions also, discussions of circuits and of testing and measurement techniques are presented clearly and simply. Solutions of sample problems are included, as are sample graphs and interpretations of these graphs. Stimulating questions, following the experimental procedure and based on the data obtained, oblige students to evaluate their data and draw the necessary conclusions. Extra credit procedures and questions are provided to stimulate student interest and resourcefulness.

The author wishes to thank the members of the Publications Review Committee of the Consumer Electronics Group (CEG) of the Electronic Industries Association (EIA) Service Committee for their guidance and encouragement; William D. Gannon, chairman (Warwick Electronics); Jack Berquist (GTE Sylvania); Frank Hadrick (Zenith); Robert Hannum (GE); Herb Horton (RCA); Andy Murnick, Jr. (Philco-Ford); and Gene Koschella (EIA-ex officio). The author also wishes to thank Prof. Edmund Turner of

Modesto Junior College for his many excellent suggestions.

Acknowledgment is also gratefully made for permission to use equipment and component photographs to: Aerovox Corporation; Centralab (Division of Globe Union Inc.); Chicago Telephone Company; Heath/Schlumberger Instruments; Hewlett Packard; Motorola Semiconductor Products; Ohmite Manufacturing Company; Radio Corporation of America; F. W. Sickles Division of General Instrument Corp.; Simpson Electric Company (Division of American Gage and Machine Company); Triplett Electrical Instruments Company; and Western Electric Company.

And last but not least, the author wishes to thank his wife May for her encouragement, inspiration, and many, many hours of proofreading.

Paul B. Zbar

SAFETY

The electronics technician works with electricity, electronic devices, motors, and other rotating machinery. He is often required to use hand and power tools in constructing prototypes of new devices or in setting up experiments. He uses test instruments to measure the electrical characteristics of components, devices, and electronic systems. He is involved in any of a dozen different tasks.

These tasks are interesting and challenging, but they may also involve certain hazards if the technician is careless in his work habits. It is therefore essential that the student technician learn the principles of safety at the very start of his career and that he practice these principles throughout his busy and exciting life.

Safe work requires a careful and deliberate approach to each task. Before he undertakes a job, the technician must understand what he is to do and how to do it. He must plan the job, setting out on the work bench in a neat and orderly fashion the tools, equipment, and instruments he will need. Extraneous items should be removed, and cables should be secured as far as possible.

When working on or near rotating machinery, loose clothing should be anchored, ties firmly tucked away.

Line (power) voltages should be isolated from ground by means of an isolation transformer. Power-line voltages can kill, so these should *not* be contacted with the hands or body. Line cords should be checked before use. If the insulation on line cords is brittle or cracked, these cords must *not* be used. TO THE STUDENT: Avoid direct contact with any voltage source. Measure voltages with one hand in your pocket. Wear rubber-soled shoes or stand on a rubber mat when working at your experiment bench. Be certain that your hands are dry and that you are not standing on a wet floor when making tests and measurements in a live circuit. Shut off power before connecting test instruments in a live circuit.

Be certain that line cords of power tools and nonisolated equipment use safety plugs (polarized 3-post plugs). Do not defeat the safety feature of these plugs by using ungrounded adapters. Do not defeat any safety device, such as fuse or circuit breaker, by shorting across it or by using a higher amperage fuse than that specified by the manufacturer. Safety devices are intended to protect you and your equipment.

Handle tools properly and with care. Don't indulge in horseplay or play practical jokes in the laboratory. When using power tools, secure your work in a vise or jig. Wear gloves and goggles when required.

Exercise good judgment and common sense and your life in the laboratory will be safe, interesting, and rewarding.

FIRST AID

If an accident should occur, shut off power immediately. Report the accident at once to your instructor. It may be necessary for you to render first aid before a physician can come, so you should know the principles of First Aid. A proper knowledge of these may be acquired by attendance at a Red Cross First Aid course.

Some First Aid suggestions are set forth here as a simple guide.

An injured person should be kept lying down until medical help arrives. He should be kept warm to prevent shock. Do not attempt to give him water or other liquids if he is unconscious. Be sure nothing is done to cause further injury. Keep him comfortable and cheerful until medical help arrives.

ARTIFICIAL RESPIRATION

Severe electrical shock may cause stoppage of breathing. Be prepared to start artificial respiration at once if breathing has stopped. The two recommended techniques are:

1 Mouth-to-mouth breathing, considered the most effective
2 Schaeffer method

These techniques are described in First Aid books. You should master one or the other so that if the need arises you will be able to save a life by applying artificial respiration.

These instructions are not intended to frighten you but to make you aware that there are hazards attendant upon the work of an electronics technician. But then there are hazards in every job. Therefore you must exercise common sense, good judgment, and safe work habits in this, as in every job.

1

Electronic Components and Their Symbols

OBJECTIVES

1. To identify 10 to 15 electrical and electronic components from their physical appearance
2. To draw the circuit symbol for each of these components

INTRODUCTORY INFORMATION

Though electronic devices are fairly complex, they are basically made up of a small number of easily identifiable components. The technician must be able to identify these components readily. Photographs of a number of these parts are shown.

Resistors are the components most frequently used in electronic circuits. They are available in different values, shapes, and sizes. Manufacturers have adopted a standard EIA (Electronic Industries Association) color-coding system for determining resistance or ohmic values of low-power resistors. You must learn this coding system. The higher-power resistors usually have the resistance value imprinted on their bodies.

There are fixed-value resistors, such as those shown in Fig. 1-1, and variable resistors such as those in Fig. 1-2.

Capacitors are also found in a wide variety of values, shapes, and sizes. Capacitors are classified as to their physical and electrical characteristics. Figure 1-3 shows some representative types. Capacitance values are given in units called microfarads (μF) or picofarads (pF). These values are identified either by EIA color code or by imprint on the body of the capacitor.

Inductors and transformers make up another classification of components. Wire-wound coils and chokes are classified as inductors. These are wound on different forms and cores. There are air-core coils, iron-core, powdered iron-core, etc. (Fig. 1-4). Transformers are coupled inductors. They serve many purposes. There are power transformers, filament transformers, radio-frequency (RF) transformers (Fig. 1-5a), horizontal-sweep transformers (Fig. 1-5b), and many others.

Almost everyone familiar with tubes knows they were the sinews of the electronics industry, but they are rapidly being replaced by transistors. Tubes vary in size from tremendously large trans-

(a)

(b)

Fig. 1-1. (a) Quarter-watt resistors (*Speer Carbon Co.*); (b) molded silicone-ceramic wirewound resistors (*Ohmite Mfg. Co.*).

Fig. 1-2. (a) Miniature composition variable resistor (*Chicago Telephone Supply Corp.*); (b) knob-operated composition trimmer (*Chicago Telephone Supply Corp.*); (c) dual-section composition variable resistor with switch (*Chicago Telephone Supply Corp.*); (d) vitreous enamel adjustable resistor (*Ohmite Mfg. Co.*).

Fig. 1-3. (a) Mylar capacitors (*Aerovox Corp.*); (b) molded cerafil capacitors (*Aerovox Corp.*); (c) molded ceramic capacitors for transistor circuits (*Aerovox Corp.*).

Fig. 1-4. Typical oscillator coil for broadcast band (*F. W. Sickles, Div. of General Instrument Corp.*).

mitting tubes to subminiature hearing-aid tubes. Tubes are described by their physical size, basing arrangement, and electrical characteristics. Figure 1-6 shows a 9-pin miniature tube and nuvistor.

Transistors (Fig. 1-7) are in the category of semiconductors. Despite their tiny size, they do a giant's job. Transistors have made possible present-day large-scale computers, inertial guidance systems for space vehicles, space communications systems, etc.

Many other components are also used in elec-

Fig. 1-5. Transformers: (a) typical *if* transformer (*F. W. Sickles, Div. of General Instrument Corp.*); (b) horizontal sweep transformer (*General Electric Company*).

Fig. 1-6. Representative tube types: (a) 9-pin miniature; (b) nuvistor (*Radio Corp. of America*).

Fig. 1-7. Representative transistor types (*Motorola Semiconductor Products, Inc.*).

tronics. You will become acquainted with these as you use them in the course of your work.

In this experiment you will examine some of the basic electronic components and become familiar with their physical appearance and circuit symbols.

Some standard electronic component symbols are shown in Fig. 1-8. They specify components which are used in a particular electronic device. The manner in which these components are connected in the device is illustrated in a drawing called

a circuit or schematic diagram. In this diagram each component is represented by its unique symbol. In the next experiment you will learn how to interpret circuit diagrams.

SUMMARY

1. Electronic devices are made from a number of identifiable components.
2. The most frequently used components are: (a) resistors, (b) capacitors, (c) inductors, (d) trans-

(a) (b) (c)
(d) (e) (f)
(g) (h) (i) (j)
(k) (l) (m) (n)
(o) (p) (q)

Fig. 1-8. Standard electronic component circuit symbols: (a) fixed resistor, (b) variable resistor, (c) potentiometer (resistor with adjustable tap), (d) capacitor (fixed), (e) polarized (electro-lytic) capacitor, (f) variable capacitor, (g) inductor (air-core), (h) inductor (iron-core), (i) transformer (air-core), (j) transformer (iron-core), (k) switch, (l) diode (solid state), (m) transistor, (n) vacuum-tube triode, (o) earphones, (p) speaker (loudspeaker), (q) fuse.

formers, (e) transistors, (f) switches, (g) fuses, and (h) tubes.

3. Any one category of component, for example, resistors, consists of many different types and sizes, with each type or size representing unique electrical characteristics.

4. Each electronic component has a unique symbol by which it is identified in a circuit diagram.

5. A circuit diagram is an electrical plan illustrating the exact components which make up a particular device and their electrical connections.

SELF-TEST

Check your understanding by answering these questions:

1. What are six of the basic components most frequently found in electronic devices? _____ ; _____ ; _____ ; _____ ; _____ ; _____

2. A drawing which shows the electrical components an electronic device consists of and how they are interconnected is called a _____ .

3. In Fig. 1-8 identify the symbol for each of the following components. *Do not refer to the component names in the figure legend.*
   Transistor _____ ; fuse _____ ; inductor (air-core) _____ ; variable resistor _____ ; transformer (iron-core) _____ ; capacitor (fixed) _____ ; switch _____ ; loudspeaker _____

MATERIALS REQUIRED

- Resistors: Some representative types (fixed and variable)
- Capacitors: Some representative types (fixed and variable)
- Inductors: Some representative types
- Tubes and transistors: Some representative types
- Miscellaneous: Transformers, tube sockets, speaker, switches, fuses, rectifiers, etc.

PROCEDURE

Study, then briefly describe in Table 1-1 the physical features of the components you receive. Draw the circuit symbol for each part. In describing the part, give its shape, approximate size, characteristic markings (including value, if given), and facilities for mounting (number of leads or lugs).

NOTE: Components prefixed with an asterisk (*) are fragile and should be handled carefully.

QUESTIONS

1. How is the resistance value of a resistor identified? Capacitance value of a capacitor?
2. What is the difference in basing between an octal tube and a 7-pin miniature?

3. What is the function of the leads extending from capacitors and resistors?
4. How many leads are there in a resistor? Capacitor?
5. Why is a power transformer so heavy?
6. Why are some resistors so much larger (physically) than others?
7. How many leads does a transistor have?
8. How do the various coils you have studied differ?

Answers to Self-Test

1. resistors; capacitors; inductors; switches; transistors; transformers
2. circuit diagram
3. transistor *m;* fuse *q;* inductor (air-core) *g;* variable resistor *b;* transformer (iron-core) *j;* capacitor (fixed) *d;* switch *k;* loudspeaker *p*

Table 1-1

Component	Circuit symbol	Description
Resistors: Fixed		
Variable		
Capacitors: Paper		
Molded		
Ceramic		
Mica		
Electrolytic		
Trimmer		
*Variable		
Transformer: Power		
Audio		
Coils and chokes: *RF choke		
*IF coil		
*Oscillator coil		
*Antenna coil		
Filter choke		
Tubes and tube sockets: *Octal		
*7-pin miniature		
*Speaker		
*Transistor		

OBJECTIVES

1. To "breadboard" a circuit from a schematic diagram
2. To trace a handwired circuit and draw its electrical diagram
3. To trace a simple printed wiring circuit and draw its electrical diagram

INTRODUCTORY INFORMATION

The Schematic Diagram

The schematic diagram is an electronic blueprint. It tells the technician where and how the electrical parts are connected in a specific circuit, such as a radio circuit. There are a few simple rules to learn:

1. Parts are shown by their electrical symbols.
2. The leads which extend from a part are shown by straight lines.
3. (*a*) If two or more parts are joined physically at a common point (terminal), this point is indicated by a heavy dot on the schematic. For three resistors which have a lead of one connected to leads of the others, we draw a simple schematic (Fig. 2-1). We can, if we wish, identify the junction point by some letter, such as *A* (Fig. 2-2).
   (*b*) If two lines in a schematic diagram cross one

another and no dot appears at the point of crossing, they are *not* connected at this point.

Breadboarding

The schematic diagram, then, is a set of instructions to the technician for assembling and connecting the parts of a circuit. Schematics are used in the experimental laboratory as well as in the production and servicing processes. The physical product created on the production line is in permanent form. The experimental circuit in the laboratory is usually assembled in temporary form so that circuit changes may be easily made until the product design is final.

Experimental circuits are assembled on a "breadboard." This is a device which ensures simple, rapid circuit connection and disassembly. Moreover, the breadboarding technique allows use of component parts over and over again, permitting a wide variety of circuit configurations for study and experimentation. The experiments in this manual will be facilitated by, and are intended for, breadboarding devices.

Wiring Methods

Handwiring

Components in electronic circuits may be connected by handwiring or printed wiring techniques.

Fig. 2-1. Three resistors meeting at a common point.

Fig. 2-2. Common point identified by letter *A*.

Fig. 2-3. Etched board ready for component insertion.

In the handwiring process, component leads are secured to terminal posts. The pigtails (leads) of parts which must be electrically connected are secured at a common terminal post. If the parts are too far apart, their terminal leads may not be long enough to reach a common post. In that case, pigtails of these remote parts (which must be joined electrically) are secured to separate terminal posts. A wire is then used to connect these terminal posts. This is the handwiring technique of interconnecting components in a chassis or on a breadboard.

Etched Wiring

An advanced means of interconnecting parts in mass-produced electronic devices utilizes etched (commonly called "printed") wiring connections. This method requires the use of a plastic, ceramic, glass, or phenolic board on which the components are mounted. Interconnection is made by means of conductive surfaces, which are etched on the board.

One technique for making etched circuit boards is as follows: The conductive circuit pattern is de-

Fig. 2-4. (a) Top and (b) bottom view of completed etched board with components soldered.

(a)

(b)

signed on paper. A clear acetate is then made of the original art. A silk screen of the circuit pattern is made. A protective material is placed on the copper foil using the silk screen of the circuit pattern. The unprotected portion of the copper foil is etched away in an acid bath. The protective coating is also removed from the circuit pattern remaining on the board. By another silk-screen process, solder resist is applied to the pattern except in the areas where solder is to be applied. In some instances, silk screen processes apply a "road map" to top and bottom of the board. The holes are punched into the boards. This requires tooling which is more accurate than drilling. A precise hole size which will produce a good solder fillet around the lead is mandatory.

Figure 2-3 shows an etched board ready for component insertion. The components are then mounted on the board by inserting their leads through the predrilled holes. The leads are then soldered to the conductive surfaces which "form" the pattern of interconnecting "printed" wiring between components. Figure 2-4a is a top view of a finished etched board. The circular blobs are points at which component leads have been soldered. The components are on the other side of the board (Fig. 2-4b).

SUMMARY

1. A schematic diagram is the electronic blueprint which a technician follows in assembling or tracing a circuit.
2. Temporary experimental circuits are assembled on breadboards. These devices are used for rapid circuit connection and disassembly.
3. In the assembly of permanent electronic devices, etched wiring techniques are often used for interconnecting components. "Printed" wiring is accomplished by etching conductive surfaces on a circuit board.
4. Handwiring is also employed in building elec-

tronic devices. Here wires serve as the conductive connections between components.
5. When interpreting a circuit diagram, follow these rules:
   (a) Parts are identified by their electronic symbols.
   (b) Terminal leads which extend from a part are shown by straight lines.
   (c) A heavy dot on a schematic indicates the common junction of two or more terminal leads.
   (d) If no dot appears where two or more lines cross, on a schematic, these lines are not connected at this crossing.

SELF-TEST

Check your understanding by answering these questions:

1. The set of instructions which a technician uses for assembling and connecting the parts of a circuit is called a _____.
2. _____ are used for temporary connection of experimental circuits.
3. A junction of two or more terminal leads is shown schematically by a _____.
4. Two wiring methods used in interconnecting components in an electronic device are: _____ and _____.
5. A fixed resistor is a _____ (how many) terminal device; a transistor is a _____ terminal device.

MATERIALS REQUIRED

- Resistors: Four or more assorted values
- Capacitors: Two or more assorted values
- Miscellaneous: A breadboarding device with terminals and connectors; mounted transistor socket; a simple prewired printed wiring circuit board

PROCEDURE

1. Breadboard the circuit shown in Fig. 2-5. When it is finished, have it approved by your instructor. Then connect the circuits of Figs. 2-6 and 2-7 in turn, and have them checked by your instructor.
2. Your instructor will give you a simple schematic diagram. Breadboard it. Exchange your circuit with that of a student near you. His circuit con-

Fig. 2-5. RC network.

Fig. 2-6. *RC* network.

figuration will be different from yours. Trace out and diagram this circuit. Have it approved by your instructor.

3. Your instructor will give you a simple printed wiring circuit board. Trace out and diagram the circuit on this board. Have your schematic checked by your instructor.

QUESTIONS

1. What is meant by the term "breadboarding"?
2. Where and when is breadboarding used?
3. How are components connected in a commercial electronic device?

Fig. 2-7. Transistor circuit.

Answers to Self-Test

1. circuit diagram
2. breadboards
3. heavy, dot
4. etched (printed) wiring and handwiring
5. two; three

EXPERIMENT 3

Familiarization with Hand Tools Used in Electronics

OBJECTIVES

1. To identify by name each of the following hand tools: diagonal cutters, long-nose pliers, soldering aid, wire stripper, soldering iron, soldering pencil, soldering gun, heat sink, knife, desoldering tool
2. To prepare hookup wire and coaxial cable for connection and for soldering

INTRODUCTORY INFORMATION

After an experimental electronic circuit has been breadboarded, its operating characteristics are tested by the technician. Design changes may be indicated and are effected as required. When the technician and engineer are satisfied that the circuit is performing as it should, it is ready for prototype assembly.

The proper chassis layout of circuit components is determined, the chassis is prepared, and the technician mounts the major components. In this phase of chassis preparation and component assembly, the technician may use chassis forming, drilling, and cutting power tools. He will also require mechanical hand tools such as scribe, punch, hammer, screwdriver, wrench, hack saw, file, etc.

Hand Tools Used in Electronics

Electrical assembly follows the preparation of the chassis and mounting of parts. This is the stage which requires electrical interconnection of components. It is concerned with the preparation and soldering of conductive wiring between parts. The common hand tools used in electrical assembly include diagonal pliers, long-nose pliers, soldering aid, wire stripper, soldering iron, soldering pencil, soldering gun, knife, and heat sink.

Diagonal pliers, commonly called dykes, cutters, or diagonals, are used for cutting soft wire and component leads. They should not be used for cutting hard metals such as iron or steel. Some diagonals have a small, notched cutting surface for stripping the insulation from a wire. This stripping hole will accommodate #22 wire, which is the standard hookup wire used in electronics.

Long-nose pliers are used for holding wire so that the stripped end may be twisted around a terminal post, or pushed through a terminal eye. Long-nose pliers sometimes include cutting edges so that the pliers can serve for both gripping and cutting wires. The 5-in. nose-pliers is a popular size, as is the 5-in. diagonal.

The needle-nose pliers is a variation of the long-nose. Its long thin jaws can get at difficult-to-reach spots.

The soldering aid is a useful tool which simplifies soldering jobs. A standard aid has a sharp metal pointer at one end and a slotted V-type grip at the other. One function of the pointer end is to help clear solder out of terminal eyes on solder lugs. The gripping end is useful in unwrapping wire and component leads, when these are being unsoldered from terminal posts.

The wire stripper removes insulation from hookup wires. There are different types of strippers, ranging from the simple type found on diagonal pliers to the automatic multisized strippers which can handle wires of different diameters. The automatic trip-type stripper is popular with electronics technicians. In addition to mechanical, there are also thermal strippers.

The soldering iron is still a standard electrical hand tool in the industry, although many technicians favor the soldering gun. A heating element inside the iron utilizes power from the power line. The heat is channeled to the tip. The tip is applied to and heats the area to be soldered.

Soldering irons are rated by the amount of power they dissipate, and thus indirectly by the amount of heat they can develop. One-hundred-watt irons are standard in handwired operations, where wires and leads are soldered to terminals. Heavier irons (250-W) are used to solder connections to large metallic surfaces (such as a metal chassis) requiring more intense heat.

The low-wattage soldering pencil is a soldering iron with a smaller heating capacity. It is used for soldering or unsoldering components from a printed wiring board, or in delicate soldering applications requiring low heat levels. Interchangeable tips in a wide variety of shapes and sizes add flexibility to the soldering pencil.

The soldering gun has gained wide popularity because of its fast heating characteristic. A trigger switch on the gun handle applies power to the heating and soldering tip and heats it in 30 s (approximately). Unlike the soldering iron, which is slow heating and must be left **on** as long as it is in use, the soldering gun is heated only at the moment it is to be used. Between soldering operations it is left **off**. One-hundred- and 125-W guns are popular sizes.

A pocket knife is useful in many small tasks requiring cutting and scraping. It can be used for scraping and cleaning terminal ends of wires and components, in preparation for soldering.

The heat sink is a small metal clip used to prevent overheating during soldering or unsoldering of heat-sensitive electronic parts. The heat sink is clipped onto the lead between the body of the part and the terminal point at which heat from a soldering iron or gun is applied. It absorbs heat and reduces the amount of heat conducted to the component.

Desoldering tools simplify the job of cleaning etched (pc) board solder holes of solder while component leads are being removed from the holes. The holes must be free of solder before the terminals of a new component may be inserted. Desoldering tools used by the technician take two forms. The first is a spring-loaded vacuum suction tool, the second a suction device in the form of a hollow rubber ball, attached to a stainless steel or plastic tube. In operation the open end of the tube is placed on a heated solder joint or solder hole. With the spring-loaded device, the spring is released and the molten solder is sucked into the tube clearing the joint or hole of solder. With the ball-type tool, the ball is squeezed and as it is released the solder is again sucked into the tube.

CAUTION: A word of caution about the use of hand tools. They should never be used in a live circuit, that is, in a circuit where electrical power is applied. These tools are made of metal, and metal is a conductor of electricity. Failure to observe this safety precaution may result in electrical shock or damage to the circuitry.

Wires

Hookup wire may come as a solid or stranded conductor. Stranded wire is the accepted standard as hookup wire. However, solid hookup wire is often used in circuits which are not subjected to movement. Solid wire is used for house wiring, bell wiring, transformer windings, etc. Component terminal leads are usually made of solid wire.

Stranded wire is made of wire threads twisted together. The number and diameter of the threads depend on the application and on the quantity of current the wire will carry. The larger the current, the larger must be the diameter of the wire. Stranded wire is flexible and may be bent and twisted without danger of breaking it.

Wires conduct electricity. To prevent adjacent wires from making electrical contact, or from touching metallic or conductive surfaces, wires are covered with an insulated coating. The industry now makes extensive use of FR-1 (fire retardent insulation) as wire covering. Older methods of insulating wires included solid-wire insulation, which used shellac or enamel into which the wire was dipped. Stranded wire was insulated by a plastic, rubber, or cloth coating.

Multiwire cables are frequently used in interconnecting electronic units or devices. For example, the several units of a radar set are interconnected by multiwire cables. A cable is made up of two or more insulated wires inside a common housing. The housing is usually a plastic or rubber sleeve, but the sleeve may also be a wire braid. Cables are designed and constructed for special applications requiring two or more separate insulated electrical conductors.

Coaxial cable is a very special type of shielded cable. It consists of an inner and outer conductor. The inner conductor may be solid or stranded. Around the inner conductor is a plastic or phenolic

Fig. 3-1. Preparation of coaxial cable for attaching microphone connector.

insulating sleeve. A braided shield is fitted around the insulating sleeve of the inner conductor. The braid acts as a second conductor around the inner conductor and is insulated from it. A rubber or plastic sleeve covers the braid and provides insulation for the coaxial assembly. Shielded cable is also manufactured without any external insulating cover.

Wire Preparation

Two or more wires that provide a conductive path for electricity must be electrically connected. This means that an uninsulated surface on one wire must be mechanically connected to an uninsulated surface on the other wire. To ensure that the wires will not separate, or the connection corrode, they are soldered at the junction.

Before wires may be connected and soldered, they must be properly prepared. This involves stripping away the insulation at the ends of the wire, thus providing terminal leads which may be connected to each other or to a terminal post. It also involves cleaning insulation or oxidation from the bare ends of the wire.

Mechanical wire strippers are usually used by the technician, though thermal strippers may also be employed. The technician selects the stripping hole which will fit the wire. He inserts the wire in the tool and squeezes the handles. The insulation is automatically removed and the stripper resets in one operation.

After the insulation is removed, the technician examines the wire. If it has a shiny look no further preparation is needed. However, a dull- or dark-appearing wire must be cleaned before it is connected and soldered. The wire may be cleaned by scraping it gently with a knife or with emery cloth. This same process is employed in removing the insulation from shellac- or enamel-covered solid wires. The technician must be careful not to "nick" the wire.

The technician may also be required to prepare coaxial or shielded wire. This type of wire is used in circuit connections where a shielded conductor is required. Coaxial cable is used for intercabling two assemblies of an electronic system and for instrument leads. The braid around the coaxial wire acts as the grounded shield for the inner conductor which is the "hot" wire.

In preparing an uncovered shielded wire for use, the braid is combed away with a soldering aid, awl, or other pointed tool. The braid is then twisted together to form the ground lead. The insulated covering is stripped from the inner conductor. The inner and outer conductors are scraped if necessary before connection to the circuit.

The processes for preparing a coaxial cable for use as an instrument lead are illustrated in Fig. 3-1. The two ends of the cable require individual treatment because one end must be terminated in a microphone connector, and the other in a probe or in two leads, depending on the use for which the cable is intended.

The preparation for mounting a microphone connector is shown in Fig. 3-1. One inch of cable insulation is first stripped away, exposing the braid. The spring shield of the connector is then placed over the braid. The braid is folded back and cut to fit the narrow neck of the spring. The insulation is stripped from the inner conductor. The knurled holding ring is placed over the spring so that the inner conductor protrudes through the eyelet. The ring is pushed back as far as it will go. The set screw is tightened. Care must be taken not to "short" the inner conductor to the braid or the wire shield.

The center conductor protruding through the eyelet is cut flush with the eyelet. This completes preparation of the cable, and the inner conductor is ready to be soldered to the eyelet.

NOTE: The cable will be soldered in the next experiment.

The insulation from the other end of the coaxial cable is stripped and the braid trimmed back. The insulation is stripped from the inner conductor. The braid is now combed and used directly, or a short piece of flexible ground wire may be soldered to it.

SUMMARY

1. Hand tools most frequently used by the electronic technician include: diagonal cutters, long-nose pliers, soldering iron, soldering pencil, soldering gun, desoldering tool, heat sink, soldering aid, wire stripper, and pocket knife.
2. Standard hookup wire in electronic assembly is #22, stranded.
3. Coaxial cable is shielded wire used for instrument leads and other applications.
4. Wire terminals must be cleaned of insulation and oxidation before soldering them in a circuit.
5. Special techniques must be followed in preparing coaxial cable for use. These include stripping the outer covering of insulation at both ends; folding back the braid just uncovered; stripping the insulation from the inner conductor. The braid serves as one conductor; the inner conductor is the other of a two-conductor coaxial cable.

SELF-TEST

Check your understanding by answering these questions:

1. Diagonal cutters are used for cutting soft _____ and component _____.
2. The wire stripper removes _____ from hookup wire.
3. Low-wattage soldering pencils are used for soldering and unsoldering components on a _____.
4. The heat sink is a _____ used to prevent _____ during soldering or unsoldering of heat-sensitive components.
5. Insulation on wires is used to prevent adjacent wires from making _____ contact.
6. Coaxial cables usually contain _____ (how many) conductors, of which the outer is a _____ or shield.
7. Wires are soldered at a junction to prevent them from _____.
8. Wire ends must be _____ and _____ before they can be soldered.

MATERIALS REQUIRED

- Tools: Diagonal cutters, long-nose pliers, soldering aid, wire stripper, soldering iron, soldering pencil, soldering gun, knife, heat sink, desoldering tool
- Miscellaneous: Short pieces of #22 and #18 stranded hookup wire; short piece of shielded wire; 3-ft piece of RG59U coaxial cable; microphone connector

PROCEDURE

Hand-Tool Familiarization

1. Examine the kit of hand tools and identify each tool.
2. Note the differences in physical appearance between the soldering iron, soldering pencil, and soldering gun.
3. Remove the tip from the soldering pencil (it unscrews) and replace it with another tip.
4. Carefully remove the tip from the soldering gun. Note how the tip ends are bent at right angles to the tip. Restore the tip. Be certain that the ends are properly secured. If they are not held tightly, the gun tip will be slow to heat, or may not heat at all.
5. Plug the soldering gun into an electrical outlet. Press and hold the trigger. Apply a piece of solder to the tip and note how quickly the tip heats. Release the trigger. Permit the tip to cool.
6. Now loosen the tip and again press and hold the trigger. Does the gun tip heat as quickly? Tighten the tip securely.
7. Plug a soldering iron into a power outlet and determine how long it takes the tip to heat for soldering.

Wire Preparation

8. Strip the insulation from the ends of a short piece of #22 stranded hookup wire. Repeat, using a short piece of #18 wire.
9. Examine the ends of the stripped wire. If necessary clean the ends.
10. Prepare a piece of shielded wire (without any insulated covering) for connection in a circuit.
11. Prepare a 3-ft piece of coaxial wire to receive a microphone connector. Mount the connector, but do not solder it. Hold assembly for use in Experiment 4.

QUESTIONS

1. List the hand tools which an electronics technician uses. State the purposes of each.
2. How does a soldering iron differ from a soldering pencil? From a soldering gun?
3. In normal use, which would be more expensive to operate, a 100-W soldering iron or a 100-W soldering gun? Why?
4. Why is hookup wire insulated?
5. Does the diameter of a wire affect its current-carrying capacity? Explain.
6. Explain the differences between ordinary hookup wire and coaxial wire.
7. Where is coaxial cable used?
8. What is a heat sink? When is it used?

Answers to Self-Test

1. wire; leads
2. insulation
3. etched (pc) boards
4. metal clip; overheating
5. electrical
6. two; braid
7. separating
8. stripped; cleaned

EXPERIMENT 4

Soldering Techniques

OBJECTIVES

1. To tin an iron
2. To connect and solder electronic components to a terminal board
3. To insert components and solder them onto a printed wiring circuit board
4. To unsolder and remove components from a handwired and printed wiring circuit using a heat sink where required to protect components against damage from excessive heat
5. To solder a microphone connector on a coaxial cable and to make needed instrument leads

INTRODUCTORY INFORMATION

Soldering is generally required to assure permanent electrical connections. Wires, or wires and terminals, are wrapped or twisted together, then solder is melted into the heated joint. When the heat is removed, the solder and wire cool making the soldered joint look like a solid piece of metal. It is not possible, after proper soldering, to separate the wires at a joint except by breaking them or by unsoldering them.

Solder is an alloy of lead and tin. It has a low melting point and comes in wire form for electronics use. Electronics solder is made up of 60 percent tin and 40 percent lead, though the composition may vary for certain applications. Rosin-core solder is used for soldering electronic components. The rosin is a flux which flows onto the surface to be soldered, assuring a more perfect union. Acid and soldering paste should *not* be used in electronics.

Proper soldering requires the following:

■ Clean metallic surfaces

■ Sufficient heat applied to the joint to melt solder when solder is applied to the heated wire surface

The solder must not come in direct contact with the tip of the soldering iron or soldering gun because it will melt readily. If the wires or terminals to be soldered have not been preheated sufficiently, the molten solder will *not* adhere to their surface. The joint may look well soldered, but the chances are that it is not. *Cold*-solder joints provide poor electrical contact and are difficult to discover when troubleshooting defects arise from poor soldering.

Soldering is not always used in the manufacturing process. There are many instances of solderless wire-wrap connections in use today.

Tinning

To ensure maximum transfer of heat from the iron to the surfaces at the joint, the tip of the iron or gun must be tinned. A new soldering tip or a tip that has been used for a long period of time must be cleaned by scraping it with a knife or emery paper, steel wool, a wire brush, or fine sandpaper. If the tip is badly pitted, it may be necessary to file it. This technique applies to *copper* tips.

Many modern tips are gold-plated or iron-bearing. A gold-plated tip should be cleaned by wiping it against a wet sponge. Iron-bearing tips can be cleaned with a wire brush. These tips should never be filed or cleaned with sandpaper or emery cloth.

After the tip is cleaned, it is heated. Solder is permitted to melt onto the tip, tinning it. If the iron is overheated and permitted to discolor before solder is applied, it will be difficult to tin.

It is not only necessary to tin the iron; the surfaces to be soldered should also be cleaned and tinned. Tinned surfaces assure good electrical connections and proper bonding when soldered. Wire may be tinned by placing it on the tip of the iron and heating it sufficiently so that the wire will melt solder. Stranded wire should be twisted together before it is tinned. Terminals should also be tinned.

The tip of a soldering iron tends to become dark and dirty when in use. To keep the iron at maximum-heat-transfer efficiency, the tip should be cleaned periodically by wiping it on a damp sponge. A wire brush will also clean the soldering tip.

Mechanical Connections

Wire and component leads should be secured by a three-quarter to a full turn around a terminal post, as in Fig. 4-1a. The tendency of the novice is to wind several turns around the post. This is wasteful of terminal area and creates difficulties if it is ever necessary to unsolder and remove the lead. In Fig. 4-1b wire is secured to a terminal strip.

If two wires are to be joined together a hook splice is used (Fig. 4-2). It is unnecessary to twist the wires together before soldering.

Heat Sinks

Heat sinks are used to avoid heat damage to heat-sensitive components during soldering. Solid-state

(a)

(b)

Fig. 4-1. Component lead secured to (a) terminal post before soldering, wrap is ¾ to 1 full turn; (b) terminal strip.

Fig. 4-2. Hook splice.

devices such as transistors and diodes are extremely sensitive to heat and can be permanently damaged if they are overheated. Any small metal clip, an alligator clip for example, may be used as a heat sink. Figure 4-3 illustrates the use of this device. It is clipped onto the lead between the component and the point at which the soldering iron is applied. The clip acts as a heat load and reduces heat transfer to the component.

After the iron is removed, the heat sink should be kept in place until the joint has cooled.

Wherever they can be used, heat sinks are a good practice. However, they are not practical in many physical applications where solid state devices have been so soldered into circuit boards that they do not allow space for heat sinks. The present trend in the manufacture of solid state devices provides for plug-in components, for example, plug-in transistors and plug-in integrated circuits.

HEAT SINK

HOT SOLDERING IRON

GERMANIUM DIODE

Fig. 4-3. Heat sink protecting germanium diode.

Soldering Aid

The pointed end of a soldering aid is used to clean terminal eyes from which wires or leads have been unsoldered and removed. Heat is applied to the terminal and the sharp tip is worked into the terminal eye until it is free of solder and is ready to receive another wire.

The slotted end of a soldering aid is used to grasp wires while they are being soldered or unsoldered.

Soldering on Etched (Printed) Wiring Circuit Boards

The network of interconnecting conductive paths on a "printed" wiring board consists of thin copper strips and pads bonded to the plastic board. Leads of components mounted on the board are inserted through holes in the board and the conductive copper. These leads are soldered to the copper at the terminus of the hole through which they emerge. If excessive heat is applied to the copper it may lift from the board, or the miniature components mounted on the board may be damaged.

A 30-W (approximately) soldering pencil is therefore used to heat the junction. This low-wattage iron provides an effective means of controlling heat. As in hand-wired circuits, component leads should be cleaned and tinned prior to soldering. Sixty-forty rosin-core solder or eutectic solder should be used. The surface of the copper bonded to the board should also be properly prepared prior to soldering. The copper can be cleaned by scraping it gently in the vicinity of the terminal hole, or an ink eraser can be used to clean the surface at the solder point.

Another factor which must be considered in working on "printed" boards is the proximity of the conductive strips and pads to each other. It is essential to avoid excess solder to prevent bridging the gap between two paths. Application of excessive solder should be avoided in handwired circuits also. Just enough solder should be melted into a junction to effect a proper bond. If tiny solder globules form in the junction area, remove them by cleaning the soldering tip on a sponge or cloth, then apply the cleaned tip to the solder globules at the junction.

Unsoldering Components

Unsoldering is the reverse of the soldering process. The terminal point is heated with a properly tinned iron. The lead to be unsoldered is grasped with a soldering aid or with a pair of long-nose pliers. When the solder melts, the lead is gently unwrapped and removed from the post. If the lead was inserted through a terminal eye, the eye is cleaned so that it will be ready to accept a new lead.

Cleaning a terminal eye, or circuit board terminal "hole," may be accomplished by manipulating a soldering aid in the heated terminal hole, or by use of a desoldering suction tool as described in the previous experiment.

Excess solder or solder globules which may have separated from the post are shaken out of the chassis or board. Wire braid and special solder "sucker" tools are especially useful in retrieving these solder globules.

SOME SAFETY PRECAUTIONS: The soldering gun or iron operates at temperatures high enough to cause serious burns. Observe these safety precautions:

(*a*) Do not permit hot solder to be sprayed into the air by shaking a hot gun or iron or a hot-soldered joint.
(*b*) Always grasp a soldering gun or iron by its insulated handle. Do *not* grasp the bare metal part.
(*c*) Do not permit the metal part of a soldering gun or iron to rest on combustible material. An iron should always rest on a soldering stand.

SUMMARY

Good soldering is part of a technician's skills. Solder connections must be mechanically strong, so that they will not shake loose with resultant loss of signal and possible damage to parts. Electrically, solder contacts must have low resistance for proper signal transfer. Some basic soldering rules are:

1. Soldering tip must be tinned and clean.
2. Metals to be soldered must be clean.
3. Support the joint mechanically where possible.
4. Pretin large surfaces before soldering them together.
5. Apply solder to joint, not to gun or iron tip. Solder must flow freely and have a shiny, smooth appearance.
6. Use only enough solder to make a solid connection.
7. Where additional flux is used, apply to joint, not to soldering tip. Only rosin flux should be used.
8. Solder rapidly and do not permit components or insulation to burn or overheat.

9. Use rosin-core solder or the equivalent. Do not use acid-core solder for any electrical work.

SELF-TEST

Check your understanding by answering these questions:

1. Soldering is needed to make good _____ _____.
2. Solder is an alloy of _____ and _____ .
3. For electronic work _____ percent _____-core solder is used.
4. An iron is _____ to ensure maximum heat transfer.
5. Wire and terminal surfaces must be _____ and _____ before soldering to ensure good electrical contact.
6. To avoid damaging heat-sensitive components during soldering a _____ _____ is used.
7. For etched board soldering a _____-W soldering iron is used.
8. Circuit board terminal points should not be _____ to avoid lifting the copper which forms the conductive paths on the phenolic board.
9. In unsoldering components from a circuit board a _____ _____ _____ is used to clean the terminal holes of solder.
10. If a junction has not been properly heated before soldering, a _____ _____ joint may result.

MATERIALS REQUIRED

- Resistors: Three or more, assorted values
- Capacitors: One, any value
- Printed wiring circuit board, with resistors and capacitors mounted on it
- Miscellaneous: 100-W soldering iron or soldering gun, 30-W soldering pencil, hookup wire, instrument lead wire, two terminal strips, breadboard, 3-in. RG59U coaxial cable assembly with microphone connector (from preceding experiment), two alligator clips, hand tools, solder

PROCEDURE

Tinning the Iron or Soldering Gun

1. Clean the tip before heating.
2. Heat the iron or gun. Apply solder to the tip as soon as it is hot enough to melt solder. Do not allow the tip to discolor by overheating before applying solder. Wipe off excess solder and flux.

Making the Soldered Connection

3. Mount two terminal boards on the breadboard or chassis parallel to each other and 2 in. apart.
4. Cut off four 3-in. lengths of wire.
5. Clean and tin the terminals.
6. Strip ½ in. of insulation from the wire at both ends. Clean and tin the wire.
7. Make the mechanical connections shown in Fig. 4-4. Loop or crimp one turn of wire around each terminal.
8. Apply the hot iron tip to the joint. Apply solder to joint, allowing it to flow freely over the joint. Do not use excess solder. If necessary, a small amount of solder may be placed at the point where the solder tip contacts the joint in order to facilitate heat transfer to joint. A solder joint should look clean and silvery.
9. Remove soldering tip and hold joint rigidly until solder has cooled and set.
10. Check the connection by pulling it with long-nose pliers. The wire should not pull away from the terminal. A "cold"-solder joint may show up when this test is made.
11. Repeat steps 8, 9, and 10 until the four wires have been soldered to the terminals.
12. Have your work checked by an instructor.

Fig. 4-4. Connections for soldering.

Fig. 4-5. Mounting resistors and capacitors on terminal board.

Opening a Soldered Connection

13. Place hot-solder tip on the joint which is to be unsoldered and hold it there during the entire process.
14. When solder at the joint starts to melt, gently uncrimp the wire from the terminal with long-nose pliers or a soldering tool and straighten it out.
15. Gently pull wire out of terminal.
16. Unsolder and remove the four wires from the terminals. Clean the terminals and prepare them for the next job.

Handwiring a Circuit on a Terminal Board

17. Mount the resistors, capacitors, and hookup wire on the terminal board, as shown in Fig. 4-5.
18. Make the required solder connections at all junction points.
19. Have your work checked by an instructor.
20. Unsolder all the parts, being careful not to damage any.

Unsoldering Components from and Resoldering Them to a "Printed" Wiring Circuit Board

21. Unsolder and remove two resistors and a capacitor from the "printed" board issued to you. *Use* a low-wattage soldering pencil. Do not damage the board or the parts.
22. Prepare the surface and holes from which the parts were removed for replacement components.
23. Place the removed components back in their original mounts.

24. Clip a heat sink onto one of the terminal leads.
25. Solder the components back on the board.
26. Remove any excess solder.
27. Have your work checked by an instructor.

Constructing Instrument Leads

Signal-generator Lead. Examine the coaxial cable and amphenol connector which you prepared in Experiment 3 to see that it is properly assembled.

Solder the center conductor to the amphenol connector eyelet. Be certain that the solder has flowed on smoothly and that the wire and eyelet are properly bonded.

Solder a 3-in. flexible ground lead to the coaxial cable braid at the other end. Solder alligator clips to the center conductor and to the ground lead. Tape the exposed braid.

This coaxial cable will be used as your signal-generator lead in future experiments.

Other Test Leads. Your instructor has prepared wire for you to construct several test leads. Construct these leads according to his instructions.

QUESTIONS

1. Why are electrical connections soldered?
2. What is a cold-solder joint?
3. What precautions must be taken to avoid a cold-solder joint?
4. What kind of flux is used in soldering electronic components? What kind of solder?
5. What precautions must be taken in unsoldering components from, and soldering them to, "printed" wiring circuit boards?
6. What is the purpose of tinning an iron? A wire? A terminal post?

Answers to Self-Test

1. electrical connections
2. lead; tin
3. 40; rosin
4. tinned
5. cleaned; tinned
6. heat sink
7. 30
8. overheated
9. desoldering suction tool
10. cold solder

EXPERIMENT 5

EVM—Electronic Voltmeter—Familiarization

OBJECTIVES

1. To identify the operating controls and write down the function of each control of an electronic voltmeter
2. To identify each of the leads of a VTVM/TVM (Vacuum Tube Volt Meter/Transistor Volt Meter) or DVM (digital voltmeter)
3. To read the voltage value at a specified point on each of the voltage ranges of an EVM
4. To set the zero and ohms adjust on the meter

INTRODUCTORY INFORMATION

Electronic Voltmeter

Voltage measurements are made primarily with an instrument called a voltmeter. There are other instruments which may be used to measure voltage, for example the oscilloscope, but voltage measurement is not the primary purpose of these devices.

Two types of voltmeters are used today, (1) the analogue and (2) the digital. A pointer on the analogue meter indicates the voltage on a calibrated scale. The voltage measured by a digital meter appears as a number on a numerical (digital) display, so that any one who reads numbers can read the voltage.

The reading errors associated with analogue meters (these are discussed in the experiment) are eliminated by the digital voltmeter (DVM). Moreover, DVMs are more accurate than analogue meters. Despite the fact that digital meters are more costly than analogue meters, their advantages are leading to their widespread adoption.

Another way of identifying voltmeters is to specify them as (1) electronic voltmeters or (2) electromechanical meters. The vacuum-tube voltmeter (VTVM), the transistorized voltmeter (TVM), and

the digital voltmeter (DVM) are examples of electronic voltmeters (EVMs). The volt-ohm-milliammeter (VOM) is an example of an electromechanical multimeter which is used to measure voltage.

Meters may be single-function devices, such as the voltmeter, which can measure only voltage. Or meters may be multipurpose devices such as the multimeter, which is used to measure a variety of electrical quantities, such as voltage, current, and resistance. The VTVM, the TVM, the DMM (digital multimeter) or the digital volt-ohm-milliammeter (DVOM), and the VOM are examples of multimeters.

In this experiment we will be concerned with electronic multimeters which we will designate simply as EVMs. When an EVM is specified in this experiment or in those which follow, any one of the three types of electronic multimeter (TVM, DVOM or VTVM) may be used.

Prior to the invention of transistors, the VTVM (vacuum-tube voltmeter) was the electronic meter used to measure dc and ac voltage and resistance. On rare occasions it also included a current measuring function. VTVMs are still in use in industry and in school laboratories. However, they are being replaced by the TVM and the DVM. The TVM and DVM include all the measuring functions of a VTVM plus current-measuring capability.

Electronic voltmeters sometimes have other specialized scales such as the decibel (dB) scale for sound level measurement.

It is evident that the transistor part of the name TVM stems from the use of transistors and other solid state devices, while in the VTVM, vacuum tubes are employed to perform similar functions. The electromechanical VOM uses neither vacuum tubes nor transistors. The general-purpose VTVM is usually a line-operated instrument—it is plugged into an electrical outlet and derives its operating power from the line. TVMs and DVMs are either battery or line operated, and frequently they contain facilities for both types of power operation.

Operating Controls (VTVM)

Figure 5-1*a* is an illustration of a typical vacuum-tube voltmeter. Note that the operating controls and meter scales are on the front panel. Of the four controls, two are labeled "Zero" and "Ohms." Though the other two are not identified by name, their functions are apparent. The lower left-hand knob might well be called the function switch. The instrument is turned **on** when it is switched from **off** to any of its four other settings. The "−Volts" and "+Volts" positions are for measurement of positive and negative direct-current volts. The "AC Volts" position is for measuring alternating-current voltage. The "Ohms" position is for measuring resistance.

The lower right-hand control is a range switch. It works in conjunction with the function switch to select the proper operating range for a specific measurement. For example, if the technician wishes to measure positive dc volts, and he knows that the voltage he will be measuring lies between 5 and 15 V, he will turn the function switch to +Volts, and the range switch to 15. Note that the range number represents the maximum voltage which can be read on that range. The meter in Fig. 5-1*a* has eight voltage ranges, 0.5 to 1,500 V.

The range switch shown in Fig. 5-1*a* acts also as a resistance-scale multiplier. Thus, the 15-V position is also the $R \times 100$ position. When measuring resistance the function switch is set on ohms. Resistance readings on the R scale must be multiplied by 100 on the $R \times 100$ position. On the $R \times 1,000$ position, they must be multiplied by 1,000, etc.

The purpose of the Zero control is to set the meter pointer on zero (at the left side) of the meter scales. When the meter is first turned to any voltage position, the pointer may go off zero during the warmup period. After the meter has been **on** for about five minutes, it should stabilize and the pointer should return to its zero setting. If it does not, it can be brought to zero by the Zero adj.

When the function switch is in the Ohms position, the meter pointer will move all the way to the right, on the R scale. The pointer should line up on the last calibration mark on the R (top) scale. If it does not, the Ohms adj. will bring the pointer into position. It should be noted that the Zero control must be properly set, *before* Ohms control is adjusted.

There are many variations among manufacturers both as to appearance of the instrument and as to operating controls. In the VTVM of Fig. 5-1*b*, for example, the function switch is on the right, and the

(*a*)

(*b*)

Fig. 5-1. (*a*) VTVM (*Radio Corp. of America*); (*b*) VTVM (*The Triplett Electrical Instrument Co.*).

Fig. 5-2. TVM—solid state VOM (*The Triplett Electrical Instrument Co.*).

range selector on the left. The two controls labeled "Zero" and "Ohms" are comparable to those on the previously discussed VTVM.

Operating Controls (TVM)

The operating controls on a TVM are very similar to those on a VTVM. In Fig. 5-2, for example, a function switch is on the lower left, a range switch is on the lower right, and there are a Zero ("0") control and an Ohms (Ω) control. The Zero and Ohms controls are adjusted like the same controls on the VTVM, and the range switch is similar in operation

to that of a VTVM. It should be noted, however, that the range switch has 12 positions, including a low 0.05 dc volts position and a 0.005 ac volts position as contrasted with the 8 positions on the VTVMs in Fig. 5-1. The function switch is also more complex, reflecting the greater versatility of the TVM. Thus, in addition to the **off**, + and − DCV, ACV, and conventional ohms (Ω) positions, there are LPΩ (low-power ohms), DC-MA, AC-MA, and Batt. positions. The low-power ohms function is used for resistance measurements in solid state circuits. DC and AC-MA positions are for measuring, respectively, dc and ac current, and the Batt. position is used to test the battery which powers this instrument. This is *not* a line-operated TVM.

In TVMs, also, there are variations among manufacturers both as to appearance of the instrument and operating controls.

Operating Controls (DVOM)

There are two operating controls on the DVOM in Fig. 5-3. On the right is a six-position function switch for selecting any one of the following measurement functions: (1) ac current, (2) dc current, (3) resistance, (4) dc volts, or (5) ac volts. The sixth position on the switch labeled "Off" shuts power off. The left-hand control is a six-position range switch. The ranges are common to each of the measurement functions. The control positions shown in Fig. 5-3 indicate that the meter is set to measure resistance and that the maximum resistance which can be measured on the range shown is 20,000 Ω.

An unusual feature of this digital instrument is an analogue scale which uses an edgewise analogue meter in addition to its digital capabilities. The ana-

Fig. 5-3. DVOM—solid state digital VOM (*Simpson Electric Co.*).

logue readings are useful in certain electronic applications.

Meter Scales

In general, the voltage scales on a TVM and VTVM are linear, the ohms scale is nonlinear.

The scale on a ruler is the most familiar example of a linear scale. On a ruler equal distances are represented by equally spaced calibrations. Thus, a 1-in. distance is represented by the same calibration interval whether it is one inch between 1 and 2 or between 9 and 10. This same relationship holds true for any linear scale, that is, the distance between consecutive corresponding calibration marks is everywhere the same on the scale.

The voltmeter scale (Fig. 5-4) is another example of a linear scale. Here the meter pointer travels equal distances along the scale *arc* for equal voltage changes. Thus, the *arc* distance between 0 and 10 V is the same as between 10 and 20 V or between 90 and 100 V.

In reading the scale in Fig. 5-4 it is necessary to supply numbers for the unnumbered calibrations. This is easily accomplished. For example, there are five equally spaced intervals between numbered voltage calibrations. Each of these distances represents one-fifth of the voltage difference between the numbered calibrations. In the case of Fig. 5-4, every calibration marker would therefore represent a 2-V interval. Thus, in Fig. 5-5, the meter pointer reads 44 V, that is, 40 V plus 2 intervals × 2 V per interval.

When the pointer falls between scale markers, the reading is approximated. For example, in Fig. 5-6 the reading is 45 V.

The voltage scales of a voltmeter are not always calibrated in the same units as the range setting. For example, the scale of Fig. 5-4 would serve admirably for a 0- to-100-V range. However, it could serve as well for a 0- to-10-V range, or a 0- to-500-V

Fig. 5-5. Meter pointer reads 44.

range. In the case of a 10-V range, every scale reading must be divided by 10. For the 500-V range, every scale reading must be multiplied by 5.

EVM Leads

A TVM, DVM, and a VTVM usually require two or three leads. The three-lead arrangement normally consists of a flexible common ground (black), a shielded coaxial lead for direct current, and a shielded coaxial lead for alternating current and ohms. The ground lead is terminated in an alligator clip which is clipped onto the common return or ground of the circuit. The two coaxial leads are terminated in probes, completely insulated except for the metallic tip which extends from the end of the probe. The technician holds the insulated probe in measuring. He does not come into contact with the conductive tip which is applied to the point of measurement. Since there are other variations of the three-lead arrangement, the technician should consult the instruction manual for the meter he is using until he is entirely familiar with the instrument.

The two-lead arrangement shown in Fig. 5-1 includes a flexible common "ground" lead and a coaxial cable. The coaxial cable is terminated in a probe. There is a two-position switch on the probe. In one position the cable serves as a dc voltage lead. In the other it is an ac voltage or Ohms lead.

The meter end of the common ground lead is terminated in a pin tip or banana plug, which plugs into an appropriate jack on the meter. The coaxial cable is connected to the meter by means of a mi-

Fig. 5-4. Linear meter scale.

Fig. 5-6. Meter pointer reads 45.

crophone-type connector. Leads should be left on the meter when it is not in use.

SUMMARY

1. The TVM (transistorized voltmeter), DVM (digital voltmeter), and VTVM (vacuum-tube voltmeter) are classified as electronic voltmeters, EVMs.
2. Electronic voltmeters are used to measure dc and ac voltages and resistance. The TVM and DVM also have current measuring ranges.
3. Electronic voltmeters are line-powered or battery-powered.
4. The meter operating controls include: (a) function switch, (b) range switch, (c) zero adjust (d) ohms adjust. The DVM does not have zero or ohms adjusts.
5. The function switch selects the desired measurement function: voltage, resistance, or current.
6. The range switch selects the proper operating range for a particular measurement.
7. The zero adjust sets the meter pointer at the zero calibration marker on the left side of the meter scales.
8. The ohms adjust is used to line up the meter pointer with the last calibration marker (infinity) on the ohms scale.
9. The voltage and current scales of an electronic voltmeter are linear.
10. The ohms scale is nonlinear.
11. The electronic voltmeter uses two or three leads. In both cases a flexible black lead serves as the common or ground lead.
12. In a three-lead meter, the other two leads are coaxial (shielded) leads, one for dc measurements and the other for ac and resistance readings.
13. In a two-lead meter, the single coaxial lead is terminated in a probe containing a two-position switch. One switch position is for dc measurements; the other position for ac and resistance measurements.

SELF-TEST

Check your understanding by answering these questions:
1. The two types of instruments used for measurement of electrical quantities are the electronic voltmeter and the _____ _____.
2. The meter which employs transistors in its circuitry is called a _____.
3. VTVMs are normally _____-powered devices.
4. Electronic voltmeters usually have four operating controls. These are: (a) _____; (b) _____; (c) _____; and (d) _____.
5. The purpose of each control listed in the answer to question 4 is: (a) _____; (b) _____; (c) _____; and (d) _____.
6. An EVM has two or three leads. These are for: (a) _____; (b) _____; (c) _____; and _____.
7. The voltage and current scales of an electronic voltmeter are linear. _____ (true/false)
8. Refer to Fig. 5-3. The range switch is set to the 10-V range. What is the meter reading? _____ V

MATERIALS REQUIRED

■ Equipment: EVM

PROCEDURE

1. Examine the meter assigned to you. Draw a panel view of the meter, showing operating controls and the functions and ranges associated with the switches. Draw also the voltage scales and the ohm scale.
2. Read the instruction manual and learn how to operate the meter.
3. Turn the EVM **on.** Set the function switch on +dc volts. Permit the meter to warm up if required. Then vary the Zero adj. control. Observe its effect on the pointer. Set the pointer on zero.

Now turn the range switch through every setting and note if the meter pointer remains on zero.
4. Check the zero setting of the meter on each range of −dc volts, and on each range of ac volts.

NOTE: A well-designed, properly adjusted meter should maintain its zero setting on every voltage function and range.

5. Set the function switch on ohms and range switch

on $R \times 1$. Be sure the lead tips do not touch each other or a metal object. Vary Ohms adj. and observe its effect on the pointer. Set this control so that the pointer rests on the maximum calibration on the resistance scale. In this position the resistance circuit is open and the meter registers infinite resistance. Check the setting of the pointer on every position of the range switch ($R \times 10$, $R \times 100$, etc.).

6. Short the ohms and ground leads (that is, touch the metal tips of these leads). The pointer should swing to zero. If it does not come to rest on zero, with the leads shorted, set it on zero with the *Zero adj*. Now open the leads and check the position of the pointer. If it is no longer on the maximum-resistance marker (infinity) set it there with the Ohms. adj.

7. Recheck the zero and infinity positions.

NOTE: Do not leave meter leads shorted together on Ohms for any length of time. In the Ohms position, an internal battery is connected in the circuit. The battery voltage will be greatly reduced in a short period of time if the leads remain shorted together.

QUESTIONS

1. List the controls on the panel of your EVM and state the purpose of each.

2. Is it possible to use the same scale on the 3-V range as on the 300-V range of your meter? Explain.

3. Draw a linear scale with number calibrations 0, 1, 2, etc., through 10. Set off each major subdivision into ten minor subdivisions. Show where *8.7* would be on your scale.

4. Is the ohm scale on your meter linear or nonlinear? Justify your answer by referring specifically to resistance calibrations.

5. Explain in detail how you would zero the meter for dc voltage readings. Identify the control or controls you would use.

6. Explain in detail how you would check to determine if the meter is properly zeroed on the $R \times 1$ range. Identify the controls you would use.

7. If in question 6 you find the meter is not properly zeroed, explain in detail how you would zero it.

Answers to Self-Test

1. electromechanical VOM
2. TVM or solid state VOM
3. line
4. (*a*) function; (*b*) range; (*c*) zero; (*d*) zero ohms
5. (*a*) Select the operating function; voltage, current or resistance; (*b*) select the appropriate range of measurement; (*c*) set the pointer at zero on the left-hand side of the scales; (*d*) set the pointer at infinity on the right-hand side of the ohms scale.
6. (*a*) common or ground; (*b*) dc; (*c*) ac and resistance
7. true
8. 2

EXPERIMENT **Resistor Color Code and Use of Ohmmeter**

OBJECTIVES

1. To determine the value of resistors from their EIA (Electronic Industries Association) color code
2. To read the resistance value at a specified point on each of the ohmmeter scales on an electronic voltmeter
3. To measure resistors of different values
4. To measure the resistance across each combination of two of the three terminals of a potentiometer and observe the range of resistance change as the shaft of the potentiometer is varied throughout its entire range

INTRODUCTORY INFORMATION

Color Code

The ohm is the unit of resistance. One symbol for ohm is Ω (Greek letter omega). Resistance values are indicated by a standard color code adopted by manufacturers. This code involves the use of color bands on the body of the resistor. The colors and their numerical values are given in the resistor color chart, Table 6-1. This code is used for $\frac{1}{4}$-W, $\frac{1}{2}$-W, 1-W, 2-W, and 3-W resistors.

The basic resistor is shown in Fig. 6-1. Note the color bands. The color of the first band tells the first significant figure of the resistor. The color of the second band tells the second significant figure. The color of the third band tells the multiplier (number of zeros to be added or the placement of the decimal point). A fourth color band is used for tolerance designation. The absence of the fourth color band means 20 percent tolerance.

In Fig. 6-1 the resistor is coded red, red, black, gold. Its value would be 22 Ω at 5 percent tolerance.

In the case of a resistor whose value is less than 1 Ω, the multiplier is silver (band or dot). In the case of a resistor whose value is greater than 1 Ω but less than 10 Ω, the multiplier is gold.

TABLE 6-1. Resistor Color Chart

Resistors, EIA and MIL		Significant figures	Color
Multiplier	Tolerance, %		
1	—	0	Black
10	—	1	Brown
100	—	2	Red
1,000	—	3	Orange
10,000	—	4	Yellow
100,000	—	5	Green
10^6	—	6	Blue
10^7	—	7	Violet
10^8	—	8	Gray
10^9	—	9	White
0.1	5	—	Gold
0.01	10	—	Silver
—	20	—	No color

RESISTOR

Figure 6-1. Resistor color code (*Centralab, a Div. of Globe-Union, Inc.*).

Fig. 6-3. A potentiometer is a three-terminal device: (*a*) circuit symbol; (*b*) end view showing shaft and terminals.

Wire-wound, high-wattage resistors usually are not color coded, but have the ohmic value and wattage rating printed on the body of the resistor.

Resistors with brown body color are insulated; those with black body color are not insulated.

In writing the values of resistors, the following designations are employed:

- k, a multiplier which stands for 1,000
- M, a multiplier which stands for 1,000,000

For example, 33 kilohms (33 kΩ) stands for 33,000 Ω; 1.2 megohms (1.2 MΩ) stands for 1,200,000 Ω.

Variable Resistors

In addition to fixed-value resistors, variable resistors are used extensively in electronics. There are two types of variable resistors, the rheostat and the potentiometer. Volume controls used in radio and the contrast and brightness controls of television receivers are typical examples of potentiometers.

A rheostat is essentially a two-terminal device whose circuit symbol is shown in Fig. 6-2. Points A and B connect into the circuit. A rheostat has a maximum value of resistance, specified by the manufacturer, and a minimum value, usually zero ohms. The arrowhead in Fig. 6-2 indicates a mechanical means of adjusting the rheostat so that the resistance, measured between points A and B, can be set to any intermediate value within the range of variation.

The circuit symbol for a potentiometer (Fig. 6-3a) shows that this is a three-terminal device. The resistance between points A and B is fixed. Point C is the variable arm of the potentiometer. The arm

is a metal contactor which moves along the uninsulated surface of the resistance element, selecting different lengths of resistive surface. Thus, the longer the surface between points A and C, the greater is the ohms resistance between these two points. Similarly, the resistance between points B and C varies as the length of element included between points B and C.

The axiom which states that the whole is equal to the sum of its parts applies to a potentiometer as well as it does to geometric figures. In this case it is apparent that the resistance R_{AC} from A to C, plus the resistance R_{CB} from C to B, make up the fixed resistance R_{AB} of the potentiometer. The action of the arm, then, is to increase the resistance between C and one of the end terminals, and at the same time to decrease the resistance between C and the other terminal, while the sum of the two resistances R_{AC} and R_{CB} remains constant.

A potentiometer may be used as a rheostat if the center arm and one of the end terminals are connected into the circuit, and the other end terminal is left disconnected. Another method of converting a potentiometer into a rheostat is to connect a piece of hookup wire between the arm and one of the end terminals, for example C connected to A. The points B and C now serve as the terminals of a rheostat. (When two points in a circuit are connected by hookup wire, these points are *shorted* together.)

Measuring Resistance

This is one of the functions of an EVM. Each manufacturer provides operating instructions for the use of his particular instrument. Hence, it will be necessary to refer to the instruction manual before using any electronic voltmeter. The student should be thoroughly familiar with the operation of the ohmmeter function before he attempts to use it in this experiment.

To measure resistance, the function switch should

Fig. 6-2. A rheostat is a variable resistor with two terminals.

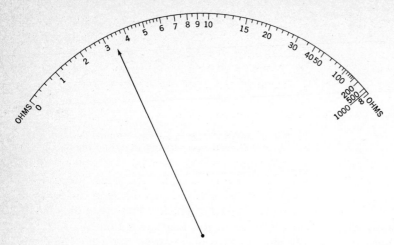

Fig. 6-4. Ohmmeter scale. On $R \times 1$ pointer shows 3.4 Ω.

be set to Ohms. Next, before using the ohms function of an electronic meter, the student should adjust to their proper settings the Ohms and the Zero controls of the meter. He is then ready to make resistance and continuity checks. Now to measure the resistance between two points, say A and B, one of the ohmmeter leads is connected to point A, the other to point B. The meter pointer then indicates, on the ohms scale, the value of resistance between A and B. If the meter reading is zero ohms, points A and B are "short-circuited." If, however, the meter pointer does not move (that is, if the indicator points to "infinity" on the ohms scale), points A and B are "open-circuited," that is, there is an infinite resistance between them.

Reading the Ohms Scale

All EVMs contain a basic ohms scale (Fig. 6-4) from which readings are made directly on the $R \times 1$ range of the meter. Figure 6-4 shows that the ohms scale is nonlinear; that is, the arc distance between consecutive graduations is not equal. Thus the arc distance between 0 and 1 is much greater than the distance between 9 and 10, though each arc represents, in this case, a change of 1 Ω.

As in the case of the voltmeter scale in Experiment 5, the operator is required mentally to supply numbers for the unnumbered calibrations. If the pointer is on the second graduation to the right of 3, between 3 and 4, as in Fig. 6-4, the corresponding ohms value on the $R \times 1$ range is 3.4 Ω.

Note that the ohms scale becomes fairly crowded to the right of the 100-Ω division. If a resistance greater than 100 Ω is to be measured with some degree of accuracy, the meter range should be switched to $R \times 10$, $R \times 100$, or $R \times 1,000$, de-

pending on the actual resistance to be measured. These three ranges, $R \times 10$, $R \times 100$, $R \times 1,000$, will usually be found on the meter. In the $R \times 10$ range, any reading made on the basic scale must be multiplied by 10. In the $R \times 100$ range, any reading must be multiplied by 100, etc.

NOTE: After switching from one ohms range to another, the settings of the Zero and Ohms adj. should be checked and reset if necessary.

SUMMARY

1. The unit of resistance is the ohm.
2. The body of a fixed carbon resistor is color-coded to specify its ohmic value and tolerance.
3. Twelve colors are contained in the color chart. These give the value of the significant figures of resistance and its tolerance. Refer to Table 6-1 for the resistor color chart.
4. The first three of four parallel color bands on the body of a carbon resistor designate its resistance, the fourth band its tolerance (see Fig. 6-1).
5. High wattage, wire-wound resistors are not color coded, but have the resistance and wattage value printed on the body of the resistor.
6. Variable resistors are of two types, the rheostat and the potentiometer.
7. A rheostat is a two-terminal device whose resistance value may be varied between the two terminals.
8. A potentiometer is a three-terminal device. The resistance between the two end terminals is fixed. The resistance between the center terminal and either end terminal is variable.
9. Ohmmeters or the Ohms functions of a VOM

or EVM are used to measure resistance and continuity.

10. The scale of an ohmmeter is nonlinear.
11. Ohmmeters or the Ohms function of a VOM or EVM have several ohms ranges ($R \times 1$, $R \times 10$, $R \times 100$, etc.).

SELF-TEST

Check your understanding by answering these questions:

1. A color code is used to indicate the _____ _____ of a carbon resistor.
2. If the color red appears in either the first or second color band on a resistor, it stands for the number (significant figure) _____.
3. If the color yellow appears in the third band on a resistor, it stands for the multiplier _____.
4. If the color _____ appears on the fourth band on a resistor, it indicates a tolerance value of 10 percent.
5. A resistor coded brown, black, black, gold has a value of _____ ohms, and a tolerance of _____ percent.
6. A high-wattage resistor is color coded in the same way as a low-wattage resistor _____ (true/false).
7. A potentiometer has _____ terminals.
8. A rheostat is a _____ _____.
9. A resistor whose value is 120 Ω and whose tolerance is 20 percent is color coded _____ _____ _____ _____.
10. If a resistor measures *infinite* ohms, the resistor is _____-circuited.

MATERIALS REQUIRED

- Equipment: EVM
- Resistors: Ten assorted values; 10,000-Ω potentiometer
- Miscellaneous: A piece of hookup wire and hand tools

PROCEDURE

1. Determine the value of each resistor supplied from its color code. Fill in the information required in Table 6-2.
2. Refer to the instruction manual of the EVM for the procedure in measuring resistance. Zero the ohmmeter. Measure each resistor with the ohmmeter, and fill in the results in the row "Measured value." The coded value and the measured value should agree within the tolerance range of the resistor. Indicate percentage of *accuracy* between coded and measured values.
3. Measure and record the resistance of a small piece of uninsulated wire. _____ Ω
4. (*a*) Measure and record the resistance of one of the resistors supplied. _____ Ω

TABLE 6-2.

| | Resistor | | | | | | | | | |
	1	2	3	4	5	6	7	8	9	10
1st color										
2d color										
3d color										
4th color										
Coded value, ohms										
Tolerance, %										
Measured value, ohms										
Accuracy										

Fig. 6-5. Resistor short-circuited.

(b) Connect the uninsulated wire across this resistor, as in Fig. 6-5. We say the resistor has been "short-circuited."

(c) Measure and record the value of this combination (i.e., the resistance from A to B). _____ Ω

5. Remove the wire from the resistor. Cut the wire in half. Place the pieces of wire near each other, but not touching. (This is an open circuit.) Measure and record the resistance between the two pieces. _____ Ω

6. Examine the potentiometer assigned to you. Orient it so that the rotatable shaft comes out toward you. Call the terminals of the potentiometer A, B, and C as in Fig. 6-3b. Measure and record in Table 6-3 the total resistance R_{AB} between A and B. Vary the arm of the potentiometer, while keeping the ohmmeter connected across AB. Does the total resistance vary? Indicate effect in Table 6-3.

7. Connect the ohmmeter terminals across AC. Turn the potentiometer control completely clockwise (CW). Measure and record the resistance R_{AC} (between points A and C), also the resistance R_{BC} (between points B and C). Compute and record the value of $R_{AC} + R_{BC}$.

8. Now observe how the resistance R_{AC} varies as the potentiometer arm is turned from its CW position to complete CCW position; how R_{BC} varies over this same range. Record the CW and CCW values for R_{AC} and R_{BC}. Compute and record $R_{AC} + R_{BC}$ in each case.

9. Set the control one-quarter of the way CW. Measure and record R_{AC} and R_{BC} in Table 6-3. Compute and record $R_{AC} + R_{BC}$.

10. Set the control three-quarters of the way CW. Measure and record R_{AC} and R_{BC} in Table 6-3. Compute and record $R_{AC} + R_{BC}$.

11. Set up an experiment and study the effect on the resistance R_{AC} when point B is shorted to C; the resistance R_{BC} when A is shorted to C. Set up a table and list the data for your experiment.

QUESTIONS

1. What resistance is in the center of your ohm scale, $R \times 1$ range?
2. At which end of the scale are resistance readings more reliable, the crowded or the uncrowded end?
3. To what range should you shift if your readings are at the crowded end of the scale?
4. What is the symbol used for ohms?
5. For carbon resistors what is the color code for (a) 0.27 Ω? (b) 2.2 Ω? (c) 39 Ω? (d) 560 Ω? (e) 33,000 Ω?
6. What is meant by a short circuit? Open circuit?
7. How much resistance would you measure across the terminals of a short circuit? Open circuit?
8. How does a potentiometer differ from a rheostat?
9. How can a potentiometer be used as a rheostat?
10. Explain the significance of your measurements in procedural steps 6 through 10.
11. State your exact procedure in procedural step 11. Include your table and readings.
12. Comment on the results of your findings in procedural step 11.

Answers to Self-Test

1. ohmic value
2. 2
3. 10,000
4. silver
5. 10 Ω; 5 percent
6. false
7. 3
8. variable resistor
9. brown; red; brown; no color
10. open

TABLE 6-3.

Step	Setting of potentiometer control	R_{AB}, ohms	R_{AC}, ohms	R_{BC}, ohms	$R_{AC} + R_{BC}$ Computed value
6	Vary over its range		X	X	X
7	Completely CW	X			
8	CW to CCW	X			
9	¼ CW	X			
10	¾ CW	X			

7

EXPERIMENT

Dry Cells and Measurement of dc Voltage

OBJECTIVES

1. To measure the voltage of a dry cell using a voltmeter
2. To measure the voltage across the combinations which result from connecting dry cells in series and in parallel

INTRODUCTORY INFORMATION

Measuring dc Voltage

Electronics technology is inconceivable without measurement of electrical quantities. Thus, circuit analysis and understanding of circuit function are facilitated by the basic measurements of voltage, current, and resistance.

The technician is the "know-how" man on the team which may include scientists, engineers, and skilled craftsmen. (Of course, he must also "know why.") He will therefore be the man primarily involved in making the measurements which will test the scientists' or engineers' theories. He will test the operation and characteristics of an experimental circuit. He will make measurements in repairing a device which is not working. It is therefore essential that the technician know the instruments of his technology, and that he understand how to use them. He must be aware of the effect an instrument may have on the circuit, the errors of measurement, and he must also know how to interpret the results of his measurements.

Measurement of dc voltage is basic in all electronics work. In this experiment we will be concerned with learning how to use the voltage function of an electronic voltmeter. We will then employ our newly acquired skill in studying the characteristics of dry-cell arrangements.

Voltage is defined as electrical pressure. It is the *difference* in electrical pressure between two points. It is also called potential difference and electromotive force (emf). The voltage *across* two points is measured with a voltmeter. An electronic voltmeter combines the functions of measuring resistance and voltage, and in some instances current. Hence its voltage-measuring functions may be used here. The student should refer to the manufacturer's operating instructions for the use of his particular meter before using it in this experiment.

Generally, the following facts apply, regardless of the make of VTVM or TVM:

1. A VTVM must be turned on and allowed to warm up. A TVM requires no warm-up.
2. The meter test leads should be plugged into the proper meter terminal jacks. The black lead is the common or ground lead, and the probe is the voltage or "hot" lead.
3. The + or − polarity switch is set to agree with the polarity of the voltage being tested. For example, when measuring voltages which are positive with respect to common, the function selector switch is set to the + dc position.
4. The dc voltage probes are shorted together and the "Zero" adjustment is used to set the meter pointer to zero. Zero is usually the left-hand side of the meter scale. (On zero-center meters, zero is the middle of the scale.)
5. The common lead is connected to the return (negative) of the circuit under test. This test lead is usually connected first.
6. The range selector switch should be set to the highest dc voltage range in measuring an unknown dc voltage. This is to avoid damaging the meter. If measurement indicates that the voltage is too small to determine its value accurately, a lower dc voltage range should be selected.
7. The dc voltage probe is then connected to the circuit under test. (The insulated probe and

Fig. 7-1. Dry cells connected in series (aiding). The negative terminal of one cell is connected to the positive terminal of the next. The battery voltage V_B is measured across the terminals *AB*.

Fig. 7-3. Dry cells connected in series-parallel.

not the metal prod should be held in using the voltmeter.)

8. The voltage value is read from the proper voltage scale. The maximum voltage for a specific voltage range is at the right end of the scale.

9. The dc voltage scale on an electronic voltmeter is separate from the *ohm* scale. The meter pointer moves in an arc from left to right.

10. The dc voltage scale is linear, with equal spacings for equal voltage changes. There may be two or three scale calibrations along the dc scale, making it possible to read voltage on all the dc ranges of the instrument. The greater the deflection toward the right, the higher the voltage that is being measured, and the greater the meter accuracy. A voltage scale should be selected that will provide maximum pointer deflection without exceeding full-scale deflection.

Dry Batteries

Dry batteries consist of arrangements of primary cells, called "dry cells." The familiar flashlight

"battery" is really a dry cell. Individual dry cells produce a low voltage.

By connecting dry cells in a *series-aiding* arrangement (Fig. 7-1), we produce a dry battery whose voltage is the sum of the dry-cell voltages. By connecting two or more dry cells in parallel (Fig. 7-2), we produce a battery whose voltage is the same as that of an individual dry cell. However, the permissible current drain of this parallel-cell battery is proportionately greater than that of any dry cell used alone in the same application. Dry batteries usually consist of a series-parallel arrangement of cells (Figs. 7-3 and 7-4). These arrangements make possible a battery whose voltage is higher than that of an individual dry cell, and one whose life is longer.

Dry cells and batteries are not always connected

Fig. 7-2. Dry cells connected in parallel. The negative terminals of the cells are connected together, and the positive terminals of the cells are connected. The output voltage V_B is measured across the terminals *AB*.

Fig. 7-4. Four dry cells connected in series-parallel (another arrangement).

Fig. 7-5. Two batteries connected in series-opposing. The negative terminal of one is connected to the negative terminal of the other. The positive terminals are free. The resultant voltage V_B is measured across AB.

in series-aiding. They are sometimes used in *series-opposing*. The student should therefore understand the effects of such a connection.

Consider the arrangement of Fig. 7-5. The two batteries are in series-opposing, not in series-aiding. That is, the resultant voltage V_B delivered across AB is equal to the difference between the battery voltages. In the illustration, terminal A is 9 V positive with respect to terminal C. Terminal B is 3 V positive with respect to C. Hence, terminal A is 6 volts positive relative to terminal B. Point A is therefore labeled "+," and point B "−."

SUMMARY

1. An electronics technician must measure voltage, current, and resistance in a circuit so that he can determine if the circuit is working properly.
2. He also makes circuit measurements in experimental circuits to determine their characteristics.
3. The term "electronic voltmeter" is understood to mean a transistorized voltmeter (TVM) or a vacuum-tube voltmeter (VTVM). Either instrument can be used to measure voltage and resistance. In addition, current can be measured with a TVM.
4. Voltage is the difference in electrical pressure between two points. Voltage is also called potential difference, or electromotive force (emf).
5. Voltage is measured by connecting a voltmeter *across* the two points where a potential difference is present.

6. In measuring an unknown voltage, the range-selector switch is first set to the highest voltage range; then, if necessary, a lower range is selected.
7. The *insulated probe* and not the metal tip is held in using a voltmeter.
8. The basic dry batteries consist of combinations of individual cells called dry cells.
9. The total voltage of a battery produced by adding dry cells in *series-aiding* is the sum of the individual cell voltages.
10. If similar dry cells are connected in parallel, the total voltage of this battery is the same as the voltage of any one cell. The advantage gained in this connection is that the life of the battery is increased (by a factor approximately equal to the *number* of parallel cells).

SELF-TEST

Check your understanding by answering these questions:
1. In an electronic circuit, three measurable quantities are _____ , _____ , and _____ .
2. A TVM can be used for measuring _____ , _____ , and _____ .
3. The man *primarily* concerned with making circuit measurements is the _____ . He must also know how to interpret these measurements.
4. The polarity switch on an electronic voltmeter may be set to _____ or _____ .
5. The _____ _____ switch of an electronic voltmeter is used to select the desired voltage range.
6. The dc voltage scale is _____ (linear/non-linear).
7. The basic unit for forming dry batteries is the _____ _____ .
8. The total voltage of dry cells connected in series depends on whether they are connected in series-aiding or series-opposing. _____ (true/false)

MATERIALS REQUIRED

- Equipment: EVM (electronic voltmeter)
- Miscellaneous: Four 1½-V dry cells, connecting wires

PROCEDURE

NOTE: Before using the electronic voltmeter, be certain that you have read and are familiar with the operating instructions. Set the meter switches properly and have them checked by an instructor. You may then proceed.

1. Connect the negative lead of the meter to the negative terminal of the cell, positive lead to positive. Measure and record in Table 7-1 the voltage of each of the dry cells supplied. Use that range of the voltmeter where you get the maximum pointer deflection without going off scale.
2. Repeat the measurements of step 1 using the next higher range of the voltmeter. Record in Table 7-2.
3. Connect cells 1 and 2 in series-aiding. Measure and record in Table 7-3 their total voltage (*A* to *B*).
4. Connect cells 1, 2, and 3 in series-aiding. Measure and record their total voltage (*A* to *B*).
5. Repeat this process, measuring and recording the voltage of four cells in series-aiding.
6. Connect two cells in parallel. Measure and record their voltage (*A* to *B*).
7. Connect three cells in parallel. Measure and record their voltage (*A* to *B*).
8. Connect four cells in series-parallel (Fig. 7-4). Measure and record their voltage (*A* to *B*).
9. Connect the cells in the series-parallel arrange-

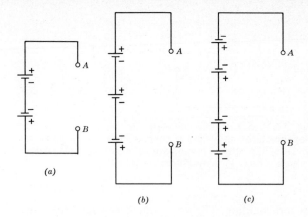

Fig. 7-6. Dry cells connected in series-opposing.

ment of Fig. 7-3. Measure and record their voltage (*A* to *B*).
10. Connect dry cells in series-opposing (Fig. 7-6*a*). Measure and record their output voltage (*A* to *B*). Repeat for Fig. 7-6*b* and *c*. Before connecting the meter, determine which is the positive terminal, *A* or *B*.

QUESTIONS

1. Name four precautions which must be observed in measuring voltage.
2. List the dc voltage ranges on your VTVM or TVM.
3. How much voltage is required to give full-scale deflection of the pointer on a 300-V range?

TABLE 7-1.

Dry-cell number	1	2	3	4
Voltage				

TABLE 7-2.

Dry-cell number	1	2	3	4
Voltage				

TABLE 7-3.

Step	3	4	5	6	7	8	9	10*a*	10*b*	10*c*
Voltage *A* to *B*										

4. In the voltage measurements of dry cells in this experiment, why must the hot lead of the meter be connected to *A* and the common lead to *B?*
5. What would happen to a dry cell or battery if the positive and negative terminals were shorted?
6. What arrangement of six dry cells gives the maximum life, regardless of voltage?
7. What arrangement of six dry cells gives the maximum voltage?
8. Draw a practical arrangement of ten 1.5-V dry cells to give a battery of 7.5 V.
9. Explain the difference in connection between two dry cells connected in series-aiding and in series-opposing.

Answers to Self-Test

1. voltage; current; resistance
2. voltage; current; resistance
3. electronics technician
4. + or −
5. range selector
6. linear
7. dry cell
8. true

EXPERIMENT 8

Familiarization with dc Power Supply

OBJECTIVES

1. To identify the controls and switches of a dc power supply
2. To turn the supply on and measure its output voltage
3. To measure the range of voltage change as the output control is varied from minimum to maximum

INTRODUCTORY INFORMATION

DC voltage and current sources are required for experiments in the laboratory. Electronic, regulated, variable power supplies satisfy this requirement and are used extensively in both school and industrial laboratories.

DC Power Supply

A variable voltage-regulated supply is one which can be manually adjusted to deliver any required voltage within its range of operation. The voltage put out by this supply remains constant despite changes in load current, within specified limits. Thus, one manufacturer states that his supply will deliver 0 to 400 V at 150 mA. This means that the output voltage can be varied from 0 to 400 V, and that the current drawn must not be greater than 150 mA.

A power supply may have facilities for providing two or more independent dc voltages, in which case separate controls and separate output terminals are contained on the instrument.

The polarity of the dc terminals on the supply is usually marked either −, +, or gnd and V⁺. A red jack is conventionally used for the positive and a black for the negative terminal of the supply.

Low-voltage, High-current DC Source

Figure 8-1 shows the front panel of a variable, regulated, low-voltage, high-current dc power source. This and equivalent sources are used for solid state experiments. This all solid state device is line-operated (117 V ac) and requires no warm-up time.

The specific model shown has two switch-selected output ranges, the first 0 to 20 V at a maximum current of 1.5 A, the second 0 to 40 V at 0.75 A. The panel meter displays, respectively, either output voltage or current. A dual concentric output control is both a coarse and a fine control of output voltage. One additional control requires mention. This is the "current" control. Maximum clockwise rotation of this control will provide the maximum current which the supply can deliver. Or the experimenter can limit the supply current by setting the current control at the required level. The purpose of the current control is to protect the experimental set up from excessive current. An on-off switch turns the equipment on and off.

There are many other types of low-voltage, high-

Fig. 8-1. Front panel of a variable, regulated, low-voltage, high-current dc power supply (*Hewlett Packard*).

current sources, including one with two sets of controls which delivers two independent voltages at two independent sets of terminals.

High-voltage, Low-current DC Sources

These sources are available for vacuum-tube and other circuits requiring higher voltages and lower currents. They normally provide variable, regulated dc voltages and fixed low-level, high-current ac voltages.

Figure 8-2 is the front panel of a supply which provides two dc and two ac outputs. One of the dc outputs is regulated and variable from 0 to 400 V at 100 mA maximum. The other dc output is non-regulated and variable from 0 to 100 V at 1 mA. The nonregulated supply *will not maintain* a steady preset output voltage when current is drawn from it. Its output voltage will drop when loaded.

The three ac output jacks at the lower left provide either 6.3 or 12.6 V, depending on which two jacks are used. The left jack and the center jack deliver 6.3 V. The right jack and the center jack provide 12.6 V.

The supply shown is line-operated and has an on-off switch on the front panel and a dc-on switch which also includes a standby position. The switch should be in the standby position when the supply is first turned on since this supply is tube-operated, and requires warm-up time before it can be used. Other controls include: voltage control for the regulated B^+ volts output; voltage control for the nonregulated C^- volts output; a meter switch which can be set to measure either the regulated or nonregulated dc voltage delivered at the respective output jacks; and a current meter to measure the output current.

Using the DC Power Supply

The following general precautions should be observed in using this device.

1. Read the operating instructions carefully and be certain that you understand them before turning a power supply on for the first time.
2. Never short the dc or ac output terminals, or you may damage the supply. To prevent the terminals from shorting, keep the leads issuing from these terminals from making contact.
3. Set the output voltage of the supply at the required level *before* connecting the supply leads to the experimental circuit. After the leads are

Fig. 8-2. Front panel of a variable, regulated, high-voltage, low-current dc power supply (*Heath Company*).

connected to the circuit, again measure the output voltage to see that it is still at the preset level.
4. If any component on the experimental circuit appears to be overheating, after power is applied, turn the supply **off** and determine the cause before proceeding with the experiment.
5. Do not grasp the uninsulated output terminals of a supply when it is on. Your body will receive the output voltage and you may experience a severe electrical shock.
6. To prevent damage to the power supply, do not switch the supply on and off excessively. If an experiment requires power to be interrupted frequently, use an external switch on the breadboard to apply and remove power from the circuit.
7. Do not operate the supply beyond its *rated* current capacity. If the current meter indicates that you are exceeding the current capability of the power supply, turn it **off** and check the experimental circuit to determine why it is drawing excessive current.

SUMMARY

1. Electronic dc voltage and current sources are used in the laboratory. They are called power supplies.
2. Electronic dc supplies derive their power from the ac line. We say they are line-operated.
3. A voltage-regulated supply is one which maintains a preset voltage level, within its range of operation, despite variations in output current and line voltage.
4. A nonregulated supply is one whose output voltage changes with changes in load current and line voltage.
5. In the school laboratory are generally found two types of supplies:

(a) Variable low-output voltage, high-output current source needed for solid state experiments

(b) Variable high-output voltage, low-output current source used in vacuum-tube experiments.

6. Power supply output terminals should *not* be shorted.

7. Power supplies should not be operated beyond their rated current capacity.

SELF-TEST

Check your understanding by answering these questions:

1. The output voltage of dc supplies found in the laboratory is made _____ to accommodate a variety of voltage needs.

2. If a power-supply voltage remains constant at a preset level, despite changes in supply current or line voltage, we say the supply is _____.

3. Variable low-voltage, high-current supplies are normally used for _____ _____ circuits.

4. The output terminals of a power supply must *not* be _____ _____ when the supply is on.

5. The _____ capacity of a supply should not be exceeded.

MATERIALS REQUIRED

- Power supply: Variable dc voltage-regulated power supply
- Equipment: Electronic voltmeter
- Resistors: 250-Ω 4-W
- Miscellaneous: SPST switch, breadboarding device

PROCEDURE

1. Draw the front panel of the power supply assigned to you. Label all switches, controls, and output terminals. Read the operating instructions and proceed only after they are clearly understood. In Table 8-1 name and describe the function of each switch, control, and terminal.

2. Power **on**. Permit the supply to warm up. Set regulated voltage control for maximum voltage as indicated on the panel meter of the power supply. In Table 8-2 indicate the panel meter voltage. Does the panel current meter show current?

TABLE 8-1.

Control, Switch, or Terminal	Function

CAUTION: Be certain that no output terminals are short-circuited.

3. Turn **on** an electronic voltmeter. After it is properly zeroed, set the meter on a voltage range equal to or just greater than the maximum rated voltage of the power supply. Connect the negative lead of the meter to the negative terminal of the regulated supply, the positive meter lead to the positive terminal. Measure and record in Table 8-2 the output voltage.

4. Reduce the supply voltage to its minimum level. Measure, both on the panel meter and on the electronic voltmeter, the lowest voltage the power supply provides. Record these measurements in Table 8-2.

NOTE: Reduce the range of the EVM when necessary to secure a more accurate measurement.

5. Set the output of the regulated supply to half its maximum voltage. Record in Table 8-2 the EVM and panel meter readings. Indicate also if panel current meter shows current.

6. Set the measured output of the regulated supply

TABLE 8-2.

Step	Position of Output Voltage Control	Output Voltage		Supply Current	
		Panel Meter Volts	EVM Volts	Yes	No
2,3	Maximum				
4	Minimum				
5	Half maximum				
6	X	X	20 V		
7	X	X			

at 20 V. Connect the circuit of Fig. 8-3. *V* is an electronic voltmeter used to measure the output voltage of the power supply. *S* is an on-off switch on the breadboard. *R* is a 250-Ω 4-W resistor. Switch *S* is **open**. Does the panel current meter show current?

7. Close switch *S*. Measure and record in Table 8-2 the output voltage of the supply. Record also if the panel current meter shows current.

8. If there is another dc voltage source on the power supply, check it in a manner similar to steps 2 through 7. Construct a new Table 8-3, similar to Table 8-2, and enter all your measurements.

QUESTIONS

1. What is the dc voltage rating of the power supply assigned to you? Current rating? If there is more than one dc output, give the ratings of each.

Fig. 8-3. Circuit to determine load, no-load characteristic of a regulated power supply.

2. What is meant by short-circuiting the output terminals of a power supply?
3. Why must we be careful to keep from short-circuiting the output terminals?
4. In addition to the voltage source (battery or power supply), is it necessary to have a complete circuit for voltage to be present? Relate your answer to the measurements in this experiment.
5. Compare your voltage measurements in Table 8-2 with the (*a*) panel meter and (*b*) EVM. Are they substantially the same?
6. How can you tell which meter is correct, the panel meter or the electronic voltmeter?
7. When is it necessary to switch the voltmeter ranges in measuring voltage?
8. Refer to your measurements in Table 8-2, steps 6 and 7. Does the voltage of the supply hold at 20 V, with and without load current?
9. Refer to Table 8-2. Under what conditions did the supply register current? What supplied this current?
10. Name at least two advantages a dc power supply has over a dry cell.

Answers to Self-Test

1. variable
2. regulated
3. solid state
4. short-ciruited (or shorted together)
5. current

OBJECTIVES

1. To identify conductive and nonconductive materials and components
2. To measure the resistance of a variety of conductive and nonconductive electrical and electronic components

INTRODUCTORY INFORMATION

Conductors and Insulators

The useful effects of electricity result from the movement of electrical charges in a circuit. This movement of electrical charges is called "current." Electrical charges move easily through paths called "conductors," but it is very difficult for these charges to move through *insulators*. Another way of saying this is to state that conductors are materials which permit current to flow easily with little electrical pressure (voltage) applied, while insulators are materials which permit very little or no current to flow. Copper is an example of a familiar conductor, while rubber is an insulator.

A complete (closed) circuit or path for dc current consists of a voltage source such as a battery or dc power supply, a conductive load such as an electric bulb, and connecting conductors (copper wire) (Fig. 9-1). In such a closed circuit there is electric current, and the bulb will light. If rubber cords were used to tie the battery to the bulb, instead of copper wires, the bulb would not light, because rubber is an insulator and does not permit current to flow through it in the circuit. The first illustration describes a *complete* or *closed* circuit, the second an *incomplete* or *open* circuit.

The tungsten filament in the electric bulb is a conductor, but it is not as good a conductor as the connecting copper wires. Tungsten is used for the filament because its properties cause it to glow and give off light when it is heated by an electric current.

It is apparent that not all conductive materials are equally good conductors of electricity. They are used in constructing electrical components because of their additional desirable characteristics.

Wire Conductors

The ability of a conductor to permit current flow is designated conductance (G), while opposition to current is called resistance (R). The ohm (Ω) is the unit of resistance. Various materials exhibit different resistivity. Thus the resistance of round wire of the same diameter increases in the following order, depending on the material of the wire: silver, copper, gold, aluminum, tungsten, nickel, iron, nichrome, etc. That is, silver has the greatest conductivity, hence the least resistance. The conductivity of copper is less than that of silver, therefore the resistance of copper wire is greater than that of silver, for the same cross-sectional area and length. However, copper wire is widely used as a conductor because of its relatively low cost and low resistance.

The resistance of round wire depends not only

Fig. 9-1. A complete (closed) electric circuit containing a battery *V*, connecting wires, and an electric bulb as load.

on the material from which it is made, but also on its length, cross-sectional area, and temperature. A formula which relates all these factors is:

$$R = \rho \frac{l}{A} \qquad (9\text{-}1)$$

where R = resistance of the wire in ohms
  l = its length in feet
  A = its cross sectional area in circular mills (CM)
  ρ = its resistivity (depends on the material) at a specified temperature

It should be noted that the area A in circular mils of a round wire whose diameter is D mils is:

$$A_{CM} = (D_{mils})^2 \qquad (9\text{-}2)$$

An example will explain how the formula is applied.

Problem

If the resistivity ρ of copper wire at 20°C equals 10.37, find the resistance R of 100 ft of such wire whose diameter $D = 0.005$ in.

Solution. A mil is defined as 0.001 in. Hence here $D = 5$ mils. Therefore, $A_{CM} = D^2 = 25$. Now since

$$R = \rho \frac{l}{A}$$

$$R = 10.37 \times \frac{100}{25} = 41.48 \ \Omega$$

Round wire is specified by gage. The Bureau of Standards has published a table of American Wire Gages (AWG) to which the technician can refer for calculations. AWG numbers decrease in diameter from gage #0000 to gage #40. The diameter of AWG 0000 is 460 mils. The diameter of AWG 40 is 3.1 mils. The diameter of AWG 36 is 5 mils. The resistance of 1,000 ft of AWG 0000 at 20°C is 0.0490 Ω, and the resistance of 1,000 ft of AWG 36 at 20°C is 414.8 Ω. If the technician wished to calculate the resistance of 100 ft of AWG 36 copper wire, he would simply refer to the AWG table. In this case, since the resistance of 1,000 ft of AWG #36 is 414.8 Ω, 100 ft would have a resistance of

$$\frac{100}{1,000} \times 414.8 = 41.48 \ \Omega$$

This is the same result as that obtained by applying Eq. (9-1).

In addition to giving the resistance of 1,000 ft of each gage wire, AWG tables also give the weight of

Fig. 9-2. A complete (closed) electric circuit containing a battery V, connecting wires, and a resistor R as load.

1,000 ft of each wire and its current capacity. Current capacity depends on the cross-sectional area A and on the type of insulation used.

Wire Components

Some components are made by winding a length of wire around a form. For example, ceramic high-power resistors consist of a wire winding around a ceramic core. Iron-core inductors (chokes) are wound around an iron core. Transformers consist of several windings around some specified core. If an ohmmeter is placed across each winding of these components, it will measure the resistance of that winding. Therefore, one way of determining if a wire winding is continuous, that is, if it is not broken, is to measure its resistance across the two end terminals. The normal resistance is usually specified by the manufacturer. If a winding measures infinite resistance, it is open.

Carbon Conductors—Resistors

In a previous experiment we learned how to determine, from their color code, the resistance of carbon resistors. In this connection we must note that carbon is also a conductor of electricity. A complete circuit which contains a battery or other dc power source, connecting wires, and a resistor (Fig. 9-2) will permit direct current to flow.

The physical construction of R will determine its resistance. Here as in circular wire the resistance of R will vary inversely with the cross-sectional area of the carbon element in the resistor, and directly with the length of the element. By controlling the diameter and length of the inner carbon element, we can control the resistance of R.

Comparing a resistor with a copper wire, we note that the resistance of the wire is *distributed* throughout its entire length. In a small carbon resistor, the resistance is lumped between the two terminals.

In the circuit of Fig. 9-2 the current can be controlled by controlling the value of R, in ohms. With V_B held constant, the higher the resistance in ohms of R, the lower will be the current. That is, as the resistance of R increases, the opposition to current increases, and the current decreases.

NOTE: Any object whose resistance can be measured with an ohmmeter is a conductor of direct current.

Insulators

All metals are conductors of electricity, though it is evident from the preceding discussion that all metals do not conduct equally. There are some materials whose resistivity is so high that they permit *very, very* little current to flow. These materials are called insulators or nonconductors. Rubber was previously mentioned as an insulator. Wood, glass, paper, mica, bakelite, and air are other examples of insulators.

If the leads of an ohmmeter are connected across an insulator, say of a rubber cord, the meter will register infinite resistance. An infinite resistance will not permit direct current to flow in a circuit. Thus Fig. 9-3 is an open circuit, because direct current will not flow through the insulator.

Insulating materials are used to cover copper wires in electricity and electronics. The purpose of insulating electrical wires is to prevent them from making electrical contact (shorting) with any other conductor or component which they may touch accidentally.

NOTE: Wires whose insulation has melted or become frayed are safety hazards and should not be used.

Before leaving the subject of insulators we should note that insulators may break down and become conductors if the voltage across them is high enough. Thus an air gap between two voltage terminals which are 1 cm apart will break down if the

Fig. 9-3. An open circuit, because the rubber insulator will not permit direct current flow.

voltage across these terminals equals or exceeds 30,000 V. Another descriptive term frequently used for this phenomenon is "arc over." For example, in the high-voltage section of a television receiver, arc-over may occur under certain abnormal conditions.

Semiconductors

This discussion would not be complete without mention of semiconductors. A transistor is an example of a device made of semiconductors. Another familiar semiconductor device is a diode.

Silicon and germanium are the materials from which semiconductors are made. These two elements, in their pure state, are insulators. However, when they have been "doped" with certain impurities such as gallium or antimony, they can be made somewhat conductive, and are then called semiconductors. The resistivity of semiconductors depends on the nature and amount of impurity injected. In general, however, the resistivity of semiconductors is said to be between that of insulators and conductors. Semiconductor devices are also referred to as solid state devices, and that branch of electronics dealing with semiconductors is known as solid state electronics.

The Human Body as a Conductor

The human body is a conductor of electricity. The most certain proof is the electric shocks that people have experienced when coming in accidental contact with the + and − terminals of a voltage source. Electric current will flow through the body, and the higher the voltage, the higher the current. Electric shock is dangerous and can be fatal. That is why the student of electricity and electronics should follow the safety precautions he is taught in order to avoid shock.

It should be mentioned that the resistance of the body at a specific point in time can be measured with an ohmmeter. The method to follow will be given under Procedure.

SUMMARY

1. Conductors are materials which permit direct current to flow through them with ease. Silver, copper, and gold are examples of conductors.
2. Insulators are materials which do not permit direct current to flow through them. Rubber, glass, and mica are examples of insulators.

3. Conductors vary in their conductivity. Thus copper is a better conductor than nichrome.
4. Round copper wire is used to connect electrical components in a circuit.
5. The resistance of wire varies directly with its length and inversely with its cross-sectional area. Resistance also depends on the material from which the conductor is made.
6. The Bureau of Standards publishes a table of American Wire Gages (AWG) which gives the diameter in mils, area in circular mils, resistance in ohms of 1,000 ft, and weight of 1,000 ft of every copper wire gage made.
7. Insulators can withstand voltages up to a critical voltage. For voltages at or higher than the critical voltage, insulators break down and permit current flow by arc-over or arc-through.
8. The breakdown voltage of an insulator depends on its material.
9. The resistance of a good conductor is very low, of an insulator very high (infinite).
10. Semiconductors are solid state devices whose resistivity lies between that of a conductor and insulator. Silicon and germanium diodes and transistors are examples of semiconductors.
11. The human body is a conductor of electricity. Body contact with electrical wires must be avoided to prevent shock.

SELF-TEST

Check your understanding by answering these questions:
 1. The following materials are examples of good electrical conductors: _____; _____; _____; _____; _____.
 2. The following materials are examples of electrical insulators: _____; _____; _____; _____; _____.
 3. For current to flow in a circuit, it must have: (*a*) _____; (*b*) _____; and (*c*) _____.
 4. A circuit in which current flows is called a _____ circuit.
 5. Current will not flow in an _____ circuit.
 6. AWG 22 round copper wire has a _____ (larger/smaller) diameter than #12 wire.
 7. The resistance of #10 wire is _____ (higher/lower) than that of #40.
 8. A 100-ft piece of #22 wire has _____ the resistance of a 50-ft piece of the same wire.
 9. An insulator will _____ _____ if it is subjected to a voltage higher than its breakdown voltage.
10. Semiconductors will, under the proper conditions, permit electric current to flow. _____ (true/false)

MATERIALS REQUIRED

- Equipment: EVM
- Resistors: One carbon resistor, any value
- Capacitors: 0.01 μF
- Silicon rectifier: 1N2615
- Miscellaneous: 12 in. length of #40 copper wire, single strand; 6-in. length of #40 nichrome wire; 12-in. length of #40 nichrome wire; piece of rubber $1/4 \times 2$ in.; piece of wood $1/4 \times 2$ in.; piece of plastic $1/4 \times 2$ in.; 8-H choke

PROCEDURE

1. Identify each of the 10 objects listed in Table 9-1.
2. Set your electronic voltmeter on Ohms and zero the meter at both ends of the scale if necessary, as in Experiment 6.
3. Measure the resistance of object 1 (resistor) and record this value in Table 9-1, column *A*. Observe which lead you connected to the right-hand terminal of the resistor and which to the left.
4. Reverse the leads of the ohmmeter and again measure the resistance of object 1. Record this value in Table 9-1, column *B*.
5. Check in the proper column if this is a conductor or insulator of direct current.
6. Repeat steps 3 through 5 for every object shown in Table 9-1. Check your meter zero if you switch to another resistance range.

NOTE: Be certain you hold the insulated probes on the meter leads when measuring resistance. This becomes *particularly important* when using the *higher* ranges of the meter ($R \times 1,000$, etc.). The reason is that the ohmmeter will read your body resistance if you hold the metal tips of the meter leads. This measurement of body resistance will

TABLE 9-1. Resistance of Conductors and Insulators

Object	Resistance (Ohms)		Conductor	Insulator
	A	B		
1. Resistor				
2. Capacitor				
3. 12-in. length #40 copper wire				
4. 12-in. length #40 nichrome wire				
5. 6-in. length #40 nichrome wire				
6. Piece of rubber $\frac{1}{4} \times 2$ in.				
7. Piece of wood $\frac{1}{4} \times 2$ in.				
8. Silicon rectifier 1N2615				
9. 8-H choke				
10. Piece of plastic $\frac{1}{4} \times 2$ in.				

affect the resistance reading of the object you are measuring, particularly if the object has high resistance.

Body Resistance

7. Set the ohmmeter on the $R \times 10,000$ or $R \times 100,000$ range and check that it is properly zeroed.
8. Measure your body resistance by grasping the metal tip of one lead with one hand, and the metal tip of the other lead with your other hand. Record. Body resistance = _____ Ω.

QUESTIONS

Refer to your experimental results, where possible, in answering these questions.

1. How can one determine whether an object is a conductor of direct current?
2. Did reversing the meter leads affect the resistance of the objects in Table 9-1? List the object(s), if any, whose resistance was affected.
3. What conclusion can you draw about the conductivity of the silicon rectifier (1N2615)?
4. Which is a better conductor of direct current, the 2-in. copper wire, or the 2-in. nichrome wire? Why?
5. Which of your measurements, if any, tend to confirm the statement that the resistance of a wire varies

directly with its length, all other factors being equal [Eq. (9-1)]?
6. Which of your measurements, if any, tend to confirm the statement that the conductivity of all metals is not the same?
7. Briefly explain how you could experimentally confirm that part of Eq. (9-1) which states that the resistance of a solid round wire varies inversely with its cross-sectional area.
8. Explain how you can zero an ohmmeter at both ends of its scale.
9. Is the ohmic resistance of your body higher or lower than that of air? How do you know?
10. Assume the wires used in the circuit of Fig. 9-1 are rubber covered. Why is it necessary to trim the rubber off the ends of the wires before connecting them in the circuit?

Answers to Self-Test

1. silver; copper; gold; aluminum; tungsten, etc.
2. wood; glass; paper; mica; air, etc.
3. (a) voltage source; (b) conductive load; (c) connecting conductive wiring
4. closed
5. open
6. smaller
7. lower
8. twice
9. arc-over (or break-down)
10. true

10 EXPERIMENT

Direct-current Measurement and Control of Current

OBJECTIVES

1. To connect a current meter to measure direct current in a circuit
2. To measure the effect of resistance in controlling current in a circuit
3. To measure the effect of voltage in controlling current in a circuit

INTRODUCTORY INFORMATION

Resistance, Voltage, and Current

In Experiments 6 and 7 the uses of the ohmmeter and voltmeter for measuring resistance and voltage were studied. From the nature of the measurements, it was apparent that there were components called resistors whose resistance could be measured directly with an ohmmeter. The quantity of ohms of resistance was not dependent on the connection of that resistor in a circuit. The characteristic of resistance was associated with the component itself.

Similarly, in the measurement of emf (electromotive force; i.e., voltage), it was evident that voltage was a characteristic of some voltage source (dry cell, battery, power supply, etc.), and that it could also exist independently without need for a complete electric circuit.

Electric current differs from voltage and resistance in that it cannot exist by itself. A view of current is that it is the result of a movement of electric charges. A voltage source by itself is insufficient to create current. A voltage source *and* a closed (complete) path are required for the movement of these charges. The movement of electric charges, then, is restricted to the closed path (circuit) within which a voltage source can act.

The quantity of current in a circuit is dependent on the amount of voltage applied by the source of emf, and on the nature of the conductive path. If the path offers little opposition, the current is larger than it would be in a circuit where the opposition to current is greater. Opposition to direct current is called resistance. Resistance is measured in ohms. Current, then, can be controlled by the amount of resistance in the circuit.

Measuring Direct Current

Direct current in a circuit may be measured by means of a dc ammeter. When the magnitude of the current to be measured is small, a milliammeter [1 milliampere (mA) = one-thousandth of an ampere] or a microammeter [microampere (μA) = one-millionth of an ampere] is used. In measuring current the circuit must be physically broken or opened and the meter inserted in series with the circuit. Suppose, for example, it is required to measure current in the circuit of Fig. 10-1. The circuit is first broken at A (Fig. 10-2). The milliammeter is then inserted in series with the circuit at the two open leads (Fig. 10-3).

Polarity must be observed, that is, the *common* meter lead (negative) must be connected to the

Fig. 10-1. Circuit where current is to be measured.

Fig. 10-2. Breaking circuit at point *A*.

Fig. 10-3. Inserting milliammeter to measure current.

point of lower potential, and the "hot" lead (positive) to the point of higher potential. When the meter is connected properly, the pointer will move in an arc from left to right. If the pointer deflects in the opposite direction, the meter leads must be reversed. In Fig. 10-3 the negative lead of the meter is connected to *B*, the positive lead to *A*. After the measurement has been made, the meter is removed from the circuit, and the original circuit connections are restored.

CAUTION: The current meter must *not* be connected across (that is, in parallel with) any component. *It must always be connected in series* with the component to measure the current in the component. Failure to observe this rule may result in serious damage to the meter.

SUMMARY

1. Opposition to current is a characteristic of a resistor. *Resistance can exist by itself.*
2. Voltage is a characteristic of an electromotive force (emf) source. *Voltage can exist by itself.*
3. *Current cannot* exist by itself. For current to exist, a voltage source and a closed path (circuit) are required.
4. In measuring direct current in a circuit, the circuit must be *broken* and a current meter must be connected in *series* with the circuit.
5. In connecting a current meter in a circuit, meter polarity must be observed.

6. Proper polarity is observed if the pointer of a current meter moves upscale, that is, from left to right, when the meter is connected in a circuit.
7. The quantity of current in a circuit can be controlled by the amount of applied voltage and by the amount of resistance in the circuit.

SELF-TEST

Check your understanding by answering these questions:
1. To measure current in a circuit, it is first necessary to _____ the circuit.
2. The current meter is then connected _____ _____ with the circuit.
3. Current can exist only in a _____ circuit.
4. _____ must be observed in connecting a current meter in a dc circuit.
5. The _____ lead of a milliammeter is the common; the _____ lead of a milliammeter is frequently called the *hot* lead.
6. A current meter must never be connected _____ a component. It must always be in _____ with the component.
7. The current in a circuit varies with the (*a*) _____ and (*b*) _____.

MATERIALS REQUIRED

- Power supply: Four 1½-V dry cells
- Equipment: 0–10 milliammeter, EVM
- Resistors: ½-W, three 1,000-Ω
- Miscellaneous: Connecting wires

PROCEDURE

Controlling Current by Resistance

1. Measure and record in Table 10-1 the resistance of one of the three 1,000-Ω resistors.
2. Connect this resistor in the circuit (Fig. 10-4).

The power source consists of four dry cells connected in series-aiding. The circuit is *incomplete* because it is *open* between points *P* and *Q*. (That is, the positive lead of the battery is not connected.)

TABLE 10-1. Controlling Current by Resistance. Voltage Source Constant

Step	Resistance in ohms	Circuit Condition	Current (mA)	Milliammeter
1, 3		Closed		A
4	Same as 3	Open		A
5		Open	X	X
6	Same as 5	Closed		A
7		Open	X	X
8	Same as 7	Closed		A
9	Same as 7	Closed		B
10	Same as 7	Closed		C
11	Same as 7	Closed		D

CAUTION: Observe polarity in connecting the milli-ammeter. The negative terminal of the milliammeter is connected to the negative end of the four-cell battery. The positive lead of the milliammeter goes to resistor R_1. *Do not complete* the circuit until the connections are checked by an instructor.

3. After the hookup is approved by an instructor, connect the positive lead of the battery, thus completing the circuit. The milliammeter now registers the current in the circuit. Read and record the current.
4. Open the circuit by removing R_1. Does the milliammeter still show current? Record in Table 10-1 the current, if any. If there is no current, record "none."
5. Now connect two 1,000-Ω resistors (R_1 and R_2) in series, outside of the circuit. Measure their resistance and record in Table 10-1.

CAUTION: The resistance of a component must *never* be measured while it is in a circuit in which there is current. The circuit must first be opened. The required resistance is then measured *outside*

of the circuit. If this rule is not observed, the ohm-meter can be *seriously* damaged.

6. Connect R_1 and R_2 in the circuit (Fig. 10-5). Measure and record the current now in the circuit.

NOTE: A battery symbol is now used to replace the four dry cells.

7. Remove R_1 and R_2 from the circuit and add a third 1,000-Ω resistor, R_3, in series. Measure and record the resistance of this series combination.
8. Connect R_1, R_2, and R_3 in the circuit (Fig. 10-6). Measure and record the current.

Fig. 10-5. Measuring current in a circuit containing two resistors connected in series.

Fig. 10-6. Measuring current in a circuit containing three resistors connected in series.

Fig. 10-4. Milliammeter connected in an open electric circuit.

Current in a Series Circuit

9. Shift the position of the milliammeter so that it is now at *B*, as in Fig. 10-7. Observe meter polarity. Measure and record the current.
10. Shift the position of the milliammeter so that it is now at *C*, as in Fig. 10-8. Measure and record the current.
11. Shift the position of the milliammeter so that it is now at *D*, as in Fig. 10-9. Measure and record the current.

Controlling Current by Voltage

12. Remove R_2 and R_3, leaving R_1 in the circuit of Fig. 10-10. The voltage source still consists of four 1½-V cells connected in series-aiding. Measure and record in Table 10-2 the voltage source *V*. Measure and record the current.
13. Remove one of the dry cells, leaving three connected in series. Measure and record the voltage source. Measure and record the current.
14. Remove another dry cell, leaving two connected in series. Measure and record the voltage source. Measure and record the current.
15. Remove another dry cell, leaving one connected in the circuit. Measure and record the voltage source. Measure and record the current.

Fig. 10-8. Measuring current in a series circuit.

16. Remove the last dry cell. Complete (close) the circuit. Measure and record current, if any.

QUESTIONS

1. Under what conditions will there be current in a circuit? Justify your answer by referring to specific data in Tables 10-1 and 10-2.
2. Explain how a milliammeter (or ammeter) is connected to measure current in a series circuit.
3. What precautions must be observed in measuring current to prevent damaging the meter?
4. Why must resistance *never* be measured in an *active* circuit, that is, in a circuit carrying current?
5. (*a*) State a conclusion showing the relationship between current and resistance in a circuit, if the voltage of the source is held constant. Refer to specific measurements in the experiment to substantiate this rule.

TABLE 10-2. Controlling Current by Voltage. Resistance in the Circuit Remains Constant

Step	No. of Dry Cells in Series	Voltage of Source	Current (mA)
12	4		
13	3		
14	2		
15	1		
16	0	0	

Fig. 10-7. Measuring current in a series circuit.

Fig. 10-9. Measuring current in a series circuit.

Fig. 10-10. Circuit for determining effect of voltage on current.

(b) Formulate this rule mathematically; using the symbols I for current and R for resistance.

6. (a) State a conclusion showing the relationship between current and voltage in a circuit if the resistance is held constant. Refer to specific measurements in the experiment to substantiate this rule.

(b) Formulate this rule mathematically, using the symbols I for current and V for voltage of the source.

7. What do your measurements in steps 9 through 11 prove concerning current in a series circuit?

Answers to Self-Test

1. break (or open)
2. in series
3. closed
4. polarity
5. negative; positive
6. across; in series
7. (a) applied voltage; (b) resistance

EXPERIMENT 11

Ohm's Law

OBJECTIVE

To determine experimentally the mathematical relationship (Ohm's law) between the current I in a resistor, the voltage V across the resistor, and the resistance in ohms R of the resistor.

INTRODUCTORY INFORMATION

Experiment 10 established that there is a definite relationship between current, voltage, and resistance in a circuit. Stated nonmathematically, one conclusion that was reached is: There is direct current I in a closed circuit containing dc voltage source V and resistance R, and the amount of current decreases as the resistance increases, if the voltage source remains constant. Another conclusion that was reached is: In a closed circuit containing resistance R and dc voltage source V, the amount of direct current I increases as the voltage increases, if the resistance remains constant.

These conclusions are important, but the form in which they are stated is qualitative (descriptive) and not quantitative (mathematical). Modern science seeks quantitative relationships between cause and effect, wherever possible. This is because a quantitative rule or formula not only states that a change will occur, if one of several related quantities is permitted to vary, but also makes it possible to predict how *much* of a change will occur.

It is possible, following the method in Experiment 10, to formulate a mathematical statement (that is, to write a formula) for the relationships among I, V, and R. To do this, it is necessary to make many precise measurements involving the two variables (V and R). Then, applying logic and mathematical reasoning, a formula is written which will fit the measured facts. Additional measurements are then made to verify the formula or to modify it, as the case may be.

An example will illustrate the process. The circuit of Fig. 11-1 was used to study the relationship between I and V, for a constant value of $R = 100\ \Omega$. A voltmeter was employed to measure the voltage applied to the circuit and an ammeter to measure the current. The following results were obtained and tabulated.

The exact relationship between V and I is quite apparent from Table 11-1, for the ratio of V/I is equal to 100 in each case. The equation which would appear to fit the facts of Table 11-1 is

$$\frac{V}{I} = 100 \qquad \text{or} \qquad I = \frac{V}{100} \qquad (11\text{-}1)$$

To test the accuracy of Eq. (11-1), the researcher would compute the value of I_1 for a voltage V_1 other than any of the voltages already used. He would then measure the current I in the circuit with V_1 applied. If I and I_1 were equal, he would check the accuracy of Eq. (11-1) for other values of voltage V_2, V_3, etc., until convinced that the equation was a valid statement of the facts.

Up to this point, it has been established that for the data presumed

$$I = \frac{V}{100}$$

is valid for a circuit where $R = 100\ \Omega$. How must this equation be modified if $R = 200\ \Omega$, $300\ \Omega$, $400\ \Omega$, etc.?

Substituting $R = 200\ \Omega$, then $R = 300\ \Omega$, etc., in the experimental circuit of Fig. 11-1, and following

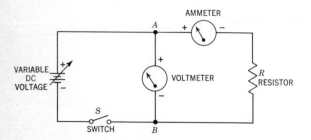

Fig. 11-1. Theoretical circuit for verifying Ohm's law.

TABLE 11-1.

R	100 Ω					Formula relating V, I, and R for $R = 100\ \Omega$	Formula test	
V volts	100	200	300	400	500		$V = 150$	550
I amperes	1	2	3	4	5	$\dfrac{V}{I} = 100$	$I = 1.5$ computed	5.5
V/I	100	100	100	100	100	or $I = \dfrac{V}{100}$	$I = 1.5$ measured	5.5

the method just outlined, new formulas are established, which show that:

for $R = 200\ \Omega$, $\qquad I = \dfrac{V}{200}$ $\qquad$ (11-2)

for $R = 300\ \Omega$, $\qquad I = \dfrac{V}{300}$ $\qquad$ (11-3)

for $R = 400\ \Omega$, $\qquad I = \dfrac{V}{400}$ $\qquad$ (11-4)

etc.

It now is intuitively obvious that formulas (11-1), (11-2), (11-3), etc., may be generalized as follows:

$$I = \frac{V}{R} \qquad (11\text{-}5)$$

Of course this generalization, Eq. (11-5), must still be verified for values of V and R not used in the experiment. Once the verification is complete, Eq. (11-5) can be stated as a law which gives the mathematical relationships among I, V, and R in a closed circuit.

NOTE: In the circuit of Fig. 11-1 it was assumed that the ammeter had no resistance, and that R was the total resistance in the circuit. In fact, the ammeter does have some resistance, but it is so low (in this case a fraction of 1 Ω) that its resistance can be ignored here and the resistance of R *can* be assumed as the total resistance in the circuit.

Ohm's Law

Now, refer again to Fig. 11-1, and to Eq. (11-5). Assume that the voltage drop measured across R is V, and that the circuit current is I. In Experiment 10 it was determined that in a series circuit the current is the same everywhere. Hence the current in R is also I. Equation (11-5) then states that the direct current I in a resistor is directly proportional to the voltage V across the resistor and inversely proportional to the resistance R, where I is measured in amperes, V in volts, and R in ohms.

Equation (11-5) is Ohm's law, first formulated by the scientist Georg Simon Ohm. This law is the foundation on which the science of electricity and electronics is based.

Measurement Errors

In the preceding discussion it was assumed that all the measurements made experimentally were 100 percent accurate. In practice, this is never so. Errors do occur and for several reasons.

One possible error results from reading a meter scale incorrectly. This is trivial and can be corrected by exercising greater care, and by taking the average of a number of the same measurement. Interpolating between calibrated markers may be another source of error.

Parallax is another source of error which can be easily corrected. It occurs when a meter reading is taken from an off-center position, that is, when the line of sight between the viewer and the meter pointer is not perpendicular to the meter scale. Figure 11-2 illustrates the error of parallax. When the

Fig. 11-2. Error of parallax occurs when line of sight of the viewer and the meter pointer are not perpendicular to the meter scale.

viewer is in position P_1, the line between his eye and the meter pointer A is perpendicular to the meter scale. This gives a correct reading of 5. However, if the viewer is in position P_2, the reading will be 7, an error due to parallax. To eliminate errors of parallax, a mirror strip is sometimes placed just below the meter scale. The correct reading position is the one in which the pointer cannot be distinguished from its reflected image in the mirror.

There are other errors which are not so obvious, and which cannot be corrected so easily. For example, there are inherent errors in the instruments used. The instrument manufacturer usually specifies the percentage of instrument error. The general-purpose EVM and other meters used in the school laboratory usually have between 2 and 5 percent error. For greater accuracy, laboratory precision instruments are required. These are highly precise instruments whose inherent error is held to a fraction of a percent.

Another source of error results from the process of inserting an instrument in a circuit to make a measurement. If the instrument alters circuit conditions in any way, incorrect readings may be obtained. Insertion errors will be discussed in greater detail in later experiments.

The fact that errors do occur is mentioned at this point because in this experiment the student will attempt to develop the equation for Ohm's law from experimental data. He can expect that his data may contain some errors of measurement.

SUMMARY

1. The relationship between the voltage V applied to a closed circuit by some electromotive force such as a battery, the total resistance R, and the current I in that circuit is given by the formula $I = V/R$.
2. The relationship between the voltage drop V across a resistor R and the current I in that resistor is given by the formula $I = V/R$.
3. The equation $I = V/R$ is a mathematical statement of Ohm's law.
4. To verify Ohm's law experimentally, many measurements must be made and the results of the measurements must be substituted in Eq. (11-5) to see that the facts actually fit the formula.
5. One set of data is obtained by measuring I, while holding the measured value of V constant and varying the measured value of R. The data obtained should fit the formula $I = V/R$.
6. Another set of data is obtained by measuring I,

while holding the measured value of R constant, and varying the measured value of V. The data obtained should also fit the formula $I = V/R$.

7. Measurement errors do occur and these must be considered in attempting to establish the accuracy of a formula such as Ohm's law.
8. Among the measurement errors that may occur are: (*a*) incorrect reading of the meter scale, (*b*) incorrect meter readings due to parallax, (*c*) errors resulting from the accuracy of the instrument used, and (*d*) errors introduced by inserting the instrument in the circuit (insertion or loading errors).

SELF-TEST

Check your understanding by answering these questions:
1. The current in a fixed resistor is _____ proportional to the voltage across that resistor.
2. If the voltage across a resistor is held constant, the current in that resistor is _____ _____ to its resistance.
3. The formula which gives the mathematical relationship between I, V, and R in a closed circuit is $I =$ _____.
4. The equation in question 4 is called _____ _____.
5. If the voltage across a 10,000-Ω resistor is 125 V, the current in the resistor is _____ A.
6. If the voltage across a resistor is 60 V, and the current in the resistor is 0.05 A, the value of the resistor is _____ Ω.
7. What is the voltage across a 1,500-Ω resistor, if the current in the resistor is 0.12 A? _____ V
8. In reading a meter, the line of sight between the viewer and the meter pointer should be _____ to the meter scale.
9. The error of parallax may be eliminated by placing a _____ _____ just below the meter scale. The correct reading position occurs when the pointer and its reflected image in the mirror _____.

MATERIALS REQUIRED

- Power supply: Source of regulated variable dc voltage
- Equipment: EVM, 0–10 milliammeter
- Resistors: ½-W 10,000-Ω; 1,000-Ω
- Miscellaneous: 10,000-Ω 2-W potentiometer, connecting wires

PROCEDURE

NOTE: In the experimental procedure which follows, use Fig. 11-3a if a 0 to 400 dc voltage supply is available. However, if a 0 to 40 V supply is employed, use Fig. 11-3b.

1. Connect the circuit of Fig. 11-3. **Power off.** Switches S_1 and S_2 are **open**, as shown. Zero the ohmmeter and connect it across R, a 10,000-Ω potentiometer connected as a rheostat (points C to D). Adjust R so that the resistance between points C and D is 2,000 Ω. *Remove* the ohmmeter from the circuit. Set the voltage control on the variable dc voltage source for 0 V, maximum counter-clockwise (CCW) position. Set the voltmeter on the 10-V range.

CAUTION: Be certain that the voltmeter and milliammeter are connected with the proper polarity as shown. Have your circuit checked and approved by an instructor before proceeding.

2. *After* circuit is approved by an instructor, turn power **on**. Close switches S_1 and S_2. Slowly adjust voltage control until 6 V is measured on the voltmeter across R (that is, V_{AB}). Measure the current I in amperes and record in Table 11-2. (Convert milliamperes to amperes before entering I in Table 11-2.)

3. Measure and record in turn, the current I (in amperes) at each of the voltage settings shown in Table 11-2, for $R = 2,000$ Ω. Reset the voltmeter to the 50-V range for voltages greater than 10 V.

4. Compute the ratio V/I for corresponding values of V and I and record in Table 11-2. (For example, if at 6 V the measured current I is 0.003

A, then $V/I = 6/.003 = 2,000$. Therefore record 2,000 in the column under 6 V, in the row "V/I".)

5. If your measurements have been correct it should now be possible for you to deduce a formula relating V and I when $R = 2,000$. Record this formula in Table 11-2, in the form $V/I = ?$ Also record this formula in the form $I = V/?$

6. Using the formula derived in step 5, calculate the current you would expect in the circuit when $R = 2,000$ Ω and $V = 9$ V; $V = 18$ V. Record the predicted (computed) values in Table 11-2 under "Formula test, I computed." Show sample calculation.

   Now measure the current in the circuit when $V = 9$ V and $V = 18$ V and record in Table 11-2.

QUESTION: If your formula is correct, how should the computed and measured values compare?

7. **Power off.** Switches S_1 and S_2 **open**. Turn voltage control on the variable dc source to zero volts (completely counter-clockwise). Place the ohmmeter across R and readjust R so that the resistance between points C and D is 4,000 Ω. Remove the ohmmeter.

8. Follow the experimental procedure set forth above for each value of voltage shown in Table 11-2 for $R = 4,000$ Ω. It is not necessary to show your other computations.

9. Follow the experimental procedure above for each value of voltage shown in Table 11-2 at $R = 6,000$ Ω, $R = 8,000$ Ω. **Power off.**

10. Draw a graph of I (measured) versus V applied at $R = 2,000$ Ω; $R = 4,000$ Ω; $R = 6,000$ Ω; $R = 8,000$ Ω. Use the same axes, with I plotted

Fig. 11-3. Experimental circuits for verifying Ohm's law: Use (a) when dc voltage source is variable from 0–400 V; use (b) when dc voltage is variable from 0–40 V.

TABLE 11-2.

R	2,000 Ω					Formula Relating V, I, and R	Formula Test			
V, volts	6	8	10	12	14	when R = 2,000	V		9	18
I, amperes						$V/I =$	I measured			
V/I						$I = \dfrac{V}{}$	I computed			

R	4,000 Ω					Formula Relating V, I, and R	Formula Test			
V, volts	8	12	16	20	24	when R = 4,000	V		10	30
I, amperes						$V/I =$	I measured			
V/I						$I = \dfrac{V}{}$	I computed			

R	6,000 Ω					Formula Relating V, I, and R	Formula Test			
V, volts	12	18	24	30	36	when R = 6,000	V		15	36
I, amperes						$V/I =$	I measured			
V/I						$I = \dfrac{V}{}$	I computed			

R	8,000 Ω					Formula Relating V, I, and R	Formula Test			
V, volts	16	20	24	28	32	when R = 8,000	V		22	40
I, amperes						$V/I =$	I measured			
V/I						$I = \dfrac{V}{}$	I computed			

along the vertical and V along the horizontal axis. Identify the value of R on each graph.

QUESTIONS

1. From the data in Table 11-2 what general conclusion can you draw about the relationship between I and V, when R is any value?
   (a) State the conclusion in words.
   (b) Write the conclusion as a formula.
2. (a) Is it necessary to confirm further the general formula set forth in answer to question 1b?
   (b) How would you confirm the general formula in question 1b?
3. What is the similarity, if any, between the graphs drawn in procedural step 10?
4. (a) From the graph drawn in step 10, what is the current I in the circuit when V = 9 V and R = 2,000 Ω?
   (b) How does this compare with the measured value?
5. (a) From the graphs in step 10, can you find the value of I when V = 15 V and R = 7,500 Ω?
   (b) If not, how can you predict the value of I for the conditions given?
6. For a constant value of R, what is the effect on I of:
   (a) Doubling V?
   (b) Tripling V?
   (c) Halving V?

7. For a constant value of V, what is the effect on I of:
   (a) Doubling R?
   (b) Halving R?
   (c) Tripling R?
8. Referring to your data, discuss the errors, if any, in your experimental results.
9. What is the accuracy of your
   (a) Ohmmeter?
   (b) Voltmeter?
   (c) Milliammeter?
10. Explain in your own words what is meant by the error of parallax in electrical measurements.

Extra Credit

11. If the ohm function of your EVM were inoperative (and you had no other ohmmeter), how could you determine the value of an unknown resistance, assuming you had the equipment listed under Materials Required? Explain in detail.
12. Plot the graph showing the relationship between I and R for a constant voltage V. What is the curve called?

Answers to Self-Test

1. directly
2. inversely proportional
3. $I = V/R$
4. Ohm's law
5. 0.0125
6. 1,200
7. 180
8. perpendicular
9. mirror strip; coincide

12

The Series Circuit

OBJECTIVES

1. To determine experimentally what the total resistance R_T is, in a circuit in which the resistors R_1, R_2, R_3, etc., are connected in series
2. From the experimental data, to formulate a mathematical rule which gives the total resistance R_T of series-connected resistors

INTRODUCTORY INFORMATION

Determining R_T for Series-connected Resistors

In electronic circuits there may be one or more resistors connected in series, in parallel, in a series-parallel arrangement, or in more complex combinations. The technician must be able to analyze such circuits so that he can determine and predict the effect of a resistor or combination of resistors in controlling current. Therefore, he must understand the laws which govern these circuits.

With the knowledge that the student has already gained it is possible for him to determine experimentally the quantitative effect of the series-connected resistors R_1, R_2, R_3, etc., on current.

Figure 12-1 shows a voltage source V applied across a resistor R_1. The student has already determined that if the values of V and R_1 are given, the

current in this circuit may be predicted (computed) by substituting the values of V and R_1 in Ohm's law:

$$I = \frac{V}{R} \qquad (12\text{-}1)$$

Now, knowing the values of V, R_1, R_2, R_3, etc., is it possible to predict the current I in the circuit containing: Two resistors R_1 and R_2 connected in series (Fig. 12-2a)? Three resistors R_1, R_2, and R_3 connected in series (Fig. 12-2b)? Any number of resistors connected in series?

What are the "facts" on which an educated guess may be made? In Experiment 10 it was established that the current I is everywhere the same in a series circuit. This, on reflection, now appears self-evident—for there is one and only one path for current in a series circuit. Moreover, in Experiment 10, measurement showed that the current I in a circuit decreased as more resistors were added in series, if the applied voltage was kept constant. It would appear, then, that the effect of adding resistors in series is to "increase" the opposition to current in a circuit. Again, this conclusion now seems evident, for since there is only one path for current, each resistor must exercise its effect in controlling current. It is the cumulative effect, the equivalent or total resistance R_T of the series combination, that is controlling current.

R_T, for a specific circuit, may be determined by measuring the current I_T in that circuit, and solving for R_T by substituting the measured values I_T and V (applied) in Ohm's law. Thus,

$$R_T = \frac{V}{I_T} \qquad (12\text{-}2)$$

Equation (12-2) suggests that a *single resistor R_T* may be substituted for two or more resistors con-

Fig. 12-1. Voltage V applied across resistor R_1.

Fig. 12-2. Voltage V applied across (a) two resistors; (b) three resistors connected in series.

nected in series, for the effect of R_T on controlling current is the same as the effect of the series-connected resistors.

An experimental procedure for determining if there is a law by which R_T may be determined readily suggests itself. Assume a group of resistors whose values are known or whose values may be determined by measurement with an ohmmeter. Connect two of these resistors in series. Apply a measured voltage V across the circuit and measure the current I_T. Compute the equivalent resistance R_T of this combination, using Ohm's law. Suppose, for the sake of discussion, that $R_T = R_1 + R_2$ in this case. Now, connect two other resistors in the circuit and experimentally determine their equivalent resistance. Repeat this procedure for two more resistors. Analyze the results. Is there a readily identifiable pattern? Can it be stated analytically by formula, as, for example, $R_T = R_A + R_B$? If it can, experimentally determine if the formula still holds for three resistors in series, four resistors in series, etc.

If the formula holds for any number of resistors, it may be accepted as a general law for resistors connected in series.

NOTE: The value of R_T may also be measured directly by disconnecting the voltage source, and measuring across the series-connected resistors with an ohmmeter. This method may be used to confirm the results of the preceding procedure.

SUMMARY

1. The effect of two or more resistors connected in series in a closed circuit containing a voltage source V is to offer more resistance to current than can any one of these resistors acting alone in the circuit.
2. The effective or total resistance R_T of series-

connected resistors R_1, R_2, etc., in a circuit containing a voltage source V may be determined experimentally by measuring the applied voltage V and the current I. Then R_T may be found by substituting the measured values of V and I in Ohm's law. Thus

$$R_T = \frac{V}{I}$$

This is method 1.
3. Two or more series-connected resistors may be replaced by a single resistor R_T, determined as in method 1, since the effect of R_T in controlling current is the same as the effect of the series-connected resistors.
4. The value R_T of series-connected resistors may also be measured directly with an ohmmeter. However, in using an ohmmeter to measure total resistance in a series circuit, the power source must be disconnected from the resistors. This is method 2.
5. A general formula for determining the value R_T of series-connected resistors may be deduced from experimental data gathered by using either method 1 or method 2. But measurements must be made involving many different combinations of resistors. In all cases the resistance of individual resistors must be measured, together with the total resistance of each series combination. The measured values must then be compared with the values predicted by the general formula to see if they are equal.

SELF-TEST

Check your understanding by answering these questions:
1. The current in a circuit containing four series-connected resistors measures 0.025 A. If the voltage applied to the circuit is 10 V, the total resistance R_T of the four series-connected resistors is: $R_T =$ _____ Ω.
2. One hundred resistors were measured with an ohmmeter, and each was found to be 100 Ω. The total resistance R_T of any two of these resistors, connected in series, measured 200 Ω. R_T of any three of these resistors connected in series measured 300 Ω; R_T of any four of these resistors connected in series measured 400 Ω. What formula is suggested by these measurements, for any number n of these resistors connected in series? $R_T =$ _____ Ω.

3. The total resistance R_T of series-connected resistors is _____ than the resistance of any one of these resistors taken by itself.
4. In a series-connected closed circuit there is just _____ path for current.
5. Current in a series circuit is the _____ everywhere.

MATERIALS REQUIRED

- Power supply: Regulated variable direct current
- Equipment: EVM, 0–10 milliammeter, ohmmeter (or ohm scale of an EVM)
- Resistors: ½-W 330-, 470-, 1,200-, 2,200-, 3,300-, and 4,700-Ω
- Miscellaneous: SPST switch

PROCEDURE

Determining R_T for Series-connected Resistors

Ohmmeter method 1

1. Measure the resistance of each of the resistors supplied and record its value in Table 12-1.
2. Connect the series circuit of Fig. 12-3 in the combinations of resistances shown in Table 12-2 and measure and record the total resistance R_T (from A to B) of each combination.
3. From Table 12-1, compute the sum of the measured values of R_1, R_2, R_3, etc., of each combination and record in Table 12-2, column labeled "Computed value $R_1 + R_2 + \cdots$."
4. In the last column of Table 12-2, write a formula for R_T which fits the measurements for each combination. For example, if measured values of $R_1 = 300$ Ω, $R_2 = 450$ Ω, and $R_T = 750$ Ω, then

$$R_T = R_1 + R_2$$

5. Write a general formula for R_T in Table 12-2, for any number of series-connected resistors.

Determining R_T for Series-connected Resistors

Milliammeter-voltmeter method 2

6. Connect the circuit of Fig. 12-4. R_1 and R_2 are the resistors in combination 1, Table 12-2. S_1 is open. Adjust the power supply for $V = 10$ V as measured with an EVM set on the proper voltage range.

7. Close switch S_1 applying power to the circuit. Check the voltmeter. The voltage V must be maintained at 10 V for this load and for each subsequent load. Record this voltage in column labeled "V applied," Table 12-3.
   Measure the current. Record this value in the column labeled "I measured" in Table 12-3.
8. Open the circuit (S_1 off). Between points A and B replace series combination 1 with series combinations 2, 3, 4, 5, 6, and 7 in turn. Measure and record in Table 12-3 the current through each combination, maintaining the measured 10 V, V applied.
9. From the measured values of I and V applied in Table 12-3, calculate R_T and record in the column labeled $R_T = V/I$ for each combination of series resistors.
10. Record from Table 12-2, in the column labeled "Computed value $R_1 + R_2 + \cdots$," the values of each combination, in the corresponding column in Table 12-3.
11. Write a formula for R_T (in the last column of Table 12-3) in terms of R_1, R_2, R_3, etc., which fits the measurements for each combination in Table 12-3.
12. Write a general formula R_T for any number of series-connected resistors, in terms of the resistors.

Design Problem (extra credit)

13. Theoretically design a series circuit with the resistors in this experiment, using first
    (a) Two resistors, then

TABLE 12-1.

Rated value, ohms	330	470	1,200	2,200	3,300	4,700
Measured value, ohms						

Fig. 12-3. Series-connected resistors.

TABLE 12-2.

Combination	Rated Value (ohms)					R_T (ohms) Measured Value	Computed Value $R_1 + R_2 + \cdots$	Formula for R_T
	R_1	R_2	R_3	R_4	R_5			
1	330	1,200	X	X	X			
2	1,200	2,200	X	X	X			
3	3,300	4,700	X	X	X			
4	330	470	1,200	X	X			
5	1,200	2,200	3,300	X	X			
6	330	470	1,200	2,200	X			
7	330	470	1,200	2,200	3,300			

General formula: $R_T =$

Fig. 12-4. Experimental circuit for determining R_T.

TABLE 12-3.

Combination	V Applied, volts	I Measured, amperes	$R_T = \dfrac{V}{I}$	Computed Value $R_1 + R_2 + \cdots$ (ohms)	Formula for R_T
1					
2					
3					
4					
5					
6					
7					

General formula: $R_T =$

TABLE 12-4.

Combination	Measured Values				V Applied, Design Value	I Measured
	R_1	R_2	R_3	R_4		
1						
2						
3						

(b) Three resistors, then

(c) Four resistors

so chosen that each combination a through c draws 5 mA, at an applied voltage V to be determined. Record in Table 12-4 the measured values of R_1, R_2, etc., selected in each combination. Show your computations. Record also in Table 12-4 the design (computed) value of V, for each combination.

14. Experimentally verify the design values for V and record your current measurements in Table 12-4.

QUESTIONS

1. Explain in your own words two methods for determining, by measurement, the total resistance R_T of resistors connected in series.
2. Why was it necessary to measure individually the resistance of each resistor used in the experiment?
3. Do the results of your measurements in Tables 12-1, 12-2, and 12-3 prove that it is possible to write a general formula in terms of R_1, R_2, etc., for total resistance of series-connected resistors R_1, R_2, etc.? Explain.

4. (a) Why must highly precise laboratory-standard instruments be used in attempting to establish or verify a scientific law?

(b) What is the accuracy of your voltmeter? Ohmmeter?

5. (a) How do the ohmmeter-measured values of R_T compare with the computed values of R_T in Table 12-2?

(b) How do the values of R_T derived from the volt-ampere method compare with the computed values in Table 12-3?

6. What effect, if any, would there be on the total resistance R_T of combination 4, Table 12-2, if we interchanged the positions of R_1, R_2, and R_3 in the circuit (Fig. 12-3)?

7. Write a formula R_T for series-connected resistors R_1, R_2, R_3 . . . , in terms of R_1, R_2, R_3, etc. Explain the formula in your own words.

Answers to Self-Test

1. 400
2. 100 n
3. greater
4. one
5. same

EXPERIMENT 13 Series-circuit Design Considerations

OBJECTIVES

1. To design a series circuit which will meet specified resistance requirements
2. To design a series circuit which will meet specified voltage and current requirements
3. To design a series circuit which will meet specified current and resistance requirements
4. To construct the circuit and check, by measurement, the design specifications

INTRODUCTORY INFORMATION

Designing a Circuit to Meet Specified Resistance Requirements

The law for total resistance of series-connected resistors can be applied to the solution of simple design problems. An example will indicate the techniques to be used in circuit design at this level.

Problem 1

A technician has a stock of the following resistors: four 56 Ω, five 100 Ω, three 120 Ω, two 180 Ω, two 220 Ω, 330 Ω, 470 Ω, 560 Ω, 680 Ω, and 820 Ω. He needs a resistance value of 1,000 Ω for a circuit he is designing. Find at least four combinations of resistors from those he has in stock, using the least possible number of components, which will satisfy the design requirement.

Solution. Experiment 12 developed the generalization that the total resistance R_T of series-connected resistances R_1, R_2, R_3, etc., is equal to the sum of these resistances. Stated as a formula,

$$R_T = R_1 + R_2 + R_3 + \cdots \qquad (13\text{-}1)$$

The technician can use Eq. (13-1) to solve problem 1. Thus

$$1,000 = R_1 + R_2 + R_3 + \cdots$$

1. By inspection it is apparent that the 820-Ω resistor and the 180-Ω resistor connected in series will add to 1,000 Ω. Hence this is one solution. It is also the solution which requires the least number of components, just two.
2. Another solution is to connect 680-Ω, 220-Ω, and 100-Ω resistors in series. Here three components are used.
3. Another solution is to connect a 560-Ω and two 220-Ω resistors in series. Again three components are used.
4. A fourth solution is to connect a 470-Ω, 330-Ω, and two 100-Ω resistors in series. Here four components are used.

There are other combinations which will add to 1,000 Ω, so that the technician has a fairly wide choice. The restriction that he use the least number of components, however, does limit his choice.

Finally, the technician should connect the resistors and measure their total resistance with an ohmmeter, to confirm his solution.

Designing a Circuit to Meet Specified Voltage and Current Requirements

Ohm's law and the law for total resistance of series-connected resistors can be applied to the solution of this type of design requirement. Again a problem will illustrate the procedures to be employed.

Problem 2

A technician has a 15-V battery and the same stock of resistors as in problem 1. He must design a cir-

Fig. 13-1. Circuit used to limit current to 0.01 A.

cuit using the least components in which current must be 0.01 A. Show the circuit arrangement he can use, including the values of all resistors.

Solution. Assume a closed series circuit is used. Two of the circuit conditions are known, namely, voltage and current. Using Ohm's law, the total resistance in the circuit can be found. Thus

$$I = \frac{V}{R_T}$$

and

$$R_T = \frac{V}{I} \qquad (13\text{-}2)$$

Substituting in Eq. (13-2) the known values of V and I, we get

$$R_T = \frac{15}{0.01} = 1,500 \ \Omega$$

This is the circuit resistance which will hold circuit current at 0.01 A. It is now necessary to find a combination of resistances from those in stock which will add to 1,500 Ω. Since the sum of 820 and 680 = 1,500, the required resistors are $R_1 = 820 \ \Omega$ and $R_2 = 680 \ \Omega$ connected as in Fig. 13-1.

As a final step the technician should connect the circuit of Fig. 13-1 and verify by measurement that there is indeed 0.01 A of current in the circuit.

The procedure to be followed, then, in this type of problem is to:

1. Solve for R_T, by substituting the known values of V and I in the formula

$$R_T = \frac{V}{I}$$

2. Find the combination of resistances whose sum will add to the given value of R_T, using the equation

$$R_T = R_1 + R_2 + R_3 + \cdots$$

3. Connect the circuit using the combination of resistors determined in 2 and verify by measurement that the circuit conditions have been met.

Designing a Circuit to Meet Specified Current and Resistance Requirements

As in the preceding design problem Ohm's law and the law for R_T of series-connected resistors are applied here. Again, a problem will illustrate the process.

Problem 3

The series circuit in Fig. 13-2 is required to draw 0.05 A. To what voltage should the technician set the variable dc source to achieve this level of current?

Solution.
1. Find the total resistance R_T.

$$R_T = 330 + 470 + 560 = 1,360 \ \Omega$$

2. Substitute the values $I = 0.05$ and $R_T = 1,360$ in Eq. (13-3) which was derived from Ohm's law

$$V = I \times R$$
$$V = (0.05)(1,360) = 68.00 \text{ V} \qquad (13\text{-}3)$$

3. Connect the circuit of Fig. 13-2 and set the power supply voltage V to 68 volts. The milli-ammeter should read 0.05 A (50 mA).

SUMMARY

1. If it is required to make up a resistance value R_T, not otherwise available, from series-connected resistors whose resistance values are R_1, R_2, R_3, etc., Ω, find the combination of resistors which will satisfy the formula

$$R_T = R_1 + R_2 + R_3 + \cdots$$

2. If it is required to design a series circuit which will meet specified voltage (V) and current (I) requirements, find the value R_T which will satisfy

Fig. 13-2. Determining voltage to which V must be set to permit 0.05 A of current in the circuit.

the given voltage and current, by substituting V and I in the equation:

$$R_T = \frac{V}{I}$$

Then select the resistors R_1, R_2, etc., whose resistance sum is R_T.

$$R_T = R_1 + R_2 + R_3 + \cdots$$

3. If it is required to design a series circuit which will meet specified current (I) and resistance (R_T) values, select first those resistors whose resistance sum is R_T. Thus

$$R_T = R_1 + R_2 + R_3 + \cdots$$

Then solve for the unknown voltage V by substituting the given values of I and R_T in the equation

$$V = I \times R_T$$

4. After the circuit has been designed theoretically, connect it and measure the unknown quantity to see that it is in fact the required design value.

SELF-TEST

Check your understanding by answering these questions:

1. The formula which gives the total resistance of series connected resistors is $R_T = $ _____.
2. The formula which gives the relationship between the applied voltage V, the current I, and the resistance R of a closed circuit is $V = $ _____.
3. To design a circuit which is powered by a battery of V volts which draws I amperes, it is necessary to find _____. This is done by substituting V and I in the formula _____.
4. To design a circuit which draws I amperes through a resistance of R ohms, it is necessary to find _____. This is done by substituting I and R in the formula _____ _____.
5. The final step in the design of a circuit, after the parameters have been determined, is to _____ the circuit and _____ the quantities involved.

MATERIALS REQUIRED

- Power supply: Regulated variable direct current
- Equipment: EVM, 0–10 milliammeter; 0–100 milliammeter, ohmmeter (or ohms scale of an EVM)
- Resistors: ½ W 330-, 470-, 1,200-, 2,200-, 3,300-4,700-Ω
- Miscellaneous: SPST switch

PROCEDURE

1. From the resistors listed under Materials Required in this experiment, select three resistors whose total color-coded resistance value R_T, when they are connected in series, is 2,000 ohms. List their color-coded resistance values in Table 13-1.
2. Connect these resistors in series and measure their combined resistance with an ohmmeter.

TABLE 13-1.

R_T Required ohms	Color-Coded Value of Resistors Whose Sum Will Satisfy R_T						R_T Measured ohms
	R_1	R_2	R_3	R_4	R_5	R_6	
2,000							
5,300							
7,500							
10,000							
11,000							

TABLE 13-2.

V Applied volts	Circuit Current I amperes		Resistor Design Values ohms					
	Required	Measured	R_1	R_2	R_3	R_4	R_5	R_6
40	0.02							
30	0.01							
5.5	0.001							
16	0.02							
11.4	0.001							

Record this measured value in Table 13-1, in column headed R_T Measured.

3. Repeat steps 1 and 2 for every value of R_T Required listed in Table 13-1. Use as many resistors as may be required.

4. Design a series circuit (like Fig. 13-1) with any combination of resistors in this experiment which will permit 0.02 A of current in the circuit when the dc supply is set at 40 V. Show your computations. List in Table 13-2 the design values of the resistors selected.

5. Connect the circuit in step 4. Set the dc supply at 40 V. Measure the circuit current and record in Table 13-2.

6. Repeat steps 4 and 5 for every value of voltage and current listed in Table 13-2.

7. Design a series circuit with the resistors in this experiment using first (a) two resistors, then (b) three resistors, then (c) four resistors so chosen that each combination a through c draws 4 mA, at an applied voltage V to be determined analytically. Record in Table 13-3 the *measured* values of R_1, R_2, etc., selected for each combination. Show your computations. Also record in Table 13-3 the design value of V, determined analytically, for each combination.

8. Experimentally verify the design values for V and record your current measurements in Table 13-3.

QUESTIONS

1. Refer to Table 13-1. Compare the R_T Required values with the R_T Measured values. If they are not the same, explain why.
2. Three ½-W resistors 1,000 Ω, 5,000 Ω, and 10,000 Ω with tolerance band color-coded gold are connected in series. What is the range of their total resistance, as measured with a meter whose error is 0%, that is, whose accuracy is 100 percent?
3. Refer to Table 13-2. Compare the required and measured values of I. Explain any difference between each set of values.
4. Consider the measured values of current I, in Table 13-2, for *each* value of applied voltage. Indicate if the R_T of the selected resistors is equal to, less than or greater than the design value, and explain why you think so.
5. Refer to Table 13-3. Compare the current I *measured*, with the design requirement of 4 mA, for every combination listed in the table. Explain any differences between the two current values for each combination.

Answers to Self-Test

1. $R_T = R_1 + R_2 + R_3 + \cdots$
2. $V = I \times R$
3. resistance; $R = V/I$
4. voltage V across R; $V = I \times R$
5. connect; measure

TABLE 13-3.

Combination	Measured Values ohms				V Applied, Design Value volts	I Measured mA
	R_1	R_2	R_3	R_4		
1						
2						
3						

14

Voltage-divider Circuits (unloaded)

OBJECTIVES

1. To develop, by analysis, a general law for computing the voltage across each resistor in an unloaded fixed resistive voltage divider
2. To confirm, experimentally, the law set forth in objective 1
3. To determine, by analysis, the voltage with respect to common, at each point in a variable resistive voltage divider
4. To confirm, experimentally, the results of objective 3

INTRODUCTORY INFORMATION

Series-connected Voltage-divider Circuits

Ohm's law finds immediate application in the analysis and design of voltage-divider circuits. Voltage dividers can be very simple circuits, or complex arrangements of resistors serving one or more "loads." In this experiment we will be concerned with "unloaded" dividers, that is, with circuits which are not required to deliver current to an external load.

The simplest dc voltage divider consists of two resistors R_1 and R_2 connected in series, across which a dc voltage V is applied (Fig. 14-1). Assume that V is 100 V and that the resistors R_1 and R_2 are 75,000 Ω and 25,000 Ω, respectively. The voltages V_1 across R_1 and V_2 across R_2, measured with an electronic voltmeter, are 75 V and 25 V respectively. The 100-V source has thus been divided by the circuit of Fig. 14-1 to produce two lower voltages.

Figure 14-1 can be modified by the addition of one or more resistors to produce any number of lower voltages, measured across individual resistors, or measured with respect to some common point such as C. An arrangement of resistors to produce specific voltages can be achieved by the "cut-and-try" method or by analysis of the circuit. The cut-and-try method is too time-consuming and inefficient. Circuit analysis is a fast and effective means of finding the values of the resistors to produce a required result.

In circuit analysis, the laws of electricity are applied to secure a "mathematical" solution of the problem. For example, in Fig. 14-1, the current I may be determined by substituting the values of V and R_T in Eq. (14-1).

$$I = \frac{V}{R_T} \qquad (14\text{-}1)$$

where V is the applied voltage

and $\qquad R_T = R_1 + R_2$

Since $V = 100$ V, and $R_1 + R_2 = 100,000$ Ω,

$$I = \frac{100}{100,000} = 0.001 \text{ A}$$

Now $\qquad V_1 = I \times R_1$

and $\qquad V_2 = I \times R_2 \qquad (14\text{-}2)$

Fig. 14-1. DC voltage divider.

Therefore $V_1 = (0.001)(75,000) = 75$ V

and $\qquad V_2 = (0.001)(25,000) = 25$ V

A general solution (formula) may be found to simplify our work. Consider Fig. 14-2. It is required to find V_1, V_2, V_3, and V_4. Assume that I is the current in this circuit. Then

$$V = I \times R_T \qquad (14\text{-}3)$$

where $\qquad R_T = R_1 + R_2 + R_3 + R_4$

Since
$$\begin{aligned} V_1 &= I \times R_1 \\ V_2 &= I \times R_2 \\ V_3 &= I \times R_3 \\ V_4 &= I \times R_4 \end{aligned} \qquad (14\text{-}4)$$

we can find the ratio of V_1, V_2, etc., to V. Thus

$$\frac{V_1}{V} = \frac{I \times R_1}{I \times R_T} = \frac{R_1}{R_T} \qquad (14\text{-}5)$$

and
$$V_1 = V \times \frac{R_1}{R_T} \qquad (14\text{-}6)$$

Similarly
$$V_2 = V \times \frac{R_2}{R_T}$$

$$V_3 = V \times \frac{R_3}{R_T} \qquad (14\text{-}7)$$

$$V_4 = V \times \frac{R_4}{R_T}$$

We have thus derived a simple formula [Eqs. (14-6) and (14-7)] for finding the voltage across any resistor in a series circuit. Stated in words, it is: The voltage V_1 across the resistor R_1 is equal to the product of the applied voltage V and the ratio of the resistance of R_1 to the total resistance R_T of the series circuit. This formula applies to a series circuit containing any number of resistors.

As an example, let us apply the formula to Fig. 14-1.

$$V_1 = V \times \frac{R_1}{R_T} = 100 \times \frac{75,000}{100,000} = 75 \text{ V}$$

Similarly $\quad V_2 = 100 \times \dfrac{25,000}{100,000} = 25$ V

Another example is given to illustrate how the formula may be used to design a voltage divider. Using a 250-V source, we wish to find the values of resistors R_1 through R_4 connected in a simple series circuit to provide 25, 50, 75, and 100 V across R_1, R_2, R_3, and R_4, respectively. Assume that the current I in this circuit must be limited to 0.001 A.

Solution.
1. First, we find R_T:

$$R_T = \frac{V}{I} = \frac{250}{0.001} = 250,000 \text{ }\Omega$$

2. Next, we can rewrite Eqs. (14-6) and (14-7) as follows:

$$R_1 = \frac{V_1}{V} \times R_T$$

$$R_2 = \frac{V_2}{V} \times R_T$$

$$R_3 = \frac{V_3}{V} \times R_T \qquad (14\text{-}8)$$

$$R_4 = \frac{V_4}{V} \times R_T$$

3. Substituting $V = 250$ volts, $V_1 = 25$ volts, $R_T = 250,000$, etc., in Eqs. (14-8) gives

$$R_1 = \frac{25}{250} \times 250,000 = 25,000 \text{ }\Omega$$

$$R_2 = \frac{50}{250} \times 250,000 = 50,000 \text{ }\Omega$$

$$R_3 = \frac{75}{250} \times 250,000 = 75,000 \text{ }\Omega$$

$$R_4 = \frac{100}{250} \times 250,000 = 100,000 \text{ }\Omega$$

These are the required values. The circuit of Fig. 14-2 may be connected using these values of R_1, R_2, etc., and V, and the required voltages may be verified by measurement with an EVM.

In analyzing the voltage-divider circuits of Figs.

Fig. 14-2. Another dc voltage divider.

Fig. 14-3. Potentiometer as a variable voltage divider.

14-1 and 14-2, we have considered the voltages developed across individual resistors of the divider. Another view of the divider is to consider the voltages at the junction points of the network, relative to a common point. In Fig. 14-1, point C is the common or ground return of the circuit. In Fig. 14-2, point G is ground. Now consider Fig. 14-2. What is the voltage at A relative to ground (G)? B to ground? C to ground? D to ground?

These voltages may be determined by a modification of Eqs. (14-6) and (14-7). First, the voltage from A to G is obviously the applied voltage V. Now, the voltage V_{BG} from B to ground is

$$V_{BG} = V \times \frac{R_2 + R_3 + R_4}{R_T}$$

and
$$V_{CG} = V \times \frac{R_3 + R_4}{R_T} \qquad (14\text{-}9)$$

and
$$V_{DG} = V \times \frac{R_4}{R_T}$$

Another method of determining V_{BG} and V_{CG} in Fig. 14-2 is to solve for the voltages V_1, V_2, V_3, and V_4 by the formulas (14-6) and (14-7). Then

$$V_{BG} = V_2 + V_3 + V_4$$
$$V_{CG} = V_3 + V_4 \qquad (14\text{-}10)$$
and
$$V_{DG} = V_4$$

Variable Voltage-divider Circuit (Unloaded)

Suppose it is required to set up a divider, as in Fig. 14-1, with a total resistance of 10,000 Ω, whose divider ratio is such that $V_1 = 69$ V and $V_2 = 31$ V. Solution of this circuit, by the method just described, yields the results

$$R_1 = 6,900 \ \Omega$$
$$R_2 = 3,100 \ \Omega$$

It would normally be expensive to secure resistors of these exact values. To overcome this difficulty we make use of a potentiometer. You will recall that a potentiometer is a three-terminal variable resistor. The resistance between the two outer terminals is fixed at the rated value of the potentiometer. The center terminal or arm is connected to a slider which makes contact with the resistive material of the potentiometer. The arm can be varied manually to select different values of total resistance. Thus, if R_1 and R_2 of Fig. 14-1 are replaced with a 10,000-ohm potentiometer, the corresponding circuit is that of Fig. 14-3.

As the arm B moves toward A, the resistance R_1 (AB) decreases, while the resistance R_2 (BC) increases. As the arm B moves toward C, R_1 increases and R_2 decreases. When B is at A, $R_1 = 0$ and $R_2 = 10,000 \ \Omega$; when B is at C, $R_2 = 0$ and $R_1 = 10,000 \ \Omega$.

Thus, by manually adjusting the position of the slider we can set the ratio R_1/R_2 as we choose, and therefore we have a means for setting the voltage V_1 at any value, between zero and the total voltage V across the potentiometer. In this process we have not changed the total resistance of the potentiometer (resistance from A to C).

In actual practice, if a potentiometer is used to develop a voltage which cannot be achieved by a combination of ordinary resistors, an EVM is connected across the arm and one of the end terminals. The arm of the potentiometer is then varied until the desired voltage is measured. This practical arrangement thus eliminates computation, while achieving the desired voltages.

It is possible to limit the range of voltage variation by placing a potentiometer in series with one or more resistors. Thus, in Fig. 14-4 the voltage variation (range) from B to C is from 25 to 75 V.

Fig. 14-4. Limiting the range of variation.

It should be noted that these results are true only if no current is drawn from any tap of the divider.

Variable voltage dividers are used as volume controls in radios, contrast controls in television receivers, speed controls in electronic motor-control circuits, voltage regulator controls, etc.

SUMMARY

1. The voltage across each resistor in a series-connected resistive voltage divider may be determined by formula. Thus in Fig. 14-2, the voltage across any resistor in the divider, say V_1 across R_1, may be found by substituting R_1, R_T (the sum of R_1, R_2, R_3, etc.), and the applied battery voltage V in

$$V_1 = V \times \frac{R_1}{R_T}$$

2. A somewhat longer method to determine the voltage across any resistor in a series-connected divider, say V_1 across R_1 in Fig. 14-2, is to find the total resistance. Thus here

$$R_T = R_1 + R_2 + R_3 + R_4$$

Next, solve for the current I in the circuit. Here

$$I = \frac{V}{R_T}$$

Now, knowing I, to find the voltage drop across R_1, simply multiply R_1 by I. Thus,

$$V_1 = I \times R_1$$

NOTE: Methods 1 and 2 are essentially the same, for if we substitute for I in the last formula, we get

$$V_1 = (I) \times R_1 = \left(\frac{V}{R_T}\right) \times R_1 = V \times \frac{R_1}{R_T}$$

3. If it is required to find the voltage to common, or to any reference, from any junction in a series-connected voltage divider, the same methods 1 or 2 may be used. For example, to find the voltage V_{CG} from C to G in Fig. 14-2, we actually find the voltage across the series-connected resistors R_3 and R_4. Applying the formula,

$$V_{CG} = V \times \frac{R_3 + R_4}{R_T}$$

4. Variable voltage divider circuits can be formed by utilizing a potentiometer across a voltage source, as in Fig. 14-3. Here the voltage from B to C can be varied all the way from V volts (when B is at A) to 0 V (when B is at C).

5. The range of voltage variation of a voltage divider can be limited to any desired value by connecting a potentiometer in series with voltage dropping resistors, as in Fig. 14-4. Here the voltage range of V_{BC} is the difference between the voltage drop V_{DC}, and the voltage drop V_{FC}, and the actual range of variation of V_{BC} is

$$V_{FC} \leq V_{BC} \leq V_{DC}$$

SELF-TEST

Check your understanding by answering these questions:
1. In Fig. 14-1, if the positions of resistors R_1 and R_2 were reversed in the circuit, the voltage across the 75-kΩ resistor would be _____ V.
2. In Fig. 14-2, $R_T = 150$ kΩ and $R_3 = 30$ kΩ. If the applied voltage V is 225 V, the voltage across $R_3 =$ _____ V.
3. In Fig. 14-2, $R_T = 150$ kΩ, $R_1 = 35$ kΩ, and $V = 300$ V. The voltage $V_{BG} =$ _____ V.
4. In Fig. 14-2, $R_T = 100,000$ Ω, $V_{BC} = 20$ V, and $V = 80$ V. The value of $R_2 =$ _____ Ω.
5. In the variable voltage divider (Fig. 14-3) $V = 35$ V. The range of variation of V_{BC} is from _____ V (maximum) to _____ V (minimum).
6. In the variable voltage divider (Fig. 14-4) the battery voltage $V = 60$ V. The values of the resistors are the same as those shown. The range of V_{BC} is from _____ V maximum to _____ V minimum.
7. In Fig. 14-2, $R_1 = 10,000$ Ω, $R_2 = 22,000$ Ω, $R_3 = 6,800$ Ω, $R_4 = 2,200$ Ω, and $V = 164$ V. The voltage $V_1 =$ _____ V, $V_2 =$ _____ V, $V_3 =$ _____ V, and $V_4 =$ _____ V.
8. For the same conditions as in question 7, $V_{CG} =$ _____ V, $V_{BG} =$ _____ V, $V_{BD} =$ _____ V.

MATERIALS REQUIRED

- Power supply: Regulated variable direct current
- Equipment: EVM, 0–10 mA meter, ohmmeter
- Resistors: ½-W 2,200-, 3,300-, 4,700-, and 5,600-Ω
- Miscellaneous: 10,000-Ω 2-W potentiometer, SPST switch

PROCEDURE

1. Connect the circuit of Fig. 14-2, using the rated values $R_1 = 2,200\ \Omega$, $R_2 = 3,300\ \Omega$, $R_3 = 4,700\ \Omega$, and $R_4 = 5,600\ \Omega$. **Power off.** M is a 0–10 milliammeter.

2. Set the power supply so that $V = 40$ V and maintain it at this level. Close switch S applying power to the circuit. Measure, and record in Table 14-1, the applied voltage V, the current I, and the voltages $V_1, V_2, V_3,$ and V_4 across $R_1, R_2, R_3,$ and R_4, respectively. Measure also and record the voltages $V_{BG}, V_{CG},$ and V_{DG}.

3. Compute, and record in Table 14-1, $I, V_1, V_2,$ etc. (as in step 2). Use Eqs. (14-6), (14-7), and (14-10). Show your computations.

4. Adjust the level of the power supply so that the measured current $I = 0.001$ A. Measure and record the applied voltages $V, V_1, V_2, V_3, V_4,$ $V_{BG}, V_{CG},$ and V_{DG}.

5. Compute and record the applied voltage V which would cause the measured current I (0.001 A). Compute and record also the voltages $V_1, V_2,$ etc. (as in step 4), which would result from application of the voltage V to the circuit. Show all your computations.

6. **Power off.** Connect the circuit (Fig. 14-3). Set the power-supply level so that the applied voltage $V = 20$ V and maintain it at this level. M is a 0–10 mA meter.

7. Close S. Measure, and record in Table 14-2, the voltage V and the current I, the voltages from B to C (V_{BC}) and from A to B (V_{AB}) as the arm of the potentiometer is varied from (a) its maximum position (B at A), to (b) three-fourths of maximum, (c) center position (approximately), (d) one-fourth of maximum, and (e) to its minimum position (B at C).

   Add and record the voltages V_{AB} and V_{BC} at each position of the arm.

8. With V still at 20 V, set the arm of the potentiometer so that the voltage, $V_{BC} = 9$ V. Measure and record in Table 14-3, $E, I, V_{BC},$ and V_{AB}.

9. Open switch S. Do not change the setting of the arm. Measure and record the resistance from the arm to C (R_{BC}), from the arm to A (R_{AB}) and across the potentiometer (R_{AC}).

10. In Table 14-3, record the computed R_{BC} and R_{AB}, knowing that $V = 20$ V, the resistance of

TABLE 14-1.

Step		V	I	V_1	V_2	V_3	V_4	V_{BG}	V_{CG}	V_{DG}
2	Measured	40								
3	Computed	—								
4	Measured		0.001							
5	Computed		—							

TABLE 14-2.

	Position of Arm	Measured Values				Computed Values
		V	I	V_{BC}	V_{AB}	$V_{AB} + V_{BC}$
(a)	Max B at A					
(b)	¾ of max					
(c)	Center					
(d)	¼ of max					
(e)	Min B at C					

TABLE 14-3.

			Measured					Computed	
V	I	V_{BC}	V_{AB}	R_{BC}	R_{AB}	R_{AC}	R_{BC}	R_{AB}
20		9						

the potentiometer = 10,000 Ω, and the measured voltage $V_{BC} = 9$ V. Show your computations.

Design Problems (extra credit)

11. (*a*) Design a circuit which will deliver a voltage which may be varied in the range 15 to 30 V (approximately). Select components from those used in this experiment. Draw the diagram showing the computed values of V, and of the required resistors. Show your computations. (*b*) Connect the circuit. Record the minimum and maximum output voltages measured from the arm of the potentiometer to ground.
   Min volts _____. Max volts _____.

12. (*a*) Design a circuit which will deliver a voltage which may be varied in the range 7.2 to 27.2 V (approximately). Select components from those used in this experiment. Draw the diagram showing the computed values of V and of the required resistors. Show your computations.

   HINT: It will be necessary to parallel the potentiometer with a resistor. To solve this problem it will be necessary to know the total resistance R_T of two resistors connected in parallel. This is given by the formula

$$R_T = \frac{R_1 \times R_2}{R_1 + R_2}$$

   where R_1 and R_2 are the two parallel resistors.

   (*b*) Connect the circuit. Record the minimum and maximum output voltages measured from the arm of the potentiometer to ground.
   Min volts _____. Max volts _____.

QUESTIONS

1. Refer to Table 14-1, steps 2 and 3. How do the measured values of V_1, V_2, V_3, and V_4 relate to the computed values? Explain any discrepancies.
2. Refer to Table 14-1, steps 4 and 5. How do the measured and computed values of V compare? Explain any differences.
3. Explain two methods by which the voltages V_1, V_2, etc., in the circuit (Fig. 14-2) may be computed.
4. Indicate, by reference to specific measurements and computations, if the results of your experiment confirm Eq. (14-10). Explain any discrepancies.
5. Refer to Table 14-2. What is true about V_{AB} and V_{BC}, regardless of the setting of the arm of the potentiometer?
6. (*a*) Refer to your measurements in Table 14-3. Find the ratios: V_{BC}/V_{AB} and R_{BC}/R_{AB}.
   (*b*) Are these ratios equal? Should they be? Why?
7. Refer to Table 14-3. What is the relationship between the measured values R_{AB}, R_{BC}, and R_{AC}?
8. How do the computed and measured values of resistance compare, Table 14-3? Explain any discrepancies.
9. What is meant by an "unloaded" voltage divider?
10. Refer to Table 14-3. *Explain* the effect on I (measured) as the arm of the potentiometer is varied.

Answers to Self-Test

1. 75
2. 45
3. 230
4. 25 kΩ
5. 35 to 0
6. 45 to 15
7. 40; 88; 27.2; 8.8
8. 36; 124; 115.2

EXPERIMENT 15

Current in a Parallel Circuit

OBJECTIVES

To determine experimentally that the total current I_T in a circuit containing resistors connected in parallel is:

1. Greater than the current in any branch
2. Equal to the sum of the currents in each of the parallel branches

INTRODUCTORY INFORMATION

Branch Currents

In considering the series circuit in Experiment 10 it was established that a closed circuit is required for current; that current ceases when the circuit is open; that current in a series circuit is the same everywhere. What are the characteristics of a parallel circuit?

Figure 15-1 shows three resistors connected in parallel and a voltage V applied across them. If the line connecting the battery to the parallel network is broken at X or at Y, and an ammeter is inserted in the circuit at X or Y (Fig. 15-2), the ammeter will measure current. This "line" current is drawn by the three resistors from the battery.

A simple experiment suggests an important characteristic of a parallel circuit. If in Fig. 15-2, resistor R_1 is removed from the circuit, the line current measured by the ammeter decreases. If R_2 is then removed from the circuit, the line current decreases further. What remains is a simple series circuit consisting of V, R_3, and the ammeter. The "line" current is now the current drawn by R_3 from V, and may be computed directly by Ohm's law.

The results of this experiment indicate that in Figs. 15-1 and 15-2 there are indeed three conductive paths for current, namely, R_1, R_2, and R_3. When all three paths are closed, there is maximum line current. When path R_1 is broken, there is less line current because only two paths remain, namely, R_2 and R_3, and so on. The individual paths are called "branches" or "legs" of the parallel circuit.

One characteristic of a parallel resistive circuit, then, is that the total current I_T in the circuit is greater than the current in any branch. The conclusion that follows is that each branch current in a parallel resistive circuit is less than the total or line current I_T.

Fig. 15-1. Voltage applied across three resistors connected in parallel.

Fig. 15-2. Measuring total current in a circuit containing parallel resistors.

Fig. 15-3. Circuit to determine the relationship between branch currents and total current in a parallel circuit.

A numerical example will illustrate this characteristic. Assume in Fig. 15-2 that the voltage source V is 6.6 V. It should be noted that 6.6 V appears across each branch resistor in the circuit, that is, the voltage across R_1 is 6.6 V, across R_2 6.6 V, and across R_3 6.6 V also. The individual currents in each branch, I_1 in R_1, I_2 in R_2, and I_3 in R_3, may be calculated by Ohm's law. Thus $I_1 = 3.3$ mA, $I_2 = 2.2$ mA, and $I_3 = 0.66$ mA. The total current I_T must be greater than the largest branch current, so in this case I_T must be greater than 3.3 mA. It will be shown in the next section that I_T is in fact equal to the sum of the branch currents, that is,

$$I_T = 3.3 + 2.2 + 0.66 = 6.16 \text{ mA}$$

Total Current in a Parallel Circuit

It is apparent that the current I_1 in R_1 combines with the current I_2 in R_2, and I_3 in R_3, to form the "line" current, and that this is the total current I_T drawn from the battery.

We say that the current "combines." Does this mean that I_T is the sum of the individual currents I_1, I_2, and I_3 drawn respectively by R_1, R_2, and R_3? This question may be answered experimentally by measuring I_1, I_2, and I_3 and the "line" current I_T in Fig. 15-3 and determining the relationship between them. I_T is the current measured by breaking the circuit at points X or Y and inserting an ammeter there. The branch current I_1 is measured by breaking the circuit at A or B, and inserting an am-

meter there; I_2 by breaking the circuit at C or D and inserting an ammeter there; I_3 by breaking the circuit at G or F and inserting an ammeter there.

If the experiment just suggested is properly performed, it should establish the validity of the formula

$$I_T = I_1 + I_2 + I_3 \cdots \qquad (15\text{-}1)$$

which states that the sum of the branch currents I_1, I_2, etc., in a parallel circuit equals the total current I_T.

SUMMARY

1. In a parallel resistive circuit the individual current in each branch is less than the total line current.
2. The total line current is greater than each individual branch current.
3. The total line current is in fact equal to the sum of all the branch currents.
4. The voltage across each branch is the same.

SELF-TEST

Check your understanding by answering these questions:

1. In Fig. 15-3, $I_1 = 1$ A, $I_2 = 2$ A, $I_3 = 3$ A. The total current I_T must therefore be greater than _____ A.
2. In Fig. 15-3, for the branch currents listed in question 1, $I_T =$ _____ A.
3. In Fig. 15-3 the voltage V_1 across R_1, the voltage V_2 across R_2, and the voltage V_3 across R_3 must all be _____ _____.
4. The voltage across each branch of a parallel network must be _____ .

MATERIALS REQUIRED

- Power supply: Regulated variable direct current
- Equipment: EVM, 0–10 milliammeter, ohmmeter (or ohm scale of an EVM)
- Resistors: ½-W 2,200-, 3,300-, 4,700-, 5,600-, and 10,000-Ω
- Miscellaneous: SPST switch

PROCEDURE

1. Measure the resistance of each of the resistors supplied and record its value in Table 15-1.
2. Connect the circuit of Fig. 15-3, using combi-

nation 1 of resistors shown in Table 15-2. Set the output of the voltage source V at 10 V and maintain it at this level under load until you

TABLE 15-1.

Rated value, ohms	2,200	3,300	4,700	5,600	10,000
Measured value, ohms					

have completed the measurements in Table 15-2.

3. Close circuit (switch **on**). Measure and record in Table 15-2 the total current I_T. Also measure and record the current I_1 in R_1, I_2 in R_2, and I_3 in R_3. Compute I_T by adding I_1, I_2, and I_3 and record.

4. Remove R_1 from the circuit. Measure and record I_T, I_2, and I_3. Compute and record I_T as in step 3.

5. Remove R_2 from the circuit and also leave R_1 *out* as in step 4. Measure and record I_T and I_3. Compute and record I_T as in step 3.

6. **Power off.** (Switch open.) Replace the resistors in combination 1 with those in combination 2. Voltage is still 10 V.

7. Repeat step 3.

8. Repeat step 4.

9. Repeat step 5.

10. **Power off.** Replace the resistors in combination 2 with those in combination 3. Voltage V is still 10 V.

11. Repeat step 3.

12. Repeat step 4.

13. Repeat step 5. **Power off.**

QUESTIONS

1. How do the individual branch currents of parallel-connected resistors compare with the total current drawn by the circuit (Fig. 15-3)?

2. Support your answer to question 1 by referring specifically to the measurements you recorded in Table 15-2.

3. What is the effect on total current of parallel-connected resistors of:
   (a) Increasing the number of resistors in parallel?
   (b) Decreasing the number of resistors in parallel?

4. Support your answers to question 3 by referring specifically to your measurements in Table 15-2.

5. Do the results of your measurements in Tables 15-1 and 15-2 prove that it is possible to write a general formula relating current in the individual branches of parallel-connected resistors with total current? If yes, write a general formula for I_T and explain it in your own words.

6. Support your answer to question 5 by comparing your computed and measured values. Comment on any discrepancies.

Answers to Self-Test

1. 3
2. 6
3. equal to V
4. equal

TABLE 15-2.

Combination	Step	Rated Value Ohms			Measured Values Volts	Measured Values Amperes				Computed Values Amperes
		R_1	R_2	R_3	V	I_T	I_1	I_2	I_3	$I_T = I_1 + I_2 + I_3$
1	3	2,200	3,300	10,000						
	4		3,300	10,000			—			
	5			10,000			—	—		
2	7	3,300	4,700	5,600						
	8		4,700	5,600			—			
	9			5,600			—	—		
3	11	4,700	5,600	10,000						
	12		5,600	10,000			—			
	13			10,000			—	—		

16

EXPERIMENT **Total Resistance of a Parallel Circuit**

OBJECTIVE

To verify experimentally that the total resistance R_T of resistors connected in parallel is given by the formula

$$\frac{1}{R_T} = \frac{1}{R_1} + \frac{1}{R_2} + \frac{1}{R_3} + \cdots$$

INTRODUCTORY INFORMATION

Total Resistance in a Parallel Circuit

The resistance R_T, which battery V "sees" in Fig. 16-1, limits the current in the circuit to the value I_T. A single resistor with value R_T could be used to replace R_1, R_2, and R_3, for R_T would also limit the current to I_T for the same value of applied voltage V. R_T, then, is the total or equivalent resistance of the parallel network.

The value of R_T for any parallel network (Fig. 16-2) may be measured directly by placing an ohmmeter across the network at points X and Y.

NOTE: If the resistors are connected in a network together with a voltage source V (Fig. 16-1) the circuit must first be broken at X or Y to remove the voltage before measuring R_T.

R_T may also be calculated by measuring I_T, as in Fig. 16-3, measuring V, the applied voltage, and substituting I_T and V in Eq. (16-1).

$$R_T = \frac{V}{I_T} \qquad (16\text{-}1)$$

One characteristic of a parallel circuit may easily be deduced from our measurements. Since the total

current I_T is greater than the currents in any of the branch circuits, the total resistance R_T must be smaller than the smallest branch resistance, for resistance varies inversely as the current according to Ohm's law.

It seems logical that there must be a more definite relationship between R_1, R_2, etc., and R_T in a parallel circuit. This relationship, however, is not immediately apparent from the measurement process. And so, reasoning from the facts which we have determined by measurement, we must call on analysis to find that relationship. Specifically, we wish to find the formula, if any, from which R_T may be calculated if the values of R_1, R_2, . . . are known.

Assume that the resistance of the wires connecting V to R_1, R_2, and R_3 in Fig. 16-1 is negligible, that is, zero. This is a valid assumption where short lengths of interconnecting wire are used. Therefore, it is apparent that electrically points B, D, and G are all the same as point Y, and that points A, C, and F are all the same as point X. Hence, if we wish to measure the voltage across R_1 or R_2 or R_3, we place our voltmeter across XY in all three cases. It follows, therefore, that the voltage across R_1, R_2, or R_3 is the same. In the case of Fig. 16-1, it is the applied voltage V.

Fig. 16-1. Voltage applied across three resistors connected in parallel.

Fig. 16-2. Three parallel resistors.

This line of reasoning can be extended to other parallel circuits. It suggests this fundamental principle which can be readily verified by measurement:

The voltage across each of the branches of a parallel circuit is the same.

Refer to Figs. 16-1 and 16-3. We can now solve for the branch currents analytically. For

$$I_1 = \frac{V}{R_1}$$

$$I_2 = \frac{V}{R_2} \qquad (16\text{-}2)$$

$$I_3 = \frac{V}{R_3}$$

Equation (15-1) states that

$$I_T = I_1 + I_2 + I_3 + \cdots$$

Now, substituting the values $I = V/R$, etc., from Eqs. (16-2) we get

$$I_T = \frac{V}{R_1} + \frac{V}{R_2} + \frac{V}{R_3} + \cdots$$

$$= V\left(\frac{1}{R_1} + \frac{1}{R_2} + \frac{1}{R_3} + \cdots\right) \quad (16\text{-}3)$$

Equation (16-1) states that

$$R_T = \frac{V}{I_T}$$

Substituting the results of Eq. (16-3) in Eq. (16-1), we find that

$$R_T = \frac{V}{V\left(\dfrac{1}{R_1} + \dfrac{1}{R_2} + \dfrac{1}{R_3} + \cdots\right)} \quad (16\text{-}4)$$

Simplifying Eq. (16-4), we get

$$R_T = \frac{1}{\dfrac{1}{R_1} + \dfrac{1}{R_2} + \dfrac{1}{R_3} + \cdots} \quad (16\text{-}5)$$

Equation (16-5) may also be written as follows:

$$\frac{1}{R_T} = \frac{1}{R_1} + \frac{1}{R_2} + \frac{1}{R_3} + \cdots \qquad (16\text{-}6)$$

Equations (16-5) and (16-6) express the analytical relationship between the total or equivalent resistance R_T, and the individual branch resistances R_1, R_2, R_3, etc., connected in parallel. Though this relationship was derived analytically, our analysis was based on experimentally observed facts. Are the results valid, and can they be verified by measurement?

A method for verifying Eq. (16-5) or (16-6) is to measure the total resistance R_T of various combinations of resistances connected in parallel by either of the two methods which have been discussed. We then substitute the measured values of the individual resistances in Eq. (16-5) or (16-6) and calculate the value of R_T. If the measured and calculated values of R_T agree, we have experimental evidence confirming the validity of the derived formulas.

Measuring Individual Resistances in a Parallel Circuit

Suppose it is required to measure the resistance R_1 in the parallel network of Fig. 16-2. How may this be achieved? Obviously it *cannot* be done by placing an ohmmeter across R_1 in the network because this procedure would give the measurement of R_T, not of R_1. We can measure R_1 by disconnecting it from the parallel network, and measuring it outside of the circuit. Or we can disconnect one lead of R_1, say at point A, thus removing the effect of the network. We can then measure R_1 by placing an ohmmeter across it.

SUMMARY

1. The voltage V across each branch (resistor in Figs. 16-1 and 16-3) of a parallel circuit is the same.

Fig. 16-3. Measuring total current in a circuit containing parallel resistors.

2. The total or equivalent resistance R_T of two or more resistors connected in parallel, as in Fig. 16-3, may be determined experimentally by measuring the total current I_T, measuring the voltage V across the parallel network, and substituting the measured values in the formula

$$R_T = \frac{V}{I_T}$$

3. Another method of determining experimentally the total resistance R_T of two or more parallel-connected resistors, as in Fig. 16-2, is to place an ohmmeter across the parallel circuit. The meter measures R_T.
4. Resistance should never be measured when there is power applied to the circuit. So if the parallel resistance of R_1, R_2, and R_3 in Fig. 16-3 is required, power must first be removed (by opening the switch S in Fig. 16-3).
5. A formula which expresses the relationship between R_T and R_1, R_2, R_3, etc., of parallel-connected resistors is:

$$R_T = \frac{1}{\dfrac{1}{R_1} + \dfrac{1}{R_2} + \dfrac{1}{R_3} + \cdots}$$

6. Another way of writing the formula for R_T is:

$$\frac{1}{R_T} = \frac{1}{R_1} + \frac{1}{R_2} + \frac{1}{R_3} + \cdots$$

7. To measure one of two or more resistors connected in parallel, say R_1 in Fig. 16-2, disconnect one lead of R_1 from the circuit. Then measure the resistance of R_1.

SELF-TEST

Check your understanding by answering these questions:
1. In the circuit of Fig. 16-1, $I_T = 0.02$ A. $V = 50$ V. The total resistance R_T equals _____ Ω.
2. For the conditions in question 1, it is possible for R_1 to equal 1,000 Ω. _____ (true/false)
3. For the conditions in question 1, the voltage V across $R_2 =$ _____ V.
4. In order to measure the resistance of R_3 in Fig. 16-2, simply place the ohmmeter leads across GF and read the resistance. _____ (true/false)
5. In Fig. 16-2 the resistance of $R_1 = 100$ Ω, $R_2 = 200$ Ω, $R_3 = 300$ Ω. $R_T =$ _____ Ω.

MATERIALS REQUIRED

- Power supply: Regulated variable direct current
- Equipment: EVM, 0–10 milliammeter, ohmmeter (or ohm scale of an EVM)
- Resistors: ½-W 2,200-, 3,300-, 4,700-, 5,600-, and 10,000-Ω
- Miscellaneous: SPST switch

PROCEDURE

1. Measure the resistance of each of the resistors supplied and record its value in Table 16-1.
2. Connect resistors R_1, R_2, etc., in parallel as in Fig. 16-2, using combination 1 of resistors shown in Table 16-2.
3. With an ohmmeter measure the total resistance of this combination and record its value in Table 16-2.
4. Repeat the procedures in steps 2 and 3 for each of the combinations shown in Table 16-2.

TABLE 16-1.

Rated value, ohms	2,200	3,300	4,700	5,600	10,000
Measured value, ohms					

5. Compute the value of R_T for each combination in Table 16-2 by substituting the measured values of R_1, R_2, etc., from Table 16-1 in the formula

$$R_T = \frac{1}{\dfrac{1}{R_1} + \dfrac{1}{R_2} + \dfrac{1}{R_3} + \cdots}$$

and record the computed values in Table 16-2. Show your computations.
6. Connect the circuit of Fig. 16-3 using the resistor values of combination 1 from Table 16-2. Just two resistors are required in combination 1. Therefore leave out R_3. Close switch and set the output of the voltage source V at 10 V and maintain it at this level, under load, until you have completed the measurements in Table 16-3.

TABLE 16-2.

Combination	Rated Value ohms					Measured Value of R_T ohms	Computed Value of R_T ohms
	R_1	R_2	R_3	R_4	R_5		
1	2,200	3,300	X	X	X		
2	2,200	3,300	4,700	X	X		
3	2,200	3,300	4,700	5,600	X		
4	2,200	3,300	4,700	5,600	10,000		

TABLE 16-3.

Combination	Measured Values		Computed Values ohms R_T
	Volts V	Amperes I_T	
1			
2			
3			
4			

7. Measure and record, in Table 16-3, the applied voltage V, and the total current I_T.
8. Repeat the procedures in steps 6 and 7 in turn for each of the combinations in Table 16-2.
9. For each combination compute the total resistance of the parallel-connected resistors by substituting the measured values of V and I_T in the formula

$$R_T = \frac{V}{I_T}$$

Record your results in Table 16-3. Show all your computations.

QUESTIONS

1. What is the effect on the total resistance of parallel-connected resistors of:

(a) Increasing the number of resistors in parallel?
(b) Decreasing the number of resistors in parallel?
2. Support your answer to question 1 by referring specifically to the measurements you recorded in Tables 16-2 and 16-3.
3. Do the measured values of R_T in Table 16-2 agree with the computed values of R_T in the same table using Eq. (16-5)? Refer specifically to your measurements.
4. Do the results of your measurements in Tables 16-1 and 16-2 prove that it is possible to write a general formula for total resistance of parallel-connected resistors? If yes, write a general formula for R_T and explain it in your own words.
5. Support your answer to question 4 by comparing your computed and measured values in Table 16-2. Comment on any discrepancies.
6. Compare the computed values of R_T in Table 16-3 with the computed values of R_T in Table 16-2, for the same combinations.
7. What was the purpose of steps 6 to 9 in this experiment?
8. What are the *three* methods you used in this experiment to determine the total resistance R_T of parallel-connected resistors?

Answers to Self-Test

1. 2,500 Ω
2. false
3. 50
4. false
5. $54^6/_{11}$ Ω

17

Parallel-circuit Design Considerations

OBJECTIVES

1. To design a parallel circuit which will meet specified voltage, current, and resistance requirements
2. To construct the circuit and check, by measurement, the design parameters

INTRODUCTORY INFORMATION

Designing a Parallel Circuit to Meet Specified Resistance Requirements

The formulas for total resistance of parallel-connected resistors can be applied to the solution of simple design problems of this type. An example will indicate the techniques to be used.

Problem 1

A technician has a stock of the following color-coded resistors: four 68 Ω, five 82 Ω, two 120 Ω, three 180 Ω, two 330 Ω, 470 Ω, 560 Ω, 680 Ω, and 820 Ω. He needs a resistance value of 37 Ω for a circuit he is designing. Find a combination of resistors, using the least possible number of components, which will satisfy the design requirement. Assume the measured values of the resistors is the same as the color-coded.

Solution. Since the value of 37 Ω is less than the resistance of the smallest resistor in stock, a parallel arrangement will be required (recall from a previous experiment that the total resistance of a parallel circuit is less than the resistance value of the smallest resistor). The formula for total resistance R_T of a parallel circuit is

$$R_T = \cfrac{1}{\cfrac{1}{R_1} + \cfrac{1}{R_2} + \cfrac{1}{R_3} + \cdots} \qquad (17\text{-}1)$$

Suppose that two resistors satisfy the design requirement. Equation (17-1) can then be written as follows:

$$R_T = \frac{R_1 \times R_2}{R_1 + R_2} \qquad (17\text{-}2)$$

that is, the total resistance of two parallel resistors is equal to the product of their resistances divided by the sum of their resistances. The use of this formula can save computation time.

Let it be assumed that two resistors will meet the design requirements and that the resistors are 68 Ω and 82 Ω. Substituting these assumed values in Eq. (17-2) gives

$$R_T = \frac{68 \times 82}{68 + 82} = \frac{68 \times 82}{150} = 37.2 \ \Omega$$

to slide-rule accuracy.

It is evident then that the two values selected meet the problem requirements, for when connected in parallel their total resistance is 37 Ω, approximately. Trial and error will show that no other combination of two resistors in stock will yield 37 Ω.

It is not always possible to select so easily the two required resistors. Another method is suggested. Suppose we assume that 68 Ω is one of the resistors. We wish to find another resistor R_x which, when paralleled with 68 ohms, will yield an equivalent resistance of 37 Ω. Substitute the known values of R_T and R_1 in Eq. (17-2):

$$37 = \frac{68 \times R_x}{68 + R_x}$$

Solving this first-degree equation yields the value

Fig. 17-1. What must be the value of R_x in parallel with 180 Ω, so that the resultant resistance $R_{AB} = 45$ Ω?

$R_x = 81$ Ω to slide-rule accuracy. It is evident then that the 82-Ω resistor in stock is the required R_x.

Problem 2

Assume the same stock as in problem 1. The technician must design a circuit requiring a 60-Ω resistor.

Solution. By inspection it is evident that 60 Ω is one-half the resistance value of 120 Ω. By manipulating Eq. (17-1), it can be shown that if n equal-valued resistors, say R_1, are connected in parallel, their total resistance R_T is equal to $(1/n) \times R_1$. That is, a general formula for the total resistance R_T of n parallel-connected resistors each of which measures R_1 ohms is

$$R_T = \frac{R_1}{n} \qquad (17\text{-}3)$$

Using Eq. (17-3) yields the solution to this problem, namely two 120-Ω resistors connected in parallel.

The technician could also use three 180-Ω resistors connected in parallel, for their total resistance is also 60 Ω. However, here three components are used rather than the two 120-Ω resistors.

Problem 3

In the circuit of Fig. 17-1, a technician measures the resistance between points A and B and finds it is 180 Ω. His circuit requires that the resistance between A and B must be 45 Ω. How can he modify

the circuit to meet the required value, assuming he has the same stock of resistors as in problem 1?

Solution. Assume that there is a resistor R_x which, when connected in parallel with 180 Ω, will bring the resistance R_{AB} down to 45 ohms. Now substitute the known values in Eq. (17-2).

$$45 = \frac{180 \times R_x}{180 + R_x}$$

The solution of this first-degree equation for R_x is

$$R_x = 60 \ \Omega$$

What is required, then, is a 60 ohm resistor. From problem 2 it is evident that two 120-Ω resistors, connected in parallel with the 180-Ω resistor, will yield the required result, namely

$$R_{AB} = 45 \ \Omega$$

The procedure, then, for designing a parallel circuit having a specified resistance value R_T, from a group of known resistors, is to apply Eqs. (17-1), (17-2), and (17-3) to the resistors at hand and find a combination which, analytically, equals or is closest to the required value. After this is done the technician should connect the resistors in parallel and measure their total resistance with an ohm-meter to confirm the solution.

Designing a Parallel Circuit to Meet Specified Resistance and Current Requirements

Ohm's law, and the formulas for total resistance in a parallel circuit, are applied in the solution of this type of problem. Again, an example will illustrate the procedure.

Problem 4

In the circuit of Fig. 17-2a determine analytically the voltage value V to which the variable dc power

(a)   (b)

Fig. 17-2. What value of V will permit 0.2-A of current in the circuit?

supply must be set to permit a total current of 0.20 A.

Solution. In order to find the applied voltage V, it is first necessary to determine the total resistance R_T in the circuit. When that is known, V can be found by using the formula $V = I \times R_T$ derived from Ohm's law.

1. *Procedure to find R_T:* The 680-Ω resistor between A and B is connected in parallel with the two series-connected resistors, 1,200 and 820 Ω. The first step is to replace the 1,200- and 820-Ω resistors with a single resistor R_1 whose resistance value is equal to the sum of the two resistors. Thus

$$R_1 = 1,200 + 820 = 2,020 \ \Omega$$

The resulting equivalent circuit is then Fig. 17-2b whose resistance and current characteristics are the same as those of Fig. 17-2a.

R_T may now be calculated by use of Eq. (17-2).

$$R_T = \frac{680 \times 2,020}{680 + 2,020} = 655 \ \Omega$$

2. *Procedure to find V:* Now

$$V = I_T \times R_T$$
$$= 0.2 \times 655 = 131 \ V$$

The required voltage is 131 V.

The technician must now connect the circuit of Fig. 17-2a and apply a measured 131 V from the dc supply. The measured current should be 0.2 A.

NOTE: Practically, the procedure the technician could have followed without attempting to find the value of V analytically, was to connect the circuit of Fig. 17-2 and adjust the output of the dc supply until the ammeter measured 0.2 A. He then could measure the dc voltage, which should be 131 V.

Designing a Parallel Circuit to Meet Specified Voltage and Current Requirements

Ohm's law and the formulas for total current and total resistance are applied in the solution of this type of problem. The following examples will illustrate the procedures to use.

Problem 5

In the circuit of Figure 17-3, a current-divider network is required which will permit 0.02 A in re-

Fig. 17-3. What are the values of the resistors R_1 and R_2 which meet the voltage and current specifications of the circuit?

sistor R_1 and 0.03 A in resistor R_2. A dc voltage source of 75 V powers the circuit. What values of R_1 and of R_2 must be used to meet the circuit requirements?

Solution. The voltage across each branch of a parallel network is the same. Therefore the voltage V_1 across R_1 equals the voltage V_2 across R_2. In this case,

$$V_1 = V_2 = 75 \ V$$

By Ohm's law

$$R_1 = \frac{V_1}{I_1} = \frac{75}{0.02} = 3,750 \ \Omega$$

$$R_2 = \frac{V_2}{I_2} = \frac{75}{0.03} = 2,500 \ \Omega$$

The required resistors, therefore, are 3,750 and 2,500 Ω. The technician should therefore connect 3,750 and 2,500 Ω in the circuit of Fig. 17-3 and measure the branch currents I_1 and I_2 to verify the solution.

Problem 6

A technician must design a circuit which will deliver 0.25 A of current from a 15-V supply. Determine analytically what combination of resistors, using the least number of components, will satisfy the design requirements. The stock of resistors which the technician has on hand includes two 120-, three 150-, three 180-, 330-, 470- and three 680-Ω resistors.

Solution. The total resistance which will satisfy the requirements of the problem is

$$R_T = \frac{15}{0.25} = 60 \ \Omega$$

A series combination will not give the required solution, but a parallel arrangement may. By inspection, it is apparent from the stock of resistors that two 120-Ω resistors in parallel will give the required value of 60 Ω. Another solution is a parallel combination of three 180-Ω resistors, but of course this requires one more component than the first combination. Hence the required solution is the combination of two parallel-connected 120-Ω resistors.

SUMMARY

1. If required to employ in a circuit a resistance value R_T which is less than the smallest resistance value of a stock of resistors, it may be possible to approximate the required value by paralleling two or more of the available resistors. The procedure is to choose a resistor R_1 whose resistance is slightly higher than the required R_T, and find from the formula

$$R_T = \frac{R_1 \times R_2}{R_1 + R_2}$$

the value of R_2 which will yield R_T. If R_2 is in stock, the two resistors have been found. Otherwise it may be necessary to repeat the process. Finally, if two resistors cannot be found whose parallel combination is R_T, it may be necessary to try three resistors, or four.

2. The equivalent resistance R_T of n equal-valued resistors R_1, connected in parallel, is

$$R_T = \frac{R_1}{n}$$

3. If it is necessary to determine analytically the voltage V to which a dc voltage source must be set to permit I amperes of current in a circuit containing R ohms of resistance, V may be found by the formula

$$V = I \times R$$

4. It is possible to determine experimentally the voltage V to which a dc voltage source must be set to permit I amperes of current in a circuit containing R ohms of resistance. A circuit is connected containing the variable dc source in series with an ammeter and the resistance R. The voltage is varied until I current is measured. Then the measured voltage V, at the terminals of the dc source, is the required voltage.

5. If it is required to determine analytically what

resistance R to connect in a circuit which will permit I amperes from a voltage source V, find R from the formula

$$R = \frac{V}{I}$$

If the value of R is available in stock, that solves the problem. Otherwise determine, using series or parallel resistors, a combination that will satisfy the required R.

SELF-TEST

Check your understanding by answering these questions:
1. What two resistors in parallel will yield a total resistance R_T of 30 Ω? Give at least two solutions. (a) _____ Ω in parallel with _____ Ω; (b) _____ ohms in parallel with _____ Ω.
2. 22-, 33-, and 47-Ω resistors are connected in parallel. Their total resistance will be _____ than 22 Ω.
3. A 10-Ω resistance may be constructed from _____ 50-Ω resistors by paralleling these resistors.
4. In the circuit of Fig. 17-4 what is the voltage V which will permit 0.2 A? _____ V
5. In the circuit of Fig. 17-3 there is 0.003 A of current. The total resistance R_{AB} between points A and B is _____ Ω.

MATERIALS REQUIRED

- Power supply: Regulated variable dc voltage source
- Equipment: Electronic voltmeter, 0–100 milliammeter
- Resistors: ½-W 560-, 1,200-, 1,800-, 3,300-, 5,600-, 10,000-Ω
- Miscellaneous: SPST switch

Fig. 17-4. What value of V will permit 0.2 A of current in this circuit?

1. Measure the resistance of each of the resistors supplied and record its value in Table 17-1.
2. Determine analytically what parallel combination of the resistors supplied will yield the value 457 Ω (approximately). Use the color-coded values of the resistors for this computation and show your computations. Record the color-coded values of R in Table 17-2.
3. Connect the parallel combination and measure the total resistance R_T with an ohmmeter. Record the measured value in Table 17-2.
4. Repeat steps 2 and 3 for each value of R_T required, listed in Table 17-2.
5. Using the resistors supplied design a parallel circuit combination, like Fig. 17-1, which will yield 0.02 A (approximately) when 40 V is applied. Show your computations. Record the color-coded values of R in Table 17-3.
6. Connect the circuit and measure the current I_T. Record this value in Table 17-3.
7. Determine, analytically, what value of voltage V will be required for a circuit similar to Fig. 17-1, if $R_1 = 1,200$ Ω, $R_2 = 1,800$ Ω, and $I_T = 0.05$ A. Show your computations. Record the computed value of V in Table 17-3.
8. Connect the circuit and adjust the supply until there is 0.05 A in the circuit. Measure the supply voltage V and record this measured value in Table 17-3.
9. Design a two-branch parallel circuit from the resistors supplied so that 0.03 A of current is in one branch, 0.02 A in the other branch. Draw the circuit, and show your computations, including the voltage V Computed. Record the color-coded values of the resistors in Table 17-4.
10. Connect the circuit. Set the dc supply at the computed value of voltage V. Measure and record the current I_1 in R_1 and I_2 in R_2.
11. Theoretically design a parallel circuit with the resistors in this experiment, using
    a. Two resistors
    b. Three resistors

c. Four resistors
so chosen that each combination **a** through **c** draws 10 mA total current I_T at an applied voltage V to be determined. Show your computations and record in Table 17-5 the measured values of R_1, R_2, R_3, and R_4 chosen, and the design (computed) value for V for each combination.
12. Experimentally verify the design values for V and record your current measurements in Table 17-5.

QUESTIONS

1. How did the measured values of the resistors in this experiment compare with their color-coded values? Refer to Table 17-1 for your answer.
2. Why was it necessary to measure the resistors supplied in this experiment?
3. In Table 17-2, how did the measured value of R_T compare with the required value? Explain any discrepancies.
4. Did the measurement in step 6 confirm the design values in step 5? If not, why not? Refer specifically to the data in Table 17-3.
5. Did the measured values in step 8 confirm the design values in step 7? If not, why not? Refer specifically to the data in Table 17-3.
6. Did the measured values in step 10 confirm the design values in step 9? If not, why not? Refer specifically to the data in Table 17-4.
7. Design a three-branch parallel circuit which will divide a total current of 0.22 A in the ratio 2:3:6. If the total resistance of the three parallel resistors is 109 Ω, find the value of each resistor and find also the applied voltage. Draw the circuit and show your computations.

Answers to Self-Test

1. (a) 60, 60; (b) 50, 75
2. less
3. 5
4. 21.4
5. 25,000 Ω

TABLE 17-1.

Rated value, ohms	560	1,200	1,800	3,300	5,600	10,000
Measured value, ohms						

TABLE 17-2.

R_T Required ohms	Combination of Parallel Resistors			Measured Value R_T ohms
	R_1	R_2	R_3	
457				
720				
315				
3,590				

TABLE 17-3.

Step	V Applied, volts			I_T, amperes		Parallel Resistors, ohms	
	Given Value	Computed Value	Measured Value	Required Value	Measured Value	R_1	R_2
5, 6	40	X		0.02			
7, 8	X			0.05		1,200	1,800

TABLE 17-4.

Branch Current, amperes				Resistance, ohms		Voltage, volts	
Required		Measured		R_1	R_2	Computed	Measured
I_1	I_2	I_1	I_2				
0.03	0.02						

TABLE 17-5.

Combination	Measured Values				V Applied Design Value	I Measured
	R_1	R_2	R_3	R_4		
1						
2						
3						

18

Total Resistance of Series-parallel Circuits

OBJECTIVES

1. To determine and verify experimentally the law for total resistance R_T of a series-parallel combination of resistors
2. To verify that the voltage across each leg in a parallel circuit is the same as the voltage across the parallel circuit
3. To design a series-parallel network which will meet specified requirements

INTRODUCTORY INFORMATION

Total Resistance of a Series-parallel Circuit

Figure 18-1 shows a series-parallel arrangement of resistors. In this circuit, R_1 is in series with the parallel circuit between points B and C, which in turn is in series with R_3. What is the total resistance between points A and D? Obviously we can measure R_T with an ohmmeter, or R_T can be found by the voltage-current method described in Experiment 12. Is it possible, however, to write a formula by which R_T of a series-parallel circuit may be computed without measurement? It is possible to determine a formula for R_T. The method we will use to

develop a formula involves *analysis* and *measurement* to verify the results of our analysis.

Refer to Fig. 18-1. If R_1 and R_3 were removed and if the parallel circuit between points B and C were left "standing alone," we could compute the total resistance R_{T2} between B and C by the formula for parallel resistors. We can therefore replace the parallel network in Fig. 18-1 by its equivalent resistance R_{T2}, and the circuit will now take the form of Fig. 18-2. It is apparent that the total resistance R_T of the series circuit (Fig. 18-2) is the same as the total resistance R_T between points A and D (Fig. 18-1).

Our analysis suggests that to find the total resistance of a series-parallel network, replace the parallel circuit by its equivalent resistance R and treat the resultant network like a simple series circuit. In the case of Fig. 18-1, the total resistance R_T between points A and D would be

$$R_T = R_1 + R_{T2} + R_3 \qquad (18\text{-}1)$$

Can this method be extended to include a series-parallel circuit involving more than one parallel network, as for example in Fig. 18-3? Observation and comparison show that the circuit between points A and D is identical with the circuit of Fig. 18-1. Hence, we can compute the resistance R_{T1} of Fig. 18-3 between points A and D using Eq. (18-1). We can now replace the section of Fig. 18-3 between A and D with R_{T1}. The resulting Fig. 18-4 is the equivalent of Fig. 18-3.

Figure 18-4 can further be simplified by replacing R_6 and R_7 with their series resistance $R_{6\text{-}7}$, where

Fig. 18-1. Series-parallel combination of resistors.

Fig. 18-2. Equivalent circuit of Fig. 18-1.

Fig. 18-3. Another series-parallel combination of resistors.

$$R_{6\text{-}7} = R_6 + R_7$$

The resultant Fig. 18-5 is the equivalent of Figs. 18-4 and 18-3. Comparison with Fig. 18-1 shows that the total resistance for Fig. 18-5 may now be computed by Eq. (18-1).

It is apparent, therefore, that series-parallel networks containing more than one parallel circuit may be simplified by replacing each parallel circuit with its equivalent resistance. Then the total resistance R_T of the network (see Fig. 18-3) is

$$R_T = R_1 + R_{T2} + R_3 + R_{T4} + R_8 + \cdots \quad (18\text{-}2)$$

where R_{T2} is the equivalent (total) resistance of the first parallel circuit, R_{T4} is the equivalent resistance of the second parallel circuit, etc., and R_1, R_3, R_8, etc., are resistors in series with the parallel circuits.

A final step in the process of establishing Eqs. (18-1) and (18-2) as a valid "law" requires experimental verification. A simple method of verification suggests itself. Measure the value of each resistance before it is connected in the circuit. Measure the resistance R_T of the total combination. Compute R_{T2}, R_{T4}, etc., for each of the parallel circuits. (Substitute the measured values of the appropriate resistors in the formula for parallel resistance and solve for R_{T2}, R_{T4}, etc.) Now add the measured values R_1, R_3, R_8, etc., to the computed values R_{T2}, R_{T4}, etc. If the measured value of

$$R_T = R_1 + R_{T2} + R_3 + R_{T4} + R_8 + \cdots$$

we have a confirmation of Eq. (18-2). Additional measurements, using other combinations, must be made to establish the validity of Eqs. (18-1) and (18-2).

Voltage across Each Leg of a Parallel Circuit

Consider the circuit of Fig. 18-6. The circuit between points B and D consists of R_2 in parallel with the series-connected combination R_4 and R_5. To determine the voltage across resistor R_2, we measure the potential difference between points B and D. To determine the voltage across the series combination of R_4 and R_5, we also measure the emf across points B and D. Moreover, the voltage across the parallel circuit between B and D is measured in the same way. This is so because a common connection joins one side of all the legs of a parallel circuit and another common connection joins the opposite sides of all the legs.

Since a voltage check across any or all legs must be made between the same two common points, the voltage across any or all legs must be the same. We reached this conclusion in Experiment 15. We will verify it in this experiment.

Fig. 18-5. Equivalent circuit of Figs. 18-4 and 18-3.

Fig. 18-4. Equivalent circuit of Fig. 18-3.

Fig. 18-6. What is the voltage across R_2?

SUMMARY

1. In a series-parallel network like that in Fig. 18-1, the total resistance R_T of the network measured across the end terminals (A and D) can be found by replacing each parallel combination by its equivalent resistance R_{T2}, leaving an equivalent series circuit (Fig. 18-2). R_T may then be computed by using the series-resistance formula:

$$R_T = R_1 + R_{T2} + R_3 + \cdots$$

2. In determining the equivalent resistance R_{T2} of one of the parallel combinations such as R_2, R_4 in Fig. 18-1, the formula for parallel resistors is used.

3. A parallel branch may sometimes have two or more series resistors, such as branch R_6–R_7 in Fig. 18-4. In that case R_6 and R_7 are combined like series resistors and their equivalent series resistance $R_{6-7} = R_6 + R_7$ is substituted in the circuit, leaving a network like that in Fig. 18-5. Then Fig. 18-5 is transformed into its series equivalent form, Fig. 18-2.

4. In a parallel network the voltage across each branch is the same. Thus in Fig. 18-6, the voltage V_2 across R_2 equals V_{BD}, and this voltage is equal to the voltage V_{4-5} across the series combination of R_4 and R_5. That is

$$V_2 = V_{BD} = V_{4-5}$$

SELF-TEST

Check your understanding by answering these questions:

1. In Fig. 18-1, $R_1 = 120$, $R_2 = 280$, $R_3 = 330$, and $R_4 = 470 \ \Omega$. What is the value of R_{T2}, measured from B to C? _____ Ω

2. For the same conditions as in question 1 what is the value of R_T, measured from A to D? _____ Ω

3. In Fig. 18-3, $R_1 = 33$, $R_2 = 56$, $R_4 = 68$, $R_3 = 100$, $R_5 = 120$, $R_6 = 47$, $R_7 = 73$, and $R_8 = 470 \ \Omega$. What is the equivalent resistance R_{T2}, between points B and C? _____ Ω

4. For the same conditions as in question 3, what is the equivalent resistance R_{6-7} in the lower branch of the parallel network between points D and F? _____ Ω

5. For the same conditions as in question 3, what is the equivalent resistance R_{T3} of the parallel network between points D and F? _____ Ω

6. For the same conditions as in question 3, what is the total resistance R_T between points A and G? _____ Ω

7. In Fig. 18-6, what is the total resistance R_T in the circuit? _____ Ω

8. In Fig. 18-6, if $V = 228$ V, what is the total current in the circuit? _____ A

9. If in Fig. 18-6 the current in R_2 is 0.0675 A, what is the voltage V_{BD} across the series combination of R_4 and R_5? _____ V

MATERIALS REQUIRED

- Power supply: Regulated variable direct current
- Equipment: EVM, 0–10 milliammeter, ohmmeter (or ohms scale of an EVM)
- Resistors: ½-W 330-, 470-, 560-, 1,200-, 2,200-, 3,300-, 4,700-, and 10,000-Ω
- Miscellaneous: SPST switch

PROCEDURE

Total Resistance of a Series-parallel Network

1. Measure the resistance of each of the resistors supplied and record its value in Table 18-1.
2. Connect the circuit of Fig. 18-1 using the values in Table 18-1. This is combination 1. With an ohmmeter measure and record in Table 18-1 the total resistance R_T (between points A and D) and the resistance R_{T2} (between points B and C).
3. Add the measured values R_1, R_{T2}, and R_3 and record their sum under column "Computed value, R_T:(a)."

Sample calculation. If the measured values of $R_1 = 300$, $R_{T2} = 300$, $R_3 = 500$, then R_T:(a) = $R_1 + R_{T2} + R_3 = 300 + 300 + 500 = 1,100$.

4. Calculate the value of R_{T2} by substituting the measured values R_2 and R_4 in the formula $1/R_{T2} = 1/R_2 + 1/R_4$ and record this value under the column "Computed value, R_{T2}."

Sample calculation. If the measured values of $R_2 = 400$ and $R_4 = 1,200$, then

$$\frac{1}{R_{T2}} = \frac{1}{R_2} + \frac{1}{R_4} = \frac{1}{400} + \frac{1}{1,200} = \frac{4}{1,200}$$

and "Computed value, R_{T2}" = 300.

TABLE 18-1.

Rated Value, ohms	R_1	R_2	R_3	R_4	R_5	R_6	R_7	R_8
	330	470	560	1,200	2,200	3,300	4,700	10,000
Measured Value, ohms								

Combination	Measured Value			Computed Value			
	R_T	R_{T2}	R_{T4}	R_T:(a)	R_{T2}	R_{T4}	R_T:(b)
1 (Fig. 18-1)			X			X	
2 (Fig. 18-1)			X			X	
3 (Fig. 18-3)							
4 (Fig. 18-3)							

5. Calculate the value of R_T by adding the measured values of R_1 and R_3 and the computed value of R_{T2} (from step 4) and record this value under the column "Computed value, R_T:(b)."

   Sample calculation. Since the measured values of $R_1 = 300$, $R_3 = 500$, and the *computed value* of $R_{T2} = 300$, R_T:(b) $= R_1 + R_{T2} + R_3 = 300 + 300 + 500 = 1,100$.

NOTE: In the sample calculations, the computed values R_T:(a) and R_T:(b) are exactly equal. Of course, we assumed perfect accuracy in our meter and in our measurements. This is the ideal situation for which the research technician strives, but never quite attains.

6. Repeat steps 2 through 5 for combination 2, using the rated values $R_1 = 470$, $R_2 = 2,200$, $R_3 = 1,200$, and $R_4 = 3,300$.
7. Connect the circuit of Fig. 18-3, using the values in Table 18-1. This is combination 3. Measure and record the total resistance R_T between points A and G; the resistances R_{T2} between points B and C; and the resistance R_{T4} between points D and F.
8. Add the measured values R_1, R_{T2}, R_3, R_{T4}, and R_8, and record their sum under column "Computed value, R_T:(a)."

9. Calculate the value of R_{T2} using the measured values of R_2 and R_4. Calculate also the value of R_{T4} by substituting the measured values of R_5, R_6, and R_7 in

$$\frac{1}{R_{T4}} = \frac{1}{R_5} + \frac{1}{R_6 + R_7}$$

   Record these values in Table 18-1.
10. Calculate the value of R_T by adding the measured values R_1, R_3, and R_8 and the computed values R_{T2} and R_{T4}, and record under the column "Computed value, R_T:(b)."
11. Repeat steps 7 through 10 for combination 4, using the rated values $R_1 = 470$, $R_2 = 2,200$, $R_3 = 1,200$, $R_4 = 3,300$, $R_5 = 10,000$, $R_6 = 560$, $R_7 = 4,700$, and $R_8 = 330$.

Voltage across Each Leg of a Parallel Circuit

12. Connect the circuit of Fig. 18-6. S is open. Set the output of the regulated power supply to 20 V and maintain it at this level.
13. Close S. With an EVM set on the proper voltage range, measure the voltage across R_1, R_2, and R_3 and the series combination of R_4 and R_5. Measure also the voltage across the parallel network V_{BD}. Record in Table 18-2.

TABLE 18-2.

V Applied	V_1 (across R_1)	V_2 (across R_2)	V_{4-5} (across R_4-R_5)	V_3 (across R_3)	V_{BD}

TABLE 18-3.

Circuit	Design Values			Measured Values		
	V	R_T	I_T	V	R_T	I_T
a						
b						
c						

Design Problem (extra credit)

14. Theoretically design a series-parallel circuit with the resistors in this experiment, using first
    (*a*) Four resistors, then
    (*b*) Six resistors, then
    (*c*) Eight resistors,
    so chosen that each combination *a* through *c* draws 5 mA total current I_T. Draw each of the three circuits, showing on the diagram the value of each resistor chosen.

NOTE: Use the measured values from Table 18-1. Show your computations and record in Table 18-3 the design (computed) values of R_T, V, and I_T in each case.

15. Connect each circuit, in turn. Determine by measurement and record in Table 18-3 (under "Measured values") V, R_T, and I_T.

QUESTIONS

1. In measuring the resistance R_{T2} in step 2, was it necessary to open the circuit at *B* or *C*? Why?
2. Why is it necessary to disconnect power before measuring resistance in a circuit?
3. Refer to Table 18-1. Are the measured value R_T and the computed values R_T:(*a*) and R_T:(*b*) the same? Should they be? Explain.
4. What conclusion can you draw from your measurements in Table 18-1 concerning the total resistance of a series-parallel circuit? Refer to specific measurements to substantiate your conclusion.
5. What is the rule for determining the total resistance R_T of a series-parallel circuit?

6. What can you say about the voltage across each leg of a parallel circuit? Refer to specific measurements to substantiate your statement.
7. What information do you need to compute the current in each leg of a parallel network?
8. Refer to Fig. 18-6. If I_2 is the current in R_2, I_{4-5} the current in R_4 and R_5, and I_T is the total current, what is the relationship between $I_2 \times R_2$, $I_{4-5} \times (R_4 + R_5)$, and $I_T \times (R_{T2})$?
9. Explain two methods you may use for measuring the total resistance R_T of a series-parallel network, such as that in Fig. 18-1.
10. Refer to Table 18-2. Does there appear to be any relationship between V applied and V_1, V_2, and V_3? If yes, what is the relationship?

Extra Credit

11. Comment on the results of your design values and measured values in the design problem. Explain any discrepancies between measured and predicted results.

Answers to Self-Test

All computations to slide-rule accuracy.

1. 176 Ω
2. 626 Ω
3. 30.8 Ω
4. 120 Ω
5. 60 Ω
6. 694 Ω
7. 2,280 Ω
8. 0.1 A
9. 148+ V

19

Kirchhoff's Voltage Law (for one generator)

OBJECTIVES

1. To determine by analysis the relationship between the sum of the voltage drops across series-connected resistors, and the applied voltage
2. To verify experimentally the relationship determined in objective 1

INTRODUCTORY INFORMATION

The solution of complex electric circuits is facilitated by the application of Kirchhoff's laws. These laws were formulated and published by the physicist Gustav Robert Kirchhoff (1824–1887), and they established the basis for modern network analysis. They are applicable to circuits with one or more voltage sources. In this experiment we will be concerned with a single voltage source.

Voltage Law

Consider the simple series circuit, Fig. 19-1. We have previously established that the series-connected resistors R_1, R_2, R_3, and R_4 may be replaced by their total or equivalent resistance R_T without affecting the current I_T in the circuit. The current I_T may then be computed by Ohm's law, for

$$I_T = \frac{V}{R_T} \qquad (19\text{-}1)$$

Fig. 19-1. $V = V_1 + V_2 + V_3 + V_4$

From Eq. (19-1), it also follows that

$$V = I_T \times R_T \qquad (19\text{-}2)$$

In the series-connected circuit of Fig. 19-1

$$R_T = R_1 + R_2 + R_3 + R_4 \qquad (19\text{-}3)$$

Substituting the equivalent value for R_T from Eq. (19-3) in Eq. (19-2) gives

$$V = I_T(R_1 + R_2 + R_3 + R_4)$$

or

$$V = I_T \times R_1 + I_T \times R_2 + I_T \times R_3 + I_T \times R_4 \qquad (19\text{-}4)$$

One of the characteristics of a series circuit is that current is the same everywhere in the circuit. I_T is, therefore, the total current in R_1, R_2, R_3, and R_4. It follows, therefore, that

$I_T \times R_1 = V_1$, the voltage drop across R_1
$I_T \times R_2 = V_2$, the voltage drop across R_2
$I_T \times R_3 = V_3$, the voltage drop across R_3

and $I_T \times R_4 = V_4$, the voltage drop across R_4

Equation (19-4) may now be rewritten as follows:

$$V = V_1 + V_2 + V_3 + V_4 \qquad (19\text{-}5)$$

Equation (19-5) is a mathematical expression of Kirchhoff's voltage law for resistors connected in series in a closed circuit.

It is apparent that Eq. (19-5) may be generalized for circuits containing one or more series-connected resistors in a closed circuit. The law also applies in the case of series-parallel circuits (Fig. 19-2). Here, $V = V_1 + V_2 + V_3 + V_4 + V_5$, where V_1, V_3, and V_5 are the voltage drops across R_1, R_4, and R_8, respectively. V_2 and V_4 are the voltages across the parallel circuits between A and B, and between C and D, respectively.

Expressed in words, Eq. (19-5) states that in a

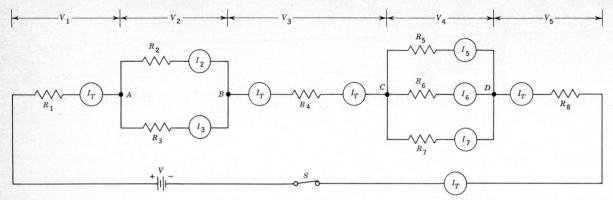

Fig. 19-2. Application of Kirchhoff's voltage law to a series-parallel circuit. $V = V_1 + V_2 + V_3 + V_4 + V_5$

closed circuit or loop, the applied voltage equals the sum of the voltage drops in the circuit.

The concept of algebraic signs or polarity is helpful in solving electrical network problems. An example will illustrate the convention employed in assigning a + or − sign to a voltage in a circuit. Consider the circuit of Fig. 19-3. Assume that current in this circuit is synonymous with electron flow, and that electrons move from a negative to positive potential. The arrow in Fig. 19-3 shows the direction of current, and the − and + signs indicate the following: point A is negative with respect to B; point B is negative with respect to C, etc. This is consistent with our assumption of current in this circuit.

Now to establish the algebraic sign for the voltages in the closed circuit, move in the direction of assumed current. Consider as positive any voltage source or voltage drop whose + (positive) terminal is reached first, and negative any voltage source or drop whose − (negative) terminal is reached first. Starting at point A in Fig. 19-3 and moving in the direction of current we have $-V_1$, $-V_2$, $-V_3$, $-V_4$ and $+V$. With this convention in mind Kirchhoff's voltage law may now be generalized as follows.

The algebraic sum of the voltages in a closed circuit equals zero.

Applying the convention on signs and Kirchhoff's law to the closed circuit of Fig. 19-3 and starting at point A, we may write as follows:

$$-V_1 - V_2 - V_3 - V_4 + V = 0 \qquad (19\text{-}6)$$

Is this equation consistent with Eq. (19-5)? Yes, for by transposing the terms on the right side of Eq. (19-5) to the left side, we get $V - V_1 - V_2 - V_3 - V_4 = 0$, a result which is identical with that in Eq. (19-6).

SUMMARY

Kirchhoff's voltage law may be expressed in two ways.
1. The sum of the voltage drops in a closed circuit equals the applied voltage; or
2. The algebraic sum of the voltages in a closed circuit equals zero.

SELF-TEST

Check your understanding by answering these questions:
1. In Fig. 19-1, $V_1 = 10$ V, $V_2 = 12$ V, $V_3 = 20$ V, $V_4 = 15$ V. The applied voltage V must then equal _____ V.
2. In Fig. 19-2, $V_1 = 15$ V, $V_2 = 20$ V, $V_4 = 27$ V, $V_5 = 9$ V, and $V = 100$ V. The voltage $V_3 = $ _____ V.

MATERIALS REQUIRED

- Power supply: Regulated variable direct current
- Equipment: EVM, 0–10 milliammeter
- Resistors: ½-W 330-, 470-, 1,200-, 2,200-, 3,300-, 4,700-, 5,600-, and 10,000-Ω
- Miscellaneous: SPST switch

Fig. 19-3. Convention for assigning polarity to voltages in a closed circuit.

PROCEDURE

Voltage Law

1. Connect the circuit of Fig. 19-1 using the values R_1, R_2, R_3, and R_4 shown in Table 19-1.
2. Adjust the output of the power supply so that $V = 40$ V. Measure and record this voltage in Table 19-2. Measure also and record the voltages V_1, V_2, V_3, and V_4 across R_1, R_2, R_3, and R_4, respectively. Add V_1, V_2, V_3, and V_4 and enter the sum in the proper column in Table 19-2.
3. Connect the circuit of Fig. 19-2 using the values R_1, R_2, etc., shown in Table 19-1.
4. Adjust the output of the power supply so that $V = 40$ V. Measure this voltage and record in Table 19-2. Measure also and record the voltages V_1, V_2, V_3, V_4, and V_5 as designated in Fig. 19-2. Add V_1, V_2, V_3, etc., and enter the sum in the proper column.

Design Problem (Extra Credit)

5. Design a series-parallel circuit like that in Fig. 19-2 using all the resistors listed in this experiment, with an applied voltage of 35 V, drawing a total current of 5 mA, approximately. Draw a diagram of the circuit, including the measured values of all resistors. Show all your computations.
6. Connect the circuit and set the power-supply voltage to 35 V. Measure the total current I_T and indicate this value on the circuit diagram.

QUESTIONS

1. How does the sum of V_1, V_2, V_3, and V_4 (step 2) in Table 19-2 compare with V? Explain the difference, if any.
2. How does the sum of V_1, $V_2 \ldots$, V_5 (step 4) in Table 19-2 compare with V? Explain the difference, if any.
3. In your own words state Kirchhoff's voltage law. Also write the equation for it.
4. Discuss the results of the design problem, step 6.
5. What made it impossible for you to design a circuit, as required, which would give *exactly* 5 mA of current?

Answers to Self-Test

1. 57
2. 29

TABLE 19-1.

	R_1	R_2	R_3	R_4	R_5	R_6	R_7	R_8
Rated value (ohms)	330	470	1,200	2,200	3,300	4,700	10,000	5,600

TABLE 19-2.

Step	V	V_1	V_2	V_3	V_4	V_5	$V_1 + V_2 + V_3 + V_4 + V_5$
2						X	
4							

OBJECTIVES

1. To determine by analysis the relationship between the sum of the currents entering any junction of an electric circuit, and the current leaving that junction
2. To verify experimentally the relationship determined in objective 1

INTRODUCTORY INFORMATION

Current Law

In Experiment 15 you verified that the total current I_T in a circuit containing resistors connected in parallel is equal to the sum of the currents in each of the parallel branches. This was one demonstration of Kirchhoff's current law, limited to a parallel network. The law is perfectly general, however, and is applicable to any circuit. It states that the *current entering any junction of an electric circuit is equal to the current leaving that junction.*

Consider the series-parallel circuit (Fig. 20-1). Designate the total current as I_T. I_T enters the junction at A, in the direction indicated by the arrow. The currents leaving the junction A are I_1, I_2, and

I_3 as shown. The currents I_1, I_2, and I_3 then enter the junction at B, and I_T leaves the junction at B. What is the relationship between I_T, I_1, I_2, and I_3? Analysis will help establish this relationship.

The voltage across the parallel circuit $V_{AB} = I_1 \times R_1 = I_2 \times R_2 = I_3 \times R_3$. The parallel network may be replaced by its equivalent resistance R_T, in which case Fig. 20-1 is transformed into a simple series circuit and $V_{AB} = I_T \times R_T$. It follows, therefore, that

$$I_T \times R_T = I_1 \times R_1 = I_2 \times R_2 = I_3 \times R_3 \quad (20\text{-}1)$$

Equation (20-1) may be rewritten as

$$I_1 = I_T \times \frac{R_T}{R_1}$$

$$I_2 = I_T \times \frac{R_T}{R_2} \qquad (20\text{-}2)$$

$$I_3 = I_T \times \frac{R_T}{R_3}$$

Adding I_1, I_2, and I_3 gives

$$I_1 + I_2 + I_3 = I_T \times R_T\left(\frac{1}{R_1} + \frac{1}{R_2} + \frac{1}{R_3}\right)$$

But $\quad \dfrac{1}{R_1} + \dfrac{1}{R_2} + \dfrac{1}{R_3} = \dfrac{1}{R_T}$

Fig. 20-1. $I_T = I_1 + I_2 + I_3$

Fig. 20-2. The algebraic sum of the currents entering and leaving a junction is equal to zero. $I_T - I_1 - I_2 = 0$

Therefore

$$I_1 + I_2 + I_3 = I_T \times R_T \times \frac{1}{R_T} = I_T$$

and
$$I_T = I_1 + I_2 + I_3 \qquad (20\text{-}3)$$

Equation (20-3) is a mathematical statement of Kirchhoff's law, applied to the circuit of Fig. 20-1. In general, if I_T is the current entering a junction of an electric circuit, and $I_1, I_2, I_3, \ldots, I_n$ are the currents leaving that junction (or vice versa), then

$$I_T = I_1 + I_2 + I_3 \cdots + I_n \qquad (20\text{-}4)$$

Kirchhoff's current law is usually stated in another way, namely, *the algebraic sum of the currents entering and leaving a junction is zero*. It will be recalled that this is similar to the formulation of Kirchhoff's voltage law, that *the algebraic sum of the voltages in a closed path or loop is zero*. Just as it was necessary to agree on a polarity convention for voltages in a loop, so it is necessary to agree on a current convention at a junction.

If the current entering a junction is considered positive (+), and the current leaving a junction is considered negative (−), then the statement that the algebraic sum of the currents entering and leaving a junction is zero can be shown to be identical with Eq. (20-4). Consider the circuit of Fig. 20-2. The total current I_T enters the junction at A and is considered +. The currents I_1 and I_2 leave the junction at A and are designated −. Then,

$$+I_T - I_1 - I_2 = 0 \qquad (20\text{-}5)$$

and
$$I_T = I_1 + I_2 \qquad (20\text{-}6)$$

It is evident that the two statements of Kirchhoff's current law lead to the same algebraic formula.

An example will show how Kirchhoff's current

law may be applied in network analysis. Suppose in Fig. 20-3 I_1 and I_2 are currents entering the junction at A and are respectively +5 and +3 A. I_3, I_4, and I_5 are currents leaving A. I_3 and I_4 are respectively 2 and 1 A. What is the value if I_5? Applying Kirchhoff's current law,

$$I_1 + I_2 - I_3 - I_4 - I_5 = 0$$

and substituting the known values of current, we get

$$5 + 3 - 2 - 1 - I_5 = 0$$
$$8 - 3 = I_5$$

and
$$I_5 = 5 \text{ A}$$

SUMMARY

1. Kirchhoff's current law states that the current entering any junction of an electric circuit is equal to the current leaving that junction.
2. An analytical statement of Kirchhoff's current law requires the assignment of polarity to current entering a junction (assume it is +) and to current leaving a junction (assume it is −).
3. With the convention of 2, Kirchhoff's current law may be stated as follows: The algebraic sum of the currents entering and leaving a junction is zero. Thus, for Fig. 20-1, at junction A

$$I_T - I_1 - I_2 - I_3 = 0$$

SELF-TEST

Check your understanding by answering the following questions:
1. In Fig. 20-1 the current entering the junction at A is 0.55 A. The current $I_1 = 0.25$ A, $I_2 = 0.1$ A. The current I_3 must therefore equal _____ A.
2. In Fig. 20-1 the current leaving the junction at $B = 1.25$ A. The sum of currents I_1, I_2, and I_3 must be _____ A.

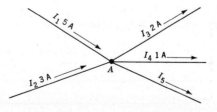

Fig. 20-3. Currents entering and leaving a junction. $I_1 + I_2 - I_3 - I_4 - I_5 = 0$

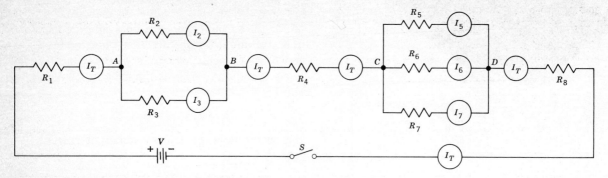

Fig. 20-4. Experimental circuit to verify Kirchhoff's current law.

3. At junction B (Fig. 20-1) the polarities which must be assigned to the currents in applying Kirchhoff's current law are: I_1 is _____; I_2 is _____; I_3 is _____; and I_T is _____.

4. The equation which describes the relationship among the currents at junction A in Fig. 20-3 is _____.

5. In Fig. 20-3, $I_2 = 10$ A, $I_3 = 4$ A, $I_4 = 4$ A, $I_5 = 4$ A. $I_1 = $ _____ A.

MATERIALS REQUIRED

- Power supply: Regulated variable dc source
- Equipment: EVM, 0–10 milliammeter
- Resistors: ½-W 330-, 470-, 2,200-, 3,300-, 4,700-, 5,600-, and 10,000-Ω
- Miscellaneous: One SPST switch

PROCEDURE

Current Law

NOTE: Before connecting the milliammeter to measure current, disconnect the power supply by opening switch S. Follow this procedure each time the meter is moved. Observe meter polarity.

1. Connect the circuit of Fig. 20-4 with $V = 40$ V. Use the values R_1, R_2, etc., in Table 20-1. Measure and record in Table 20-2 the currents I_T at A; I_2, I_3, and I_T at B; I_T at C; and I_5, I_6, I_7, and I_T at D. Add the currents I_2 and I_3 and record the sum in the proper column. Add also the currents I_5, I_6, and I_7, and record the sum in the proper column.

2. Connect the circuit of Fig. 20-1, using the values $R_1 = 330$ Ω, $R_2 = 470$ Ω, and $R_3 = 1,200$ Ω. Adjust the output of the power supply so that $V = $

40 V. Measure and record in Table 20-3 the currents I_T at A; I_1, I_2, I_3; and I_T at B. Add the currents I_1, I_2, and I_3 and record their sum in Table 20-3.

Design Problem (extra credit)

3. Design a series-parallel circuit with three legs in the parallel circuit, such that the currents in the legs I_1, I_2, and I_3 are in the ratio 1:2:3 (approximately) and the total current in the circuit is 6 mA. Use resistors listed in this experiment which will give the closest approximation to the desired ratio. Draw a diagram of the circuit showing the values of resistance chosen and the design voltage V. Show all your computations. Measure the required currents and record them

TABLE 20-1.

	R_1	R_2	R_3	R_4	R_5	R_6	R_7	R_8
Rated value ohms	330	470	1,200	2,200	3,300	4,700	10,000	5,600

TABLE 20-2.

	I_T at A	I_2	I_3	I_T at B	I_T at C	I_5	I_6	I_7	I_T at D	$I_2 + I_3$	$I_5 + I_6 + I_7$
mA											

TABLE 20-3.

	I_T at A	I_1	I_2	I_3	I_T at B	$I_1 + I_2 + I_3$
mA						

in Table 20-4. Record also the ratio of the currents $I_1:I_2:I_3$.

QUESTIONS

1. How does the sum of I_2 and I_3 in step 1 compare with I_T at A? I_T at B? Explain any discrepancies.
2. How does the sum of I_5, I_6, and I_7 in step 1 compare with I_T at C? I_T at D? Explain any differences.
3. Comment on the results of your measurements in step 2. Explain any unexpected result.
4. In your own words state Kirchhoff's current law and write its equation.
5. What information do you need to compute I_2 and I_3 in Fig. 20-4?
6. Discuss the results of the design problem (step 3).

7. What made it impossible for you to design a circuit which would give an *exact* current ratio 1:2:3?

Answers to Self-Test

1. 0.2
2. 1.25
3. I_1 is +; I_2 is +; I_3 is +; I_T is −
4. $I_1 + I_2 - I_3 - I_4 - I_5 = 0$
5. 2

TABLE 20-4.

	I_1	I_2	I_3	I_T	$I_1:I_2:I_3$
mA					

EXPERIMENT 21

21

Voltage-divider Circuits (loaded)

OBJECTIVES

1. To determine analytically the effects of a load on the voltage relationships in a resistive voltage-divider circuit
2. To confirm experimentally the results of objective 1

INTRODUCTORY INFORMATION

In the simple dc voltage-divider circuits studied in Experiment 14, no load current was drawn. The only current was in the divider network itself. In electronics, divider networks are frequently used as a source to supply voltage to a load, which draws current. When this is the case, the divider-voltage relationships which were obtained for no-load conditions no longer hold. The actual changes depend on the amount of current drawn and on the circuit connections. Consider the circuit of Fig. 21-1. Suppose V is a constant-voltage source, which maintains 90 V across the divider network without load and under load. Without load, points A, B, and C are 30, 60, and 90 V, respectively, with respect to G. Moreover, a 3-mA bleeder current I_1 is drawn. Now if we add a load resistor R_L (at point B) which draws 2 mA of load current I_L, the voltages at A and B will be different from the condition of no load. See Fig. 21-2 for the load circuit. We can compute the voltages from the circuit information.

Assume that there is bleeder current I_1 when a load current of I_L is drawn by R_L. By Kirchhoff's law we can set up the equation

$$I_1(R_2 + R_3) + (I_1 + I_L)(R_1) = 90 \quad (21\text{-}1)$$

Solving for I_1 (since this is the only unknown), we get

$$I_1(R_1 + R_2 + R_3) = 90 - (R_1)I_L \quad (21\text{-}2)$$

$$I_1 = \frac{90 - (I_L \times R_1)}{R_1 + R_2 + R_3} \quad (21\text{-}3)$$

Substituting 10,000 for R_1, for R_2, and for R_3, and $I_L = \frac{2}{1,000}$ in Eq. (21-3), we get

$$I_1 = \frac{7}{3}\text{mA} \quad (21\text{-}4)$$

It now follows that the voltage from A to G is

$$V_{AG} = I_1(10,000) = \frac{70}{3} = 23\frac{1}{3} \text{ V} \quad (21\text{-}5)$$

and the voltage from B to G is

$$V_{BG} = 46\frac{2}{3} \text{ V} \quad (21\text{-}6)$$

It becomes apparent that both the voltages and bleeder current are affected when a load is added to a voltage-divider circuit.

To verify this effect experimentally it is necessary to have a voltage source across the divider which will remain constant with and without load. In Fig. 21-2 the divider network consists of R_1, R_2, and R_3 connected to a voltage source. When a load resistor (R_L) is added as shown, the voltage source

Fig. 21-1. Voltage divider without load.

Fig. 21-2. Voltage divider with fixed load.

Fig. 21-3. Voltage divider with variable load.

is readjusted if necessary to maintain the same voltage across the network as under no-load conditions. The required measurements are made under no-load and load conditions.

SUMMARY

1. A series-resistive circuit, such as that in Fig. 21-1, is a simple voltage divider without load.
2. The voltage, say V_1 across R_1, in such a divider may be computed by use of the formula

$$V_1 = V \times \frac{R_1}{R_T}$$

where V is the applied battery voltage and R_T is the sum of the series-connected resistors. The voltages at A, B, and C with relation to G can be determined in this manner.

3. If a load R_L drawing current I_L is added to such a circuit, Fig. 21-2, the voltages at A and B with relation to G, will change, as will the "bleeder" current I_1.
4. To determine analytically the new voltages in this voltage divider, Kirchhoff's current and voltage laws must be applied. The equation or equations which result can then be solved to give the required voltages.

SELF-TEST

Check your understanding by answering these questions:

1. In the circuit of Fig. 21-1, assume $R_1 = 10$ kΩ, $R_2 = 18$ kΩ, and $R_3 = 22$ kΩ. $V = 150$ V. Then the (a) voltage V_1 across $R_1 = $ _____ V; (b) V_{BG} (across $R_2 + R_3$) = _____ V; (c) V_{AG} (across R_3) = _____ V.
2. In question 1, bleeder current $I_1 = $ _____ mA.
3. In the circuit of Fig. 21-2, $R_1 = 10$ kΩ, $R_2 = 18$ kΩ, $R_3 = 22$ kΩ, and $V = 150$ V. $I_L = 5$ mA. The bleeder current $I_1 = $ _____ mA.
4. For the conditions in question 3, (a) $V_1 = $ _____ V; (b) $V_{BG} = $ _____ V, (c) $V_{AG} = $ _____ V.

MATERIALS REQUIRED

- Power supply: Regulated, variable direct current
- Equipment: EVM, 0–10 milliammeter, ohmmeter
- Resistors: ½-W, three 1,200-Ω
- Miscellaneous: 10,000-Ω 2-W potentiometer, SPST switch

PROCEDURE

1. Connect the circuit of Fig. 21-3. Maintain a constant voltage V of 10 V in the output of the power supply.

2. With zero-load current (i.e., with the lead to the 10,000-Ω rheostat open) measure the bleeder current I_1 (mA) and record in Table 21-1. Mea-

TABLE 21-1.

	Measured Values					Computed Values			
V	I_L (load current), mA	I_1, mA	V_{BG}, volts	V_{AG}, volts	R_L, ohms	I_1, mA	V_{BG}, volts	V_{AG}, volts	R_L, ohms
10	0								
10	2								
10	4								
10	6								

sure also and record the voltages V_{BG} and V_{AG}.

3. Adjust the rheostat to draw 2 mA of load current, while maintaining $V = 10$ V, measured. Measure and record the bleeder current, the voltages V_{BG} and V_{AG}. Open the load resistor R_L, but do not vary the setting of the arm. Measure and record the resistance to which it was set to draw 2 mA of load current. Reconnect R_L after this measurement.

4. Repeat step 3 for conditions of 4 mA and 6 mA of load current.

5. Compute, and record in Table 21-1, the bleeder current I_1, the voltages V_{BG} and V_{AG}, and the load resistance R_L, for each of the load conditions in the experiment. Show your computations.

QUESTIONS

1. Refer to the data in Table 21-1. How does load current vary with load resistance R_L? Explain why.
2. Refer to Table 21-1. What is the effect on bleeder current I_1 as the load current increases? Explain why.
3. What is the effect on the voltages V_{AG} and V_{BG} at the divider taps, as the load current increases (Table 21-1)? Explain why.
4. Compare the computed values in Table 21-1 with the measured values. Explain any differences.

Answers to Self-Test

1. (a) 30; (b) 120; (c) 66
2. 3
3. 2
4. (a) 70; (b) 80; (c) 44

22

EXPERIMENT **22** Voltage- and Current-divider Circuit Design Considerations

OBJECTIVES

1. To design a voltage divider which will meet specified voltage and current requirements
2. To design a current divider which will meet specified current and voltage requirements
3. To construct the circuits, and confirm, by measurement, the design parameters

INTRODUCTORY INFORMATION

Designing a Voltage Divider for Specific Loads

The electronics technician may be called upon to design a voltage divider which will meet specified requirements of voltage, load current, and bleeder current. The process he will follow will be illustrated by an example.

Example

Design a voltage-divider circuit for a 300-V power supply. The loads which must be served are 0.05 A at 300 V, and 0.04 A at 250 V. The bleeder current under these conditions is 0.01 A.

NOTE: A "rule of thumb" value of bleeder current is 10 percent (approximately) of load current.

Solution. Draw a diagram, as in Fig. 22-1. Label the known load currents and voltages. Designate the *unknown resistors* as R_1 and R_2. (The loads are shown as resistors drawing the required currents.)

Apply Kirchhoff's current law and show the currents entering and leaving the junctions A, B, and C. The total current I_T which must be supplied is $I_T = 0.05 + 0.04 + 0.01 = 0.1$ A. I_T is shown enter-

ing the junction at C and leaving the junction at A. The current I_1 in R_1 is the bleeder current. Now

$$I_1 \times R_1 = 250 \text{ V} \tag{22-1}$$

or

$$R_1 = \frac{250}{I_1} \, \Omega \tag{22-2}$$

Substituting the value of bleeder current (0.01) for I_1 gives

$$R_1 = \frac{250}{0.01} \, \Omega = 25,000 \, \Omega \tag{22-3}$$

The current in R_2 is 0.05 A, the sum of the bleeder current and load 1 current. Moreover, as is evident from Fig. 22-1, the voltage from A to B is 50 V. Therefore

$$(0.05)(R_2) = 50 \text{ V} \tag{22-4}$$

and

$$R_2 = \frac{50}{0.05} \, \Omega = 1,000 \, \Omega \tag{22-5}$$

We have thus determined the proper values of R_1 and R_2 which will sustain a load of 0.04 A at

Fig. 22-1. Voltage-divider circuit for illustrative example.

250 V, a load of 0.05 A at 300 V, a bleeder current of 0.01 A, from a 300-V power supply.

Designing Current-divider Circuits

It is sometimes necessary for the technician to design current-divider circuits. To avoid trial and error methods, analysis is again suggested. The actual steps in the analytical method will depend on the nature of the design problems. But in general certain steps must be taken to facilitate the solution. These are:

1. Draw a circuit diagram labeling all the known values and the unknowns.
2. Apply Ohm's and Kirchhoff's laws and set up one or more equations which will reflect the electrical relationships involved.
3. Solve these equations for the unknown component values.
4. Construct the circuit from the computed design values and by measurement determine if the design requirements have been met.

Several examples will illustrate this process.

Example 1

It is required to design a three-branch parallel network, like that in Fig. 22-2, such that a 100-V source V will deliver a total current I_T of 0.9 A. A further requirement is that the current I_T be distributed among R_1, R_2, and R_3, such that $I_1:I_2:I_3 = 2:3:4$. What values of R_1, R_2, and R_3 will satisfy these conditions?

Analysis. We know V, I_T, and the ratios $I_1:I_2:I_3$. If we could find the actual values of I_1, I_2, and I_3,

Fig. 22-2. Current-divider circuit. If $I_T = 0.9$ A, what are the values of I_1, I_2, and I_3?

we would then know two of the Ohm's law values for each of the branches, namely V and I. From V and I, R can be found, since

$$R = \frac{V}{I} \qquad (22\text{-}6)$$

Solution. We can set up two equations for I from the current ratios which will help us.

$$\frac{I_1}{I_2} = \frac{2}{3}$$
$$\frac{I_1}{I_3} = \frac{2}{4} \qquad (22\text{-}7)$$

Solving for I_2 and I_3 in terms of I_1 yields

$$I_2 = \frac{3}{2}I_1$$
$$I_3 = \frac{4}{2}I_1 \qquad (22\text{-}8)$$

Now applying Kirchhoff's current law,

$$I_T = I_1 + I_2 + I_3 = 0.9 \qquad (22\text{-}9)$$

and substituting in Eq. (22-9) the values of I_2 and I_3 from Eq. (22-8) gives

$$I_1 + \frac{3}{2}I_1 + \frac{4}{2}I_1 = 0.9 \qquad (22\text{-}10)$$

Solving,

$$\frac{9I_1}{2} = 0.9 \qquad (22\text{-}10)$$

$$I_1 = 0.2 \text{ A} \qquad (22\text{-}11)$$

Now substituting 0.2 for I_1 in Eqs. (22-8) gives

$$I_2 = 0.3 \text{ A}$$
$$I_3 = 0.4 \text{ A} \qquad (22\text{-}12)$$

Now we can find the values of R_1, R_2, and R_3. Thus

$$R_1 = \frac{V}{I_1} = \frac{100}{0.2} = 500 \ \Omega$$

$$R_2 = \frac{V}{I_2} = \frac{100}{0.3} = 333.3 \ \Omega \qquad (22\text{-}13)$$

$$R_3 = \frac{V}{I_3} = \frac{100}{0.4} = 250 \ \Omega$$

Example 2

Design a divider network which will supply three resistive loads from a 125-V source such that loads 1, 2, and 3 receive 0.02 A, 0.05 A, and 0.06 A, re-

Fig. 22-3. A voltage and current-divider network designed to supply three loads from a 125-V source.

spectively. All three loads are in parallel and the resistance of load 1 is 3,000 Ω.

Solution. Draw the three parallel loads R_1, R_2, and R_3. Since the current and resistance of load 1 are given, we can find the voltage across load 1, and so across each of the other two parallel loads. Thus

$$V_1 = I_1 \times R_1 = 0.02(3,000) = 60 \text{ V} \quad (22\text{-}14)$$

We have a 125-V source. Therefore we will have to drop 65 V before we reach the three parallel loads. A series-parallel circuit, such as that in Fig. 22-3, will serve our requirements. It is now a simple matter to find the values of R_4 and loads R_2 and R_3.

First, solving for R_2 and R_3:

$$R_2 = \frac{V_{AB}}{I_2} = \frac{60}{0.05} = 1,200 \ \Omega$$

$$R_3 = \frac{V_{AB}}{I_3} = \frac{60}{0.06} = 1,000 \ \Omega \quad (22\text{-}15)$$

Finally, the total current I_T in R_4 may be found by Kirchhoff's current law:

$$I_T = I_1 + I_2 + I_3 = 0.02 + 0.05 + 0.06 = 0.13 \text{ A}$$

Therefore

$$R_4 = \frac{65}{0.13} = 500 \ \Omega \quad (22\text{-}16)$$

SUMMARY

1. In designing either voltage- or current-divider circuits, an analytical solution is found first. Ohm's and Kirchhoff's laws are applied in the analysis.
2. The general steps in analyzing a design problem are:

(a) Draw a diagram of the proposed circuit, labeling all the knowns and unknowns.
(b) Set up one or more equations which reflect the electrical relationships involved.
(c) Solve these equations for the values of the unknown components.
(d) Construct the circuit from the computed design values and determine by measurement if the design requirements have been fulfilled.

3. In designing voltage-divider circuits for specified load currents, a rule-of-thumb value for bleeder current is 10 percent (approximately) of load current.

SELF-TEST

Check your understanding by answering these questions:

1. It is required to design a voltage-divider circuit like that in Fig. 22-1, so that a 200-V source V can supply two loads, a 0.1-A load 1 at 50 V and a 0.4-A load 2 at 200 V. Assume a bleeder current I_1 of 0.05 A. The values of R_1 and R_2 which will accomplish this result are:
   (a) $R_1 = $ _____ Ω
   (b) $R_2 = $ _____ Ω
2. For the same conditions as in question 1, the resistances of the two loads are:
   (a) Resistance of load 1 is _____ ohms
   (b) Resistance of load 2 is _____ ohms
3. In a divider circuit similar to that in Fig. 22-3, it is required to supply three equal-load resistors, drawing a total current of 0.15 A, at a load voltage of 20 V. The supply source V is 50 V. The values of R_1, R_2, R_3, and R_4 which will achieve this are:
   (a) $R_1 = $ _____ Ω
   (b) $R_2 = $ _____ Ω
   (c) $R_3 = $ _____ Ω
   (d) $R_4 = $ _____ Ω

MATERIALS REQUIRED

- Power supply: Regulated, variable dc source
- Equipment: EVM, 0–10 milliammeter
- Resistors: ½ W, as required for procedural steps 1 through 6
- Miscellaneous: SPST switch, and potentiometer(s) as required for procedural steps 1-6

PROCEDURE

1. Design a voltage-divider circuit for a 35-V regulated power supply which must feed a 9-mA load at 20 V. The bleeder current should be 1 mA (approximately). Draw the circuit diagram, showing all values of voltage, current, and resistance. Show your computations and have your instructor check the solution before connecting the circuit.

2. Select the required resistors from your kit. If your kit does not contain a design-value resistor, adjust a potentiometer connected as a rheostat to the desired value, or make up the resistor from a combination of other resistors.

3. Connect the circuit. Measure the required voltages and currents and record in Table 22-1 which you will construct for the required data.

4. Design a circuit which will feed three parallel loads from a 40-V source. The total current drawn by the three loads must equal 7 mA and the load currents must be in the ratio 1:2:4. The load voltage is 20 V. Draw the circuit diagram, showing all values of voltage, current, and resistance. Show your computations and have an instructor check your solution before connecting the circuit.

5. Select required resistors from your kit. If your kit does not contain a design-value resistor, adjust a potentiometer connected as a rheostat to the desired value, or make up the resistor from a combination of other resistors.

6. Connect the circuit. Measure the required voltages and currents and record in Table 22-2, which you will construct for the required data.

QUESTIONS

1. In the circuit of Fig. 22-1 what would be the effect on the voltage V_{BC} of:
   (a) Increasing load 1 current? Why?
   (b) Increasing load 2 current? Why?
2. In the circuit of Fig. 22-1, for the design requirements shown, will the value of current in load 2 affect the values of R_1 and R_2? Explain.
3. How did the measured values of current in step 3 compare with the design values? Refer specifically to your measurements in Table 22-1. Explain any discrepancies.
4. How did the measured values of current in step 6 compare with the design values? Refer specifically to your measurements in Table 22-2. Explain any discrepancies.
5. Write a list of Materials Required for procedural steps 1-6.

Extra Credit

6. Write a general formula for R_1, and another for R_2 (Fig. 22-1). Designate the supply voltage as V; the voltage for load 1 as V_1; the bleeder current as I_1; load 1 current as I_{L1}; load 2 current as I_{L2}.

Answers to Self-Test

1. (a) 1,000
   (b) 1,000
2. (a) 500
   (b) 500
3. (a) 400
   (b) 400
   (c) 400
   (d) 200

23

Defect Analysis by Voltage, Current, and Resistance Measurement

OBJECTIVES

1. To investigate an analytical system of trouble-shooting electric circuits by voltage, current, and resistance measurements
2. To apply these troubleshooting techniques in locating a defective part in a:
   (a) Series circuit
   (b) Parallel circuit
   (c) Series-parallel circuit

INTRODUCTORY INFORMATION

The job of an electronics technician has many aspects. He must be competent in many areas. One of the interesting and challenging tasks he faces is the repair of electronic equipment which has become defective.

Modern equipment is too complex to follow hit-or-miss troubleshooting procedures. What is required is a logical, analytical, scientific trouble-shooting system which will assure locating a defective component efficiently and quickly. Some of the elements of such a system will be considered in this experiment.

Electronic equipment is made up of direct-current and alternating-current circuits. Direct-current circuits consist of one or more sources of power (voltage and current), a network of resistors, and other elements with which you will become familiar in due time. If trouble has been isolated to a dc network, how is the defective component found? One way is to replace the entire network of components with known good parts. This, however, is too time-consuming and costly. A better method is to find the defective part and replace it. Our attention, then, will be directed here to an analytical system

of locating a defect in a dc circuit containing a power source, resistors, and interconnecting wiring.

Defects in a DC Circuit

Consider the circuit (Fig. 23-1). A power source V supplies 100 V through a switch S to three resistors, $R_1 = 2,000 \ \Omega$, $R_2 = 3,000 \ \Omega$, and $R_3 = 5,000 \ \Omega$, connected in series. When S is open, as shown, there is no complete path for current. There is *no* current I. Since there is no current, there is no voltage drop ($I \times R$) across any of the resistors. A current meter M, connected in the circuit under these conditions, would measure zero current. A voltmeter connected across R_1, R_2, or R_3 would measure 0 V in each case.

When S is closed, there is a complete circuit. Under normal conditions, the measured current is 10 mA. The measured voltage drop across R_1 is 20 V. Across R_2 it is 30 V, and across R_3 it is 50 V. These values of voltage and current can be computed before measurement by applying the laws of electricity. Thus

$$I = \frac{V}{R_T} = \frac{100}{10,000} = \frac{10}{1,000} = 10 \text{ mA} \quad (23\text{-}1)$$

Fig. 23-1. Troubleshooting a series dc circuit.

The voltage V_1 across R_1 is

$$R_1 = I \times R_1 = \frac{10}{1,000} \times 2,000 = 20 \text{ V} \quad (23\text{-}2)$$

Similarly, the other voltages, V_2 and V_3, can be evaluated.

In Fig. 23-1, if any one of the components changes value or becomes defective, circuit conditions are altered. Measurement would show values of current and voltage different from the expected (computed) values. This condition would affect the operation of any circuit or device dependent on normal operation of Fig. 23-1. For example, if M were a small motor which required 10 mA to make it turn, a change in the circuit which appreciably reduced the current below 10 mA would result in the inability of the motor to run.

Changes in circuit operation can be caused by changes in the characteristics of individual components.

Resistors. The resistance of a resistor can increase or decrease. A resistor may become *open* or short-circuited. An open resistor acts like an open switch. It results in an *incomplete* or open circuit. A short-circuited resistor acts like an interconnecting wire which has *zero* resistance.

Switch. Mechanical troubles in a switch may cause the switch to remain *open* at all times, *shorted* at all times (that is, it acts as though it were always closed), or intermittent in operation (working properly sometimes, but not always).

Interconnecting Wires. The inner conductive wire, covered by insulation, may break, opening the circuit as an open switch would. An interconnecting wire may also become intermittent in use.

Power Source. A battery develops high internal resistance, with time. The terminal voltage, under load (that is, when it is supplying current to a circuit), will therefore drop to a lower value than that for which the battery is rated.

The output of an electronic power supply which is not regulated will also drop under load. That is, the no-load voltage will be higher than the voltage under load. Though this is a normal characteristic, it must be taken into consideration.

Finally, a power source may become totally inoperative and produce no voltage.

Checking Components

Resistors, switches, and connecting wires, out of the circuit, may be checked with an ohmmeter. The resistance of a resistor is measured to determine if it equals its rated value. Tolerance must be considered. A 1,200-Ω $\pm$ 10 percent resistor is considered within tolerance if it measures within the range 1,080 to 1,320 Ω.

The "continuity" of a switch may be checked by measuring resistance across its terminals. When the switch is open, the ohmmeter should show *infinite* resistance. When the switch is closed, the reading should be *zero* ohms.

A continuity check of an interconnecting wire will show whether it is open (infinite resistance) or closed (zero ohms). Flexing the wire while the meter is connected across it may reveal an intermittent condition.

The resistors and interconnecting wires in the series circuit of Fig. 23-1 may be checked individually or altogether, without removing them from the circuit. Switch S must be open, however, because resistance must *never* be measured in a circuit to which power is applied. The resistance between points A and B must measure 2,000 Ω; between B and C, 3,000 Ω; and between C and D, 5,000 Ω. Measuring the circuit in this manner checks the resistors and interconnecting wires.

If a resistance check between points A and D shows 10,000 Ω, we may assume that the components and interconnecting wires between A and D are good.

The power source may be checked with a voltmeter. The terminal voltage of the supply (or battery) must be measured under load, that is, S must be closed, permitting the load (R_1, R_2, and R_3 in this case) to draw current.

Dynamic Troubleshooting Measurements

The resistance of resistors and the continuity of switches and interconnecting wires may measure "good" out of the circuit. Yet these parts may break down or change under load. A *dynamic* test of their operation under load is therefore necessary.

Dynamic measurements may be required for another reason. It may be inconvenient to remove parts from a circuit or even to unsolder one end of a component to free it for resistance measurement. In this case, also, circuit measurements under load are indicated.

In a dynamic test, voltage and current checks are made to determine if the measured values equal the computed values. If they are different, analysis of the measurements should provide the clue to the defective part.

Troubleshooting Axioms and Troubleshooting Assumption

The conclusions which may be drawn from dynamic tests are predicated on two basic troubleshooting axioms. They are:

1. In a resistive circuit which is operating normally, the voltage across individual linear resistors and the current in every part of the circuit are distributed according to Ohm's and Kirchhoff's laws.
2. Conversely, if the voltage across individual linear resistors or the current in any part of the circuit is not distributed according to Ohm's and Kirchhoff's laws and the indicated values of resistance, then the circuit is not operating normally, that is, there is a defect in the circuit.

In addition to the two axioms, this basic troubleshooting assumption is helpful in an evaluation of circuit measurements: The trouble in a resistive circuit which is not operating normally is due to *one* and *only one* defective component.

This assumption greatly simplifies troubleshooting analysis and, from a statistical standpoint, is relatively accurate. Of course, if the circuit still operates abnormally after *one* defective component is found and is replaced with a known good component, we proceed to look for a second *single* possible defective component, etc. In all cases the technician should determine the cause for the initial failure and correct it.

Troubleshooting a Series Circuit

A circuit illustration will serve to demonstrate the troubleshooting process. Consider Fig. 23-1. Voltage V measures 100 V. Current in the circuit measures 4 mA, rather than 10 mA. Voltage V_1 across R_1 is 8 V, V_2 across R_2 is 12 V, V_3 across R_3 is 80 V. Since current is everywhere the same in a series circuit, by applying Ohm's law we can compute the value of each resistor, from the measured values. The computed values are: $R_1 = 2,000 \ \Omega$, $R_2 = 3,000 \ \Omega$, and $R_3 = 20,000 \ \Omega$. It is apparent, therefore, that R_3 has increased in value to 20,000 Ω. Replacing R_3 with a 5,000-Ω resistor will restore the circuit to normal operation.

The measured values can be interpreted in another way. Since current has decreased from 10 mA to 4 mA, it follows that the total resistance of the circuit has increased. The troubleshooting assumption states that there is just one defect in the circuit. Hence, just *one* resistor has increased in value, namely, R_3, because R_3 is the only resistor across which the voltage has increased despite a decrease in circuit current.

The following conclusion can be applied to any series circuit: In a series circuit a *decrease* in current results from an increase in total resistance. The defective resistor is the one across which an *increased* voltage is measured.

It can be readily demonstrated that if the current in a series circuit *increases*, the change results from a *decrease* in total resistance. The resistor whose value has *decreased* is the only one across which a *decreased* voltage is measured. Across each of the other resistors, the voltage has increased.

These conclusions may be drawn about a series circuit only if the applied voltage V measures the same as the indicated circuit value. If the measured value of V is not the same, it must be determined why, before any other step is taken. Hence, *one* measurement which must be made is a voltage check of the source V.

If the current in Fig. 23-1 measures zero, it is because the circuit is open and there is no complete path for current. A logical procedure to follow is to check the voltage of the source V first. If that is normal, power is removed and an ohmmeter may be used to check the continuity of the switch, connecting wires, and the resistance of the resistors.

There is another method to find the discontinuity (open) in the circuit. The negative lead of the voltmeter is connected to the (−) terminal of the battery. Switch S is closed. The positive lead is then used to probe the circuit. First place the positive lead at A. If the battery and interconnecting leads are good, the full voltage V, 100 V in this case, will be measured. Next, shift the positive lead to B. If R_1 and the interconnecting wires are good, the meter will read 100 V. However, if either R_1 or the interconnecting wires is open, the voltage will be zero.

We continue probing the circuit from points A to B to C to D until we find a point where the voltage is zero. The trouble lies between that point and the point immediately preceding it, where the voltage read the full value V.

Troubleshooting a Parallel Circuit

Consider the parallel circuit, Fig. 23-2. By Kirchhoff's laws $I_T = I_1 + I_2 + I_3$, and $I_1 = V/R_1$, $I_2 = V/R_2$, $I_3 = V/R_3$.

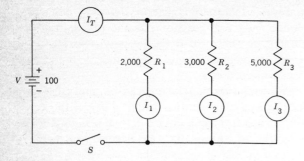

Fig. 23-2. Troubleshooting a parallel dc circuit.

It is therefore, a simple matter to compute the values I_1, I_2, I_3, and I_T from the circuit voltage V and the rated values of R_1, R_2, and R_3. In Fig. 23-2, the computed values are $I_1 = 50$ mA, $I_2 = 33\frac{1}{3}$ mA, $I_3 = 20$ mA, and $I_T = 103\frac{1}{3}$ mA.

Suppose V measures 100 V as required and the total current I_T measures $83\frac{1}{3}$ mA. It is apparent that one of the branch currents has decreased, and that this decrease results from an increase in the resistance of that branch. Measurement shows I_2 and I_3 are normal, but $I_1 = 30$ mA. R_1 is therefore defective and its resistance has increased from 2,000 to 3,333 Ω. Replacing R_1 with a 2,000-Ω resistor will restore the circuit to normal operation.

If V measures 100 V, and I_T has increased to 170 mA, we conclude that one of the branch currents has increased, and that this increase results from a decrease in resistance in that branch.

Measurement shows that I_1 and I_3 are normal, but that $I_2 = 100$ mA. Therefore, R_2 is defective because its resistance has decreased from 3,000 to 1,000 Ω.

The following conclusions may be drawn about a parallel circuit, if the source voltage measures the required voltage V, and the measured value of I_T is:

1. *Smaller* than the computed value. The decrease in I_T results from an *increase* in resistance of a resistor in one of the parallel branches. That resistor is defective whose branch current is *smaller* than the computed value.
2. *Greater* than the computed value. The increase in I_T results from a *decrease* in resistance of a resistor in one of the parallel branches. That resistor is defective whose branch current is *greater* than the computed value.

As in the case of the series circuit, these conclusions are valid only if the applied voltage V measures the same as the indicated circuit value.

Hence, one measurement which must be made is a voltage check of V. The other measurements in the troubleshooting process include I_T, and measurement of the branch currents until the defective resistor is found.

Troubleshooting a Series-parallel Circuit

In the series-parallel circuit (Fig. 23-3), computation reveals that the normal resistance between points B and C, $R_{BC} = 2,000$ Ω. R_T of the circuit = 9,000 Ω, $I_T = 10$ mA, V_{AB} (across R_1) = 20 V, $V_{BC} = 20$ V, $V_{CD} = 50$ V, $I_2 = 6\frac{2}{3}$ mA, and $I_3 = 3\frac{2}{3}$ mA.

Suppose V measures 90 V as required, and I_T measures 8 mA. It is apparent that R_T has *increased* and that this increase is due to an increase in the resistance of one of the resistors. The procedure for finding the defective resistor is the same as in the case of a series circuit. The voltage is measured across each of the resistors and across each parallel circuit. The defective resistor is the one, or is one of the branch resistors of the parallel circuit, across which an *increased* voltage is measured.

For example, in Fig. 23-3 if the measured values are $I_T = 8$ mA, $V = 90$ V, $V_{AB} = 16$ V, $V_{BC} = 34$ V, and $V_{CD} = 40$ V, the trouble must be in R_2 or R_3. Now a measurement of branch current I_2 and I_3 will reveal the defective resistor. Thus, if $I_2 = 2\frac{1}{3}$ mA and $I_3 = 5\frac{2}{3}$ mA, the trouble is in R_2, which has changed value from 3,000 to 14,570 Ω.

If the current in a series-parallel circuit *increases*, the change results from a *decrease* in total circuit resistance R_T. The resistor whose value has decreased, or the parallel circuit whose voltage has *decreased*, is the only one across which a *decreased* voltage is measured. Trouble in a branch of a parallel circuit is found in the same manner as in the parallel circuit in Fig. 23-2.

The method of troubleshooting a series-parallel circuit, then, involves:

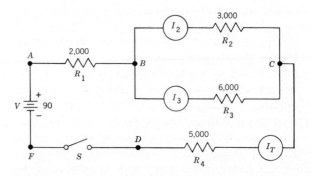

Fig. 23-3. Troubleshooting a series-parallel dc circuit.

1. Computation of the normal circuit values of current and voltage
2. Measurement of the actual values of current and voltage
3. Analysis of the measurements, applying the principles of both series- and parallel-circuit troubleshooting techniques

SUMMARY

1. In a dc circuit, troubles in operation may result from any defective component in the circuit. Thus resistors may open or change value; switches may open, short, or become intermittent; interconnecting wires may open or become intermittent; power sources may develop high internal resistance or become totally inoperative.
2. Components can be checked with an ohmmeter for resistance and continuity. Accordingly a resistor should measure within the tolerance of its color-coded value. An open switch will measure infinite resistance across its terminals; a shorted or closed switch will measure zero resistance. Interconnecting wires, if open, will measure infinite resistance.
3. The output voltage of a power source should be checked under load. If it is lower than the rated value, it either has developed a high internal resistance, is defective for other internal reasons, or is being excessively loaded (that is, too much current is being drawn).
4. Ohmmeter checks of components are considered static checks. Dynamic tests of all components made when they are operating under load, that is, with power applied, are more reliable than ohmmeter tests.
5. In a dynamic test, voltage and current checks are made in a circuit to determine if the measured values are the same as the computed values. If they are different, analysis of the measurements should help identify the defective part.
6. One troubleshooting assumption is that if the voltage across linear resistors or the current in such a circuit is not distributed according to Ohm's and Kirchhoff's laws, then there is a defect in the circuit.
7. In the troubleshooting process it also simplifies matters to assume that there is just *one* defect in the circuit.
8. In a *series* circuit a *decrease* in current results from an *increase* in total resistance. The de-

fective resistor is the one across which an *increased* voltage is measured.
9. In a *series* circuit an *increase* in current results from a *decrease* in total resistance. The resistor whose value has *decreased* is the one across which a *decreased* voltage is measured.
10. In troubleshooting any circuit, check the output voltage of the power source to determine if it is the same as its rated value.
11. If the current in a *series* circuit is *zero*, some component is open, or the output of the power source is zero volts.
12. In a *parallel* circuit if the source voltage is normal and the measured value of I_T is:
(a) *Smaller* than the computed value, the decrease in I_T results from an *increase* in *resistance* of a resistor in one of the parallel branches. *That resistor is defective whose branch current is smaller than the computed value.*
(b) *Greater* than the computed value, the increase in I_T is the result of a *decrease in resistance* of one of the resistors in the parallel branches. *That resistor is defective whose branch current is greater than the computed value.*
13. If the current in a *series-parallel* circuit has decreased, there has been an *increase* in the *total resistance* of the circuit.
14. If the current in a series-parallel circuit has *increased*, there has been a *decrease* in the *total resistance* of the circuit.
15. In troubleshooting a series-parallel circuit, the same procedures are followed as in troubleshooting series circuits and parallel circuits, including voltage and current measurements and analysis of these measurements, applying the principles of both series- and parallel-circuit troubleshooting techniques.

SELF-TEST

Check your understanding by answering these questions:
1. In the circuit of Fig. 23-1, the color-coded values of R_1 is 2,200 Ω, R_2 is 4,700 Ω, and R_3 is 8,200 Ω. $V = 150$ V. When the switch is closed, milliammeter M reads 4.5 mA. The voltage V_1 (measured across R_1) = 10.0 V, V_2 = 103 V, V_3 = 37.0 V.
   (a) If the circuit were normal, the values of current and voltage would be:
      (1) $I =$ _____ mA
      (2) $V_1 =$ _____ V

(3) $V_2 =$ _____ V

(4) $V_3 =$ _____ V

(b) What has happened to the total resistance in the circuit? _____ (increased/decreased)

(c) The defective component is _____ .

2. In the circuit of Fig. 23-1, the resistors are color coded the same as in question 1 and $V = 150$ V. When the switch is closed milliammeter M reads *0* current. The measured voltages $V_{AF} = 150$ V, $V_{BF} = 150$ V, $V_{CF} = 150$ V. The components which are possibly defective (open) are: _____ or _____ or _____ .

3. In the circuit of Fig. 23-2, the color-coded values of the resistors are: $R_1 = 1,200$ Ω, $R_2 = 1,800$ Ω, $R_3 = 3,000$ Ω, and $V = 300$ V. When the switch S is closed, the measured currents are as follows: $I_T = 0.4$ A, $I_1 = 0.133$ A, $I_2 = 0.167$ A, and $I_3 = 0.1$ A.

(a) If the circuit were normal, the total current I_T would measure _____ A.

(b) The total resistance in the circuit has therefore _____ (increased/decreased).

(c) The normal currents should be:

(1) $I_1 =$ _____ A

(2) $I_2 =$ _____ A

(3) $I_3 =$ _____ A

(d) The defective resistor is _____ whose resistance has _____ (increased/decreased).

4. In the circuit of Fig. 23-2, when switch S is closed, $I_T = 0$, $I_1 = 0$, $I_2 = 0$, $I_3 = 0$, $V = 100$ V.

With switch S open, an ohmmeter check from the positive terminal of V to the right-hand terminal of S shows continuity. Assuming all the wiring is good, the defective component must be (a) _____ which is (b) _____ (open/shorted).

5. In the circuit of Fig. 23-3 the measured voltages are as follows: $V_{AB} = 18$ V, $V_{BC} = 27$ V, $V_{CD} = 45$ V. The defective component is (a) _____ which is (b) _____ (open/decreased resistance).

MATERIALS REQUIRED

- Power supply: Regulated variable direct current
- Equipment: EVM, 0–10 milliammeter, 0–100 milliammeter
- Resistors: ½-W 2,200-, 2,700-, 3,300-, 4,700-, 5,100-, 5,600-, 6,800-, 8,200-, 10,000-, and 12,000-Ω

NOTE: Your instructor will assign resistors as required from the coded values shown above. One or more of these resistors will be defective.

- Miscellaneous: SPST switch, interconnecting wires

NOTE: These parts may also be either good or defective.

PROCEDURE

1. Connect a series circuit, similar to that in Fig. 23-1, using the components assigned to you by your instructor.

NOTE TO INSTRUCTOR: Resistors R_1, R_2, and R_3 should be selected so that their normal total resistance is no greater than 10,000 Ω. The smallest normal valued resistor should be no less than 2,200 Ω.

Apply 40 V across the circuit and maintain the voltage V at this level. One of the components will be defective, but it will not be possible to determine which, simply by observation.

Draw a circuit diagram and on it show the color-coded value of each resistor. Also record the rated values in Table 23-1. Compute the total resistance R_T, the value of voltage which should appear across each resistor under normal operation, and the circuit current I, and record in Table 23-1. Show your computations.

2. Troubleshoot the circuit, entering in Table 23-1 each measurement you actually make. Analyze your measurements and under "Analysis" in Table 23-1, record the defect which you believe exists. State the reasoning which led to your conclusion.

Verify your conclusion by replacing the defective component with a known good part and measure the current I. Indicate under "Verification and measurements" the measured current I and whether the defect was found. If you were incorrect in your analysis, reevaluate your conclusion and find the defective part. List the defective part in the column headed "Defect."

3. Connect a parallel circuit similar to that in Fig. 23-2, using the components assigned by your instructor.

TABLE 23-1.

	Resistance			Voltage			Current		Defect
	Rated	Computed		Computed	Measured		Computed	Measured	
R_T	—		V	40		I_T			
R_1		—	V_1			I_1	—	—	
R_2		—	V_2			I_2	—	—	
R_3		—	V_3			I_3	—	—	

Analysis:

Verification and measurements:

TABLE 23-2.

	Resistance			Voltage			Current		Defect
	Rated	Computed		Computed	Measured		Computed	Measured	
R_T	—		V	30		I_T			
R_1		—	V_1			I_1			
R_2		—	V_2			I_2			
R_3		—	V_3			I_3			

Analysis:

Verification and measurements:

TABLE 23-3.

	Resistance			Voltage			Current		Defect
	Rated	Computed		Computed	Measured		Computed	Measured	
R_T	X		V	40		I_T			
R_1		X	V_{AB}				X	X	X
R_2		X	V_{BC}			I_2			
R_3		X	X	X	X	I_3			
R_4		X	V_{CD}				X	X	X

Analysis:

Verification and measurements:

NOTE TO INSTRUCTOR: The normal values of R_1, R_2, and R_3 should be so selected that the total current I_T is no greater than 30 mA. The smallest normal-valued resistor should be no smaller than 2,200 Ω.

Apply 30 V across the circuit and maintain the voltage V at this level. Draw a circuit diagram and on it show the color-coded value of each resistor. Also record these rated values in Table 23-2. Compute the currents I_1, I_2, I_3, and

I_T which should be found under normal operation, and record in Table 23-2. Show your computations.

4. Troubleshoot the circuit as in step 2. Enter the measurements you actually make, your analysis, suspected defect, verification, etc., in Table 23-2.
5. Connect a series-parallel circuit similar to that in Fig. 23-3, using the components assigned by your instructor.

Apply 40 V across the circuit and maintain the voltage V at this level. Draw a circuit diagram and on it show the color-coded values of each resistor. Also record the rated values in Table 23-3. Compute the total resistance R_T, total current I_T, the voltages V_{AB}, V_{BC}, and V_{CD}, the currents I_2 and I_3 and record in Table 23-3. Show your computations.
6. Troubleshoot the circuit, as in steps 2 and 4. Enter the measurements you actually make, your analysis, suspected defect, verification, etc., in Table 23-3.

QUESTIONS

1. State in your own words the two troubleshooting axioms.
2. Is the troubleshooting assumption valid? Why is it used?
3. In a series circuit what is the effect of an increased resistance on:
   (a) Current in the circuit?
   (b) Voltage across each of the resistors?

NOTE: Assume voltage source remains constant.

4. In a parallel circuit, assuming that the voltage source remains constant, what is the effect of one increased resistance on:
   (a) Total current in the circuit?
   (b) Each branch current?
5. Same as question 3, except discuss the effect of a *decreased* resistance.
6. Same as question 4, except discuss the effect of a *decreased* resistance.
7. In a series-parallel circuit, assuming the voltage source remains constant, what is the effect of an *increased* series resistance on:
   (a) Total current?
   (b) Individual branch currents?
   (c) Voltage across each resistor and across each parallel circuit?
8. Same as question 7, except discuss the effect of a *decreased* parallel resistance.
9. How can you check whether the following parts are good:
   (a) Resistor?
   (b) Switch?
   (c) Connecting wire?
   (d) Battery?
10. Explain a dynamic check for finding an open resistor in a:
   (a) Series circuit
   (b) Parallel circuit

Answers to Self-Test

1. (a) (1) 9.93; (2) 21.8; (3) 46.7; (4) 81.5
   (b) increased
   (c) R_2
2. R_3; S; M
3. (a) 0.517
   (b) increased
   (c) (1) 0.25; (2) 0.167; (3) 0.1
4. (a) S; (b) open
5. (a) R_3; (b) open

24

OBJECTIVES

1. To specify analytically the conditions for maximum power transfer from a dc source to a load
2. To verify experimentally the condition for maximum power transfer, indicated in objective 1

INTRODUCTORY INFORMATION

Power has been defined as the rate of doing work. Electrically, the unit of power is the watt (W). The relationships among dc power (W) in a resistance R, the voltage V across R, and the current I in R are given by the equations

$$W = V \times I = I^2R = \frac{V^2}{R} \qquad (24\text{-}1)$$

Delivery of electric power to a load R_L involves a source of power V, and a circuit network between V and the load resistance R_L. Let us consider the power relationships in a simple circuit containing a generator V with internal resistance R, delivering power to a load resistance R_L.

The current I in R_L is given by

$$I = \frac{V}{R + R_L} \qquad (24\text{-}2)$$

Applying the I^2R formula for power, it is evident that the power developed in R_L is

$$P_L = \left(\frac{V}{R + R_L}\right)^2 \times R_L = \frac{V^2R_L}{(R + R_L)^2} \qquad (24\text{-}3)$$

If V is a constant source with a fixed internal resistance R, for what value of R_L will there be a maximum power transfer from V to R_L? A mathematical analysis to find this value requires the use of the

calculus. However, an experimental method will be helpful in determining the required R_L.

Suppose the voltage V is 100 V, the internal resistance $R = 100\ \Omega$. Let R_L take a series of values as shown in Table 24-1. The power W, calculated by using Eq. (24-3), is also shown in Table 24-1.

The table indicates that as R_L increases from 0 to 100 Ω, the number of watts dissipated by R_L increases from 0 to a maximum of 25. As R_L increases from 100 to 100,000 Ω, the number of watts transferred to R_L decreases from 25 to 0.099.

It would appear that in this circuit maximum transfer of power occurs when the load resistance equals the internal resistance of the generator. Does this conclusion hold for other generators V', with internal resistance R', transferring power to a load R_L? The answer to this question may be determined by taking at random any voltage V', an internal resistance R', varying the load R_L, and then calculating the power in R_L. Again, the result will be that maximum power will be delivered to the load when $R_L = R'$, the internal resistance of the generator.

We may therefore state the law governing maximum transfer of power to a load in a dc circuit:

A generator transfers a maximum power to a load when the resistance of the load equals the internal resistance of the generator.

SUMMARY

1. The power W, in watts, dissipated by a resistor R (ohms), across which there is a dc voltage V (volts), is given by the equation $W = V^2/R$.
2. The power W, in watts, dissipated by a resistor

TABLE 24-1.

R_L, ohms	$R + R_L$, ohms	$W = \dfrac{V^2 R_L}{(R + R_L)^2}$ (watts)
0	100	0
10	110	8.26
20	120	13.9
30	130	17.7
40	140	20.4
50	150	22.2
60	160	23.4
70	170	24.2
80	180	24.7
90	190	24.9
100	200	25
110	210	24.9
120	220	24.8
130	230	24.6
140	240	24.2+
150	250	23.9+
200	300	22.2
400	500	16
600	700	12.25
800	900	9.87
1,000	1,100	8.26
10,000	10,100	0.98
100,000	100,100	0.099

R (ohms) in which there is dc current I (amperes) is given by the equation $W = I^2 R$.

3. If the dc voltage is V (volts) across a resistor and current is I (amperes), the wattage dissipated by the resistor is given by the equation $W = V \times I$.

4. When power is delivered by a power source V, whose internal resistance is R_{in}, to a load R_L, maximum transfer of power occurs when the load resistance equals the internal resistance of the source, that is, when $R_{in} = R_L$.

SELF-TEST

Check your understanding by answering these questions:

1. The current in a 120-Ω resistor is 0.1 A. The power in watts dissipated by the resistor is $W =$ _____ W.
2. The voltage across a resistor is 12 V, and the current in the resistor is 0.05 A. The power dissipated by the resistor is $W =$ _____ W.
3. The voltage across a 220-Ω resistor is 16.0 V. The power dissipated in the resistor is $W =$ _____ W.
4. A power supply with an internal resistance of 25 Ω delivers power to a 50-Ω load connected across its terminals. If the voltage delivered by the supply without load is 15 V, the power dissipated by the load is $W =$ _____ W.
5. A power supply with an internal resistance of 25 Ω delivers power to a resistive load. If the no-load voltage at the output of the supply is 50 V, the maximum power would be delivered to a load whose resistance $R_L =$ _____ Ω.
6. The power delivered by the supply in question 5 is $W =$ _____ W.

MATERIALS REQUIRED

- Equipment: Regulated variable dc supply; EVM
- Resistors: ½-W 1,000-Ω (330-, 470-, 1,000-, and 2,200 Ω for extra-credit question)
- Miscellaneous: Resistance decade box or 10,000-Ω potentiometer (10,000-Ω potentiometer for extra-credit question); one DPST switch

PROCEDURE

1. Connect the circuit of Fig. 24-1. V is a voltage-regulated power supply, R a 1,000-Ω resistor, R_L a resistance decade box or a 10,000-Ω potentiometer connected as a rheostat. A double-pole, single-throw switch (two on-off switches ganged together) S_{1A}, S_{1B} is used to open the load so that the resistance of R_L, completely isolated from V, may be adjusted and measured as required. Set the voltage $V = 10$ V and maintain it at this level during the remainder of the experiment.

Fig. 24-1. Experimental circuit for determining maximum power transfer to a load.

2. Open Switch S_1. Adjust the rheostat so that R_L is 0 Ω, as measured with an ohmmeter. Close S_1. With an EVM, measure the voltage V_L across R_L, and record in Table 24-2.
3. Open Switch S_1. Adjust the rheostat so that R_L measures 100 Ω. Close S_1. With a VTVM, measure the voltage V_L across R_L, and record in Table 24-2.

4. Repeat step 3 for every value of R_L shown in the table.
5. From the corresponding measured values of V_L and R_L, calculate the power in R_L using the formula $W = V_L^2/R_L$. Convert W into milliwatts and record in Table 24-2. Show a sample calculation.
6. Now calculate W_T using the formula $W_T = V^2/(R + R_L)^2$. Convert W_T into milliwatts and record in Table 24-2. Show a sample calculation.
7. Draw a graph of W versus R_L. Let R_L be the horizontal, W the vertical axis.
8. Draw a graph of W_T versus R_L.

Extra Credit

9. Describe in detail the experimental procedure you would use to determine the value of R_L for which there is a maximum transfer of power from the circuit of Fig. 24-2 to R_L. Draw a diagram of the experimental circuit. Tabulate your

TABLE 24-2.

R_L, ohms	$R + R_L$, ohms	V_L, volts	$W = \dfrac{V_L^2}{R_L}$ (milliwatts)	$W_T = \dfrac{V^2}{(R + R_L)^2}$ (milliwatts)
0				
100				
200				
400				
600				
800				
850				
900				
950				
1,000				
1,100				
1,200				
1,500				
1,700				
2,000				
4,000				
6,000				
8,000				
10,000				

Fig. 24-2. Determining value of R_L for which there is maximum transfer of power to R_L.

Fig. 24-3. Equivalent circuit for Fig. 24-2.

measurements, including a column for R_L and for W. When you have determined the value of R_L experimentally, compare it with the computed value of R_L, determined by transforming the circuit of Fig. 24-2 into an equivalent circuit like that in Fig. 24-3.

HINT: Connect the circuit of Fig. 24-2, with switch S open and the power supply disconnected. Short points C and D. Experimentally measure the resistance between points A and B. If the measured value is R_T, then the maximum power transfer should occur when $R_L = R_T$. Select a value for V when you are conducting the experiment that will give readings in the center region of the voltmeter scale when measuring V_L, the voltage across R_L.

QUESTIONS

1. In your experiment, for what value of R_L is there maximum power transfer?

2. Do the measurements and computations in Table 24-2 confirm the maximum-power-transfer law? Discuss any unexpected results.
3. In Fig. 24-1, how does the voltage across R_L vary with R_L? The current in R_L?
4. In Fig. 24-1, how does W_T vary with R_L?
5. When is the *percentage* of power dissipated by the load greatest in the circuit of Fig. 24-1? (See Table 24-2 and compute $W/W_T \times 100$.)
6. Does maximum power transfer in Fig. 24-1 occur at the same value of R_L for which there is maximum efficiency?

NOTE: Maximum efficiency occurs when most of the total power (W_T) is dissipated by the load.

Answers to Self Test

Answers to slide-rule accuracy.

1. 1.2
2. 0.6
3. 1.16
4. 2.0
5. 25
6. 25

25

Balanced-bridge Circuit

OBJECTIVES

1. To find by analysis the mathematical relationship among the resistors in a balanced-bridge circuit
2. To apply experimentally the principle of the balanced bridge in measuring an unknown resistance

INTRODUCTORY INFORMATION

The volt-ohm milliammeter and the electronic voltmeter (except the DVOM) are direct reading instruments utilizing a current-actuated meter movement. The measured quantity, volts, milliamperes, or

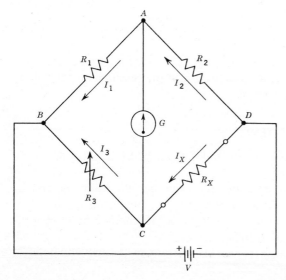

Fig. 25-1. Wheatstone bridge for measuring unknown resistance R_x.

ohms, is read directly from a calibrated scale along which the pointer travels.

For measuring resistance with greater accuracy than that possible with an ohmmeter, a bridge is used. A resistance bridge employs a highly sensitive galvanometer as an indicating device, together with a calibrated variable-resistance standard and a voltage source in a suitable circuit arrangement. The galvanometer serves as a null indicater which signals a balance condition. Resistance is read from a calibrated scale associated with the standard variable resistance and range multiplier.

Figure 25-1 illustrates the most common type, the Wheatstone bridge. The form shown is the diamond arrangement, which the four resistors in the circuit diagram resemble. The unknown resistance to be measured, R_x, is connected between terminals C and D. R_1 and R_2 are fixed-value precision resistors called "ratio arms," and R_3 is a variable resistance known as the "standard arm." The indicator is a highly sensitive zero-centered galvanometer G. Current will flow through G when there is a difference of potential between points A and C. When there is no potential difference between A and C, that is, when $V_{AC} = 0$, the pointer of the galvanometer will return to zero. When $V_{AC} = 0$, the bridge is said to be balanced.

Consider the relationships in the circuit of Fig. 25-1. Let I_1 be the current in R_1, I_2 in R_2, I_3 in R_3, and I_x in R_x. I_1 times R_1 is the voltage from A to B, that is,

$$V_{AB} = I_1 R_1$$

Similarly

$$V_{AD} = I_2 R_2$$
$$V_{CB} = I_3 R_3 \qquad (25\text{-}1)$$

and

$$V_{CD} = I_x R_x$$

To obtain balance ($V_{AC} = 0$), the voltage from A to B must equal the voltage from C to B. That is,

$$V_{AB} = V_{CB} \qquad (25\text{-}2)$$

Therefore, $I_1 R_1 = I_3 R_3$, and

$$\frac{I_1}{I_3} = \frac{R_3}{R_1} \qquad (25\text{-}3)$$

Similarly $\qquad V_{AD} = V_{CD}$

and $\qquad \dfrac{I_2}{I_x} = \dfrac{R_x}{R_2} \qquad (25\text{-}4)$

At balance, there is no current in G. Therefore

$$I_1 = I_2$$
and $\qquad I_3 = I_x \qquad (25\text{-}5)$

Substituting the results of Eqs. (25-5) in Eqs. (25-3) and (25-4), we find that

$$\frac{I_1}{I_x} = \frac{R_3}{R_1}$$
$$\frac{I_1}{I_x} = \frac{R_x}{R_2} \qquad (25\text{-}6)$$
and

Therefore $\qquad \dfrac{R_x}{R_2} = \dfrac{R_3}{R_1} \qquad (25\text{-}7)$

and $\qquad R_x = \left(\dfrac{R_2}{R_1}\right) \times R_3 \qquad (25\text{-}8)$

The fundamental equation (25-8) states that at balance the unknown resistance R_x equals the product of the ratio of the ratio arms (R_2/R_1) and the standard arm R_3.

Maximum accuracy and sensitivity occur when $R_1 = R_2$, that is, when the ratio $R_2/R_1 = 1$. For this condition $R_x = R_3$. If R_3 is a highly accurate decade-type rheostat, the value R_x can be read directly from the calibrated scale of the rheostat when R_3 is adjusted for balance.

The condition $R_2/R_1 = 1$ would limit the range of measurement of R_x to the range of variation of the rheostat, so that if the maximum resistance of $R_3 = 10,000\ \Omega$, a resistor whose value is larger than $10,000\ \Omega$ could not be measured. To overcome this limitation a range switch permits selection of different ratio arms. Thus, if $R_2/R_1 = 3$, the maximum value of R_x which may be measured is $3R3$, etc. It should be apparent, though, that any small error of galvanometer imbalance which the eye may not detect is multiplied by the ratio R_2/R_1, 3 in this case, thus increasing the error.

There are other bridge circuits, such as the slide-wire bridge, for measuring resistance. In this experiment, however, we will be concerned with the diamond-type Wheatstone bridge. There will be no attempt to build a bridge of great accuracy. The experiment will simply demonstrate the principle of a balanced bridge.

SUMMARY

1. A Wheatstone bridge is used for measuring resistance values with greater accuracy than that possible with the volt-ohm milliammeter or electronic voltmeter.
2. The ratio arms R_1 and R_2 in Fig. 25-1 are precision resistors which are switched in to give the desired resistance-range multiplier of the bridge.
3. The standard arm R_3 is a variable calibrated precision resistance by means of which the bridge is *balanced*, when measuring an unknown resistance R_x. The value of R_x is read, at balance, from the calibrated dial of the standard arm, and multiplied by the setting of the ratio arms.
4. A highly sensitive zero-center galvanometer is used as the indicator. A voltage source powers the bridge.
5. When the voltage across the galvanometer (V_{AC} in Fig. 25-1) is zero, there is no current in the galvanometer, which then reads zero. This is the balance setting at which the resistance of the unknown resistor is read.
6. At balance in a Wheatstone bridge, the product of the resistors of opposite arms is equal, that is, in Fig. 25-1

$$R_x \times R_1 = R_2 \times R_3$$

and from this condition the value of R_x is

$$R_x = \frac{R_2}{R_1} \times R_3$$

7. R_x, therefore, is equal to the product of the ratio of the ratio arms and the resistance of the standard arm.

SELF-TEST

Check your understanding by answering these questions:
1. A Wheatstone bridge has the following adjustable controls: _____ and _____.
2. Power for a portable Wheatstone bridge is provided by a _____.
3. At balance, the current in the galvanometer is _____.
4. A galvanometer is a _____.
5. In Fig. 25-2, the circuit of the experimental

Wheatstone bridge, balance occurs when R_3 is set for 3,750 Ω. The value of R_x is _____ Ω.

6. In Fig. 25-1, at balance $R_2/R_1 = 10$ and $R_3 = 1,950$. The resistance value of R_x is _____ Ω.

7. A Wheatstone bridge is known to be in good operating order and can measure resistors in the range 0.1 to 100,000 Ω. In measuring a resistor whose value is about 50,000 Ω, the bridge cannot be balanced for a specific setting of the multiplier (ratio arms). In order to balance the bridge it is necessary to change the setting of the _____ .

Fig. 25-2. Experimental Wheatstone bridge.

MATERIALS REQUIRED

■ Power supply: Voltage-regulated variable dc source
■ Equipment: Zero-center galvanometer or a 20,000-Ω/V VOM; electronic voltmeter
■ Resistors: ½-W, two 5,100-Ω (5 percent); assortment of six different values of resistors (minimum value 500, maximum value less than 10,000 Ω)
■ Optional: Other resistors as required in step 5

■ Miscellaneous: Decade resistance box, variable from 1 to 10,000 Ω, or 10,000-Ω potentiometer; two SPST switches

PROCEDURE

1. Connect the circuit (Fig. 25-2). S_1 and S_2 are open. Adjust regulated power supply for $V = 6$ V. Set R_3, the resistor decade box, to maximum resistance. (If a decade box is not available, use a 10,000-Ω potentiometer.) M is a zero-center galvanometer. (If one is not available, use the most sensitive current range on a 20,000-Ω/V VOM, e.g., 50-μA or 60-μA range.) The purpose of R_4 is to reduce the sensitivity of M during the initial stages of adjustment.

2. Connect R_x, the resistor to be measured, between C and D. (R_x must be no smaller than 560 Ω, or greater than 10,000.) Close switch S_1.

S_2 is still open. The meter pointer should deflect upscale. If it does not, check to see that the standard arm R_3 is set at maximum.

3. Reduce the resistance of R_3 gradually until M approaches zero. Now close S_2 and readjust R_3 if necessary until M reads zero. Read the value of R_x from the calibrated resistance decade box and record in Table 25-1. Record also the color-coded value of R_x and its rated tolerance.

NOTE: If R_3 is a 10,000-Ω potentiometer connected as a rheostat instead of the resistance decade, remove R_3 from the circuit without varying its setting

TABLE 25-1. Bridge Measurement of R_x: $R_2/R_1 = 1$

Resistor No.	1	2	3	4	5	6
Rated (code) value						
Percent tolerance						
Measured value						

and measure it with an ohmmeter. Record this as the value R_x in Table 25-1.

4. Open S_1 and S_2. Reset R_3 at maximum resistance. Remove the first R_x from the circuit. Using the bridge-circuit procedure in steps 2 and 3, measure in turn each of six resistors supplied to you. Record the measured and coded values in Table 25-1.

Extra Credit—Design Problem

5. Using the same R_3 as standard arm, design a bridge circuit which will measure resistances up to
   a. 50,000 Ω (approximately)
   b. 250,000 Ω (approximately)
   Draw the circuit diagram, for a and b, showing required values of all components in the bridge. Secure the necessary parts. If 5 percent resistors are not available, measure the ratio arm resistors with an ohmmeter and compute the multiplier ratio from the measured values. With bridge a measure three resistors greater than 10,000 Ω, but less than 50,000 Ω. With bridge b measure three resistors greater than 50,000 Ω, but less than 250,000 Ω.
   Record the results of your measurements in separate tables similar to Table 25-1. Identify the multiplier ratio in each case.

QUESTIONS

1. What determines the accuracy of measurements made with the experimental resistance bridge?
2. Why is it necessary to use a very sensitive galvanometer as an indicator in this experiment?
3. What is the purpose of the voltage source V? Is the voltage level ($V = 6$ V) of the voltage source critical?
4. Were the results of your measurements within the rated values of the resistors measured? If not, explain why not.

Extra Credit

5. In Fig. 25-1 at balance, $R_1 = R_2 = 5{,}000$ Ω; $R_3 = R_x = 6{,}000$ Ω; $R_m = 3{,}000$ Ω. Find the total resistance which the power supply (V) sees. Show your computations.

Answers to Self-Test

1. standard arm; ratio arms or multiplier
2. battery
3. zero
4. very sensitive zero-center dc meter movement
5. 3,750
6. 19,500
7. multiplier

EXPERIMENT 26   Thévenin's Theorem

OBJECTIVES

1. To determine by analysis the values V_{TH} (Thévenin generator voltage) and R_{TH} (Thévenin generator resistance) in a dc circuit containing a single voltage source
2. To confirm experimentally the values V_{TH} and R_{TH} proposed by Thévenin's theorem in the solution of unbalanced bridge circuits

INTRODUCTORY INFORMATION

Ohm's and Kirchhoff's laws have served us in the solution of dc circuits. In the study of the complex electric and electronic circuits, however, more

powerful analytical tools are sometimes required. Thévenin's theorem is one of a group of network theorems (including the superposition theorem and Norton's theorem) which provides the means for simplifying analysis of complex circuits. The technique employed involves reducing a network to a simple equivalent circuit which acts like the original network.

Thévenin's Theorem

Thévenin's theorem states that any *two* terminal networks may be replaced by a simple equivalent circuit consisting of a Thévenin generator, whose voltage V_{TH}, acting in series with an internal resistance R_{TH}, causes current to flow through the load. Thus, Fig. 26-1d is the Thévenin equivalent

Fig. 26-1. Solution of a network by Thévenin's theorem.

Fig. 26-2. Unbalanced-bridge circuit.

for the circuit of Fig. 26-1a considering R_L as the load. If we knew the values V_{TH} and R_{TH}, the process of finding the current I in R_L would be a simple application of Ohm's law.

The rules for determining V_{TH} and R_{TH} are as follows:

1. The voltage V_{TH} is the voltage "seen" across the load terminals in the original network, with the load resistance removed (open circuit voltage).
2. The resistance R_{TH} is the resistance seen from the terminals of the open load, looking into the original network when the voltage sources in the circuit are replaced by their internal resistance.

The development of the Thévenin equivalent circuit for Fig. 26-1a may be readily followed from Fig. 26-1b, c, and d. In Fig. 26-1b, R_L has been open circuited and V_{TH} is the voltage which appears across the terminals AB. In this case it is obviously 50 V. In Fig. 26-1c, we see that the original voltage source V has been replaced by its internal resistance R_1, which equals 5 Ω. The Thévenin re-

sistance R_{TH} is the resistance measured across AB in Fig. 26-1c. R_{TH} here is 100 Ω. Finally, Fig. 26-1d, shows the Thévenin equivalent generator V_{TH} in series with R_{TH}, which replaces the original network. Applying Ohm's law to the circuit of Fig. 26-1d, we see that

$$I = \frac{V_{TH}}{R_{TH} + R_L} = \frac{50}{450} = 0.111 \text{ A}$$

This is the required current in R_L.

It might appear, at first, as though this method unnecessarily complicates the problem, for the circuit of Fig. 26-1a can be easily solved by Ohm's and Kirchhoff's law. Although this is true in this single instance, the value of Thévenin's theorem becomes apparent if the problem in Fig. 26-1a is slightly modified. Suppose it is required to find the current in R_L, in Fig. 26-1a, for different values of R_L, as for example $R_L = 20$ Ω, 50 Ω, 100 Ω, 1,200 Ω, etc., while the rest of the circuit remains the same. It would be quite laborious to apply Ohm's and Kirchhoff's laws for each individual value of R_L in solving for I.

On the other hand, a single determination of the

Thévenin equivalent circuit (Fig. 26-1d) would suffice for each value of R_L, because V_{TH} and R_{TH} are independent of the value of R_L. Now it would be much simpler to solve I for the succession of values of R_L, using the Thévenin equivalent generator of Fig. 26-1d.

Solving Unbalanced-bridge Circuit by Thévenin's Theorem

Figure 26-2a is an unbalanced-bridge circuit. It is required to find the current I in R_5. Thévenin's theorem lends itself readily for a solution of this problem.

Consider R_5 the load. The problem is to transform the circuit into a Thévenin equivalent generator supplying current to R_5, as in Fig. 26-2b.

We open circuit R_5 to determine V_{TH}, the voltage V_{BC} across BC, Fig. 26-2c. To determine the voltage V_{BC}, we can find the voltage V_{BD} from B to D, and V_{CD} from C to D. Then

$$V_{BD} - V_{CD} = V_{BC} = V_{TH} \qquad (26\text{-}1)$$

The voltage V_{BD} in Fig. 26-2c is

$$V_{BD} = \frac{160}{200} \times 60 = 48 \text{ V} \qquad (26\text{-}2)$$

The voltage V_{CD} is

$$V_{CD} = \frac{120}{180} \times 60 = 40 \text{ V} \qquad (26\text{-}3)$$

Therefore

$$V_{BC} = 48 - 40 = 8 \text{ V} = V_{TH} \qquad (26\text{-}4)$$

The resistance R_{TH} of the Thévenin generator V_{TH} is found by replacing V by its internal resistance (assumed to be zero in this case), and solving for R_{BC}. Figure 26-2d is the result of shorting A to D and redrawing the circuit in a form which lends itself readily for computation. Hence

$$R_{BC} = R_{TH} = \frac{R_1 R_4}{R_1 + R_4} + \frac{R_2 R_3}{R_2 + R_3}$$

$$R_{TH} = \frac{40 \times 160}{40 + 160} + \frac{60 \times 120}{60 + 120} \qquad (26\text{-}5)$$

$$= 32 + 40 = 72 \ \Omega$$

Substituting these values in the Thévenin equivalent circuit (Fig. 26-2b) and solving for I, we find

$$I = \frac{V_{TH}}{R_{TH} + R_5} = \frac{8}{172} = 0.0465 \text{ A}$$

Experimental Verification of Thévenin's Theorem

It is possible to determine the values of V_{TH} and R_{TH} for a load R_L in a specific network by measurement. Then we can experimentally set the output of a voltage-regulated power supply to V_{TH} by connecting a resistor whose value is R_{TH} in series with V_{TH} and with R_L. We can measure I in this equivalent circuit. If the measured value of I_L in R_L in the original network is the same as the I measured in the Thévenin equivalent, we have one verification of Thévenin's theorem. For a more complete verification, this process would have to be repeated many times with random circuits.

SUMMARY

1. Thévenin's theorem proposes that any two terminal network may be replaced by a simple equivalent circuit, which acts like the original circuit at the load connected to the two terminals.
2. The equivalent circuit consists of a generator or power source called the Thévenin generator (V_{TH}) in series with the Thévenin internal resistance (R_{TH}) in series with the load across the two terminals in question. Thus for the complex circuit Fig. 26-2a, the Thévenin equivalent at the terminals BC is Fig. 26-2b.
3. To determine the Thévenin voltage V_{TH}, open the load at the two terminals in the original network and calculate the voltage at these two terminals. This *open-load* voltage is V_{TH}.
4. To determine the Thévenin resistance R_{TH}, keep the load open at the two terminals in the original network and replace the original power source with its own internal resistance. Then calculate the resistance at the *open-load* terminals, looking back into the original circuit.
5. Thévenin's theorem is applicable to a circuit with one or more power sources.
6. Once the complex network has been replaced by the Thévenin equivalent circuit, solution of the current in the load may be accomplished by Ohm's law. Thus in Fig. 26-1d current in the 350-Ω load equals V_{TH} divided by the sum of R_{TH} and R_L.

$$I_{\text{load}} = \frac{V_{TH}}{R_{TH} + R_L}$$

7. In this experiment we will verify Thévenin's

Fig. 26-3. Experimental circuit for verifying Thévenin's theorem, applied to an unbalanced bridge circuit.

theorem experimentally for one set of values of an unbalanced bridge circuit.

SELF-TEST

Check your understanding by answering these questions:

1. Consider the circuit of Fig. 26-1a. $V = 160$ V, $R_1 = 30$ Ω, $R_2 = 270$ Ω, $R_3 = 500$ Ω, and $R_L =$ 560 Ω. In the equivalent Thévenin circuit, assuming the internal resistance of the supply is zero,

   (a) $V_{TH} =$ _____ V
   (b) $R_{TH} =$ _____ Ω
   (c) $I_{load} =$ _____ A

2. Consider the circuit of Fig. 26-2a. $V = 120$ V, $R_1 = 200$ Ω, $R_2 = 500$ Ω, $R_3 = 300$ Ω, $R_4 = 600$ Ω, and $R_5 = 112.5$ Ω. In the equivalent Thévenin circuit, assuming the internal resistance of the supply is zero,

   (a) $V_{TH} =$ _____ V
   (b) $R_{TH} =$ _____ Ω
   (c) $I_5 =$ _____ A, where I_5 is the current in R_5.

MATERIALS REQUIRED

- Equipment: Regulated variable dc supply; EVM; milliammeter (or VOM) with a 5- or 10-mA range
- Resistors: ½-W 2,700-, 3,300-, 5,600-, 10,000-, and 15,000-Ω; and others as required in step 7 (optional)
- Miscellaneous: Resistance decade box or, if not available, a 10,000-Ω 2-W potentiometer; two SPST switches

PROCEDURE

1. Connect the circuit of Fig. 26-3. S_1 and S_2 are both open. M is a milliammeter set on the 1-mA range. Close S_1. Adjust V so that the voltage measured with an EVM is 40 V. Close S_2. Measure the current in R_L and record in Table 26-1 under column "I_L Measured Original Circuit."

2. Open S_2. Switch S_1 remains closed. With an EVM, measure the voltage V_{BC} across BC, and record in Table 26-1, under the heading "V_{TH} Measured."

3. Remove the voltage supply V from the circuit. S_2 is still open. Short points A to D.

NOTE: It is assumed that the internal resistance of the voltage-regulated supply is very low, and hence can be neglected.

TABLE 26-1.

R_L	V_{TH} (volts)		R_{TH} (ohms)		I_L (mA)		
	Measured	Computed	Measured	Computed	Measured		Computed
					Thév.	Original Circuit	
2,700							

S_2 is still open. Measure the resistance between points B and C, and record under "R_{TH} Measured."

4. Adjust the output of the power supply to V_{TH} measured. Set the resistance of the decade box to the value R_{TH} and connect between points P and B. If a decade box is not available, connect a 5,600-Ω resistor in series with a 10,000-Ω potentiometer serving as a rheostat, and adjust the rheostat so that the resistance between points PC (Fig. 26-4) is equal to the measured value of R_{TH}. Now connect R_L, the 2,700-Ω load resistor, in series with the milliammeter M between points C and B as in Fig. 26-4.

5. Close switch S. Monitor V_{TH} to see that it has not changed in value. Measure the current in R_L and record under column "I_L Measured Thév."

6. **Power off.** In the circuit of Fig. 26-3 compute the value of V_{TH} for the Thévenin equivalent generator for load R_L and record in Table 26-1 under column "V_{TH} Computed." Show your computations. Compute also the value of R_{TH} and record under column "R_{TH} Computed." Show your computations.

Now find the value of I_L in R_L using the Thévenin equivalent circuit with the computed values. Show your computations. Record in Table 26-1 under column "I_L Computed."

Extra Credit

7. Verify that Thévenin's theorem is true for the circuit of Fig. 26-3 for other values of R_L, substituting for R_L, in turn, each of three resistors whose values are no less than 1,200 Ω nor greater than 10,000 Ω. Measure the current I_L in R_L in the original circuit and also in the Théve-

Fig. 26-4. Experimental Thévenin equivalent for Fig. 26-3.

nin equivalent circuit. Record in Table 26-1 the values of R_L used, and the other pertinent data. Explain in detail the method you used.

QUESTIONS

1. How does the value of I_L measured in the circuit of Fig. 26-3 compare with that in the Thévenin equivalent circuit of Fig. 26-4? Should they be the same? Explain.
2. Discuss the data in Table 26-1 and explain any unexpected results.
3. Does the experiment confirm the validity of Thévenin's theorem for the circuit of Fig. 26-3? Explain.
4. Does the experiment "prove" Thévenin's theorem for any circuit? Explain.
5. What specific advantages, if any, do you see in using Thévenin's theorem in the solution of dc circuits?

Answers to Self-Test

Answers to slide-rule accuracy.

1. (*a*) 100; (*b*) 188; (*c*) 0.134
2. (*a*) 45; (*b*) 338; (*c*) 0.1

27

EXPERIMENT   Norton's Theorem

OBJECTIVES

1. To determine by analysis the values I_N (Norton's generator current) and R_N (Norton's generator resistance) in a dc circuit containing one or two voltage sources
2. To verify experimentally the values I_N and R_N proposed by Norton's theorem in the solution of complex dc networks containing two voltage sources

INTRODUCTORY INFORMATION

Norton's Theorem

Thévenin's theorem simplifies analysis of complex networks by reducing the original circuit to a simple equivalent circuit involving a constant voltage source, the Thévenin generator (V_{TH}) in series with an internal resistance R_{TH}. This generator delivers current to the circuit load resistance we wish to study. It is then a simple matter to determine the

(a)   *(b)*

Fig. 27-1. (a) Network terminated by a load resistance R_L; (b) Norton equivalent generator.

current in the load and the voltage across the load. Norton's theorem utilizes a similar technique of simplification. The Norton generator, however, is a constant current source.

Norton's theorem states that any two terminal network may be replaced by a simple equivalent circuit consisting of a constant current generator I_N, shunted by an internal resistance R_N. Figure 27-1a shows the original network as a *block* terminated by a load resistance R_L. Figure 27-1b shows the Norton equivalent circuit. The Norton current I_N is distributed between the shunt resistance R_N and the load R_L. The current I_L in R_L may be found from the equation

$$I_L = \frac{I_N \times R_N}{R_N + R_L} \qquad (27\text{-}1)$$

The rules for determining the constants in the Norton equivalent circuit are as follows:

1. The constant current I_N is the current that would flow in the short circuit between the load resistance terminals if the load resistance were replaced by a short circuit.
2. The Norton resistance R_N is the resistance seen from the terminals of the *open* load, looking into the original network, when the voltage sources in the circuit are replaced by their internal resistance. Thus R_N is defined in exactly the same manner as is R_{TH} in Thévenin's theorem.

Applications

Consider the circuit, Fig. 27-2a. We wish to find the current I_L in R_L. We can solve for I_L by simple application of Ohm's and Kirchhoff's laws or by

Fig. 27-2. Applying Norton's theorem to solution of a dc network.

Thévenin's theorem. Let us, however, find I_L by Norton's theorem.

The development of the Norton equivalent circuit for Fig. 27-2a may be readily followed from Fig. 27-2b, c, and d. In Fig. 27-2b, R_L has been short circuited. As a result, R_3 is also "shorted." The Norton current I_N in the short circuit is readily found to be

$$I_N = \frac{V}{R_T} = \frac{100}{195 + 5} = \frac{100}{200} = 0.5 \text{ A} \quad (27\text{-}2)$$

In Fig. 27-2c we see that the original voltage source V has been replaced by its internal resistance R_1, which equals 5 Ω. The Norton resistance R_N is the resistance measured across AB in Fig. 27-2c. R_N is 100 Ω. Finally, Fig. 27-2d shows the Norton equivalent generator I_N in shunt with R_N, across the load resistance R_L. Applying Eq. (27-1) we have

$$I_L = \frac{I_N \times R_N}{R_N + R_L} = \frac{0.5(100)}{100 + 350}$$

$$I_L = \frac{50}{450} \quad (27\text{-}3)$$

and
$$I_L = 0.111 + \text{A} \quad (27\text{-}4)$$

This is the same value obtained in Experiment 26 when we used Thévenin's theorem to find I_L.

It would now be possible to find the current in any value R_L in the circuit of Fig. 27-2a by using the Norton equivalent generator with the values in Fig. 27-2d.

Solution of DC Networks with Two or More Generators

The solution of complex dc networks containing two or more voltage sources is possible using Kirchhoff's laws, Thévenin's theorem, and Norton's theorem. Application of the two latter theorems frequently simplifies analysis and computations. So that the student will have all methods at his disposal, we will solve the circuit of Fig. 27-3 by all three.

By Kirchhoff's laws

Let us recall Kirchhoff's two laws. These are:

1. The sum of the voltage drops across series-connected resistors in a closed circuit is equal to the

Fig. 27-3. Applying Kirchhoff's voltage law to a closed circuit.

applied voltage. This law may be stated in another form, namely, *the algebraic sum of the voltages around any closed circuit loop is zero*.

2. The current entering any junction of an electric circuit is equal to the current leaving that junction. Again, this law may be restated in another form, namely, *the algebraic sum of the currents at any point in a circuit is zero*.

In applying the second form of Kirchhoff's voltage law, certain conventions must be followed. Consider the circuit of Fig. 27-3. The formulation "algebraic sum," etc., requires careful attention to the *signs* of voltage sources and voltage drops. Thus, starting at A in Fig. 27-3 and moving in the loop $ABCDA$, we have first the voltage V_S. This is positive at A. Hence, the first term in our equation is $+V_S$. Next is the voltage drop V_1 across R_1. Assuming electron flow in the direction of the arrow, the polarity of voltage V_1 is as shown in Fig. 27-3, namely, $-$ to $+$. Hence the next term in our equation is $-V_1$. Similarly proceeding in the direction CDA the polarity of voltages are, respectively, $-V_T$ and $-V_2$. The equation satisfying Kirchhoff's voltage law may now be written as follows:

$$+V_S - V_1 - V_T - V_2 = 0 \qquad (27\text{-}5)$$

NOTE: The opposite *signs* of V_S and V_T show that these batteries are bucking, not aiding, each other,

a fact which is quite evident from the circuit diagram.

Now let us apply Kirchhoff's laws to the circuit of Fig. 27-4. It is necessary to set up as many equations as there are independent paths for current (loops). Loop 1 is $ABCDA$. We assume electron current I_1 in this loop in the direction of the arrow as shown. Another complete and independent loop, 2, is $DFGCD$ and electron current I_2 is assumed in the direction shown. I_1 and I_2 constitute the unknowns in this system. Let it be required to find I_2 in R_3.

Start at A in loop 1 and write the equation for voltage:

$$+V_1 - (I_1 + I_2)R_2 + V_2 - I_1 \times R_1 = 0 \quad (27\text{-}6)$$

Substituting the circuit values in Eq. (27-6), we have

$$100 - 2{,}200(I_1 + I_2) + 150 - 1{,}000\, I_1 = 0 \quad (27\text{-}7)$$

Collecting terms and simplifying, we have

$$3{,}200\, I_1 + 2{,}200\, I_2 = 250 \qquad (27\text{-}8)$$

Now start at D in loop 2. The equation for voltage is

$$-I_2 R_3 - (I_1 + I_2)R_2 + 150 = 0 \qquad (27\text{-}9)$$

Substituting the circuit values in Eq. (27-9),

Fig. 27-4. Solution of a complex network by Kirchhoff's laws and Thévenin's theorem.

(a)

(b)

(c)

(d)

Fig. 27-5. Solution of a complex network by Norton's theorem.

$$-1,200\,I_2 - 2,200(I_1 + I_2) + 150 = 0 \quad (27\text{-}10)$$

Collecting terms and simplifying, we get

$$2,200\,I_1 + 3,400\,I_2 = 150 \quad (27\text{-}11)$$

Equations (27-8) and (27-11) may now be solved simultaneously for I_1 and I_2, yielding the results

$$I_1 = 0.0861 \text{ A}$$
$$I_2 = -0.0115 \text{ A}$$

The fact that the sign before I_2 is minus simply means that I_2 flows in a direction opposite to that assumed in Fig. 27-4. However, the numerical value is correct. I_2 is the required current in R_3.

Solution by Thévenin's theorem

Let us solve for I_2 in R_3 (Fig. 27-4a) by applying Thévenin's theorem. First we must find V_{TH}, the open circuit voltage across FG, Fig. 27-4b. The open circuit voltage across FG is $V_{FG} = V_{TH} = -150 + I_3R_2$. We must therefore solve for I_3 before we can determine V_{TH}. Kirchhoff's voltage law applied to Fig. 27-4b in the loop $ABCDA$ gives

$$100 - I_3(2,200) + 150 - I_3(1,000) = 0 \quad (27\text{-}12)$$

and

$$3,200\,I_3 = 250$$

$$I_3 = \frac{250}{3200} = 0.0781 \text{ A}$$

Now $I_3R_2 = (0.0781)(2,200) = 171.82 \text{ V}$

The voltage

$$V_{TH} = -150 + I_3R_2 = 21.82 \text{ V} \quad (27\text{-}13)$$

Hence, V_{TH} in Fig. 27-4d is 21.82 volts.

To find R_{TH}, replace V_1 and V_2 by their internal resistances, which we will assume to be zero, and solve for the resistance across FG (Fig. 27-4c).

$$R_{TH} = \frac{2,200 \times 1,000}{2,200 + 1,000} = 687.5 \ \Omega \quad (27\text{-}14)$$

Now to solve for I_2 in R_3 of the original network, find I in Thévenin's equivalent circuit (Fig. 27-4d). Thus

$$I = I_2 = \frac{V_{TH}}{R_{TH} + R_3} = \frac{21.82}{1887.5} = 0.0115 + \text{A} \quad (27\text{-}15)$$

We see that the current I_2 in R_3 is 0.0115 A, the same as in the solution of the original circuit using Kirchhoff's laws.

Solution by Norton's theorem

Let us now solve for I_2 in R_3 (Fig. 27-5a, which is the same as Fig. 27-4a) by Norton's theorem. The development of Norton's equivalent circuit (Fig. 27-5d) may be seen from Fig. 27-5b and c. In Fig.

27-5b, R_3 is replaced by a short circuit. I_N, the current in the short circuit between terminals F and G, is the current in the Norton equivalent generator. To solve for I_N, apply Kirchhoff's laws to loops 1 and 2 of Fig. 27-5b. Loop 1 is the path $ABCGFDA$, while loop 2 is the path $DFGCD$.

In loop 1,

$$+100 - I_4(1,000) = 0$$

$$I_4 = \frac{100}{1,000} = 0.1 \text{ A} \qquad (27\text{-}16)$$

In loop 2,

$$-2,200\, I_5 + 150 = 0$$

$$I_5 = \frac{150}{2,200} = 0.0682 \text{ A} \qquad (27\text{-}17)$$

Figure 27-5b shows that $I_N = I_4 - I_5$, since the directions of I_4 and I_5 are opposite. Therefore

$$I_N = 0.1000 - 0.0682 = 0.0318 \text{ A} \quad (27\text{-}18)$$

The Norton generator resistance R_N is the resistance measured across terminals FG in Fig. 27-5c. This is the same as the Thévenin resistance R_{TH}, previously computed for Fig. 27-4c, and is

$$R_N = 687.5 \ \Omega \qquad (27\text{-}19)$$

Now, to find I_2 in Fig. 27-5d, use the formula

$$I_2 = \frac{I_N \times R_N}{R_N + R_L} = \frac{0.0318 \times 687.5}{687.5 + 1,200}$$

$$= \frac{0.0318 \times 687.5}{1887.5} = 0.0115 + \text{A}$$

$$(27\text{-}20)$$

Again we find that the value of current in R_3 is the same as that obtained using Kirchhoff's laws and Thévenin's theorem.

SUMMARY

1. Norton's theorem offers another method of solving complex circuits. By Norton's theorem a complex two-terminal network may be replaced by a simple equivalent circuit which acts like the original circuit at the load connected to the two terminals.
2. The equivalent circuit is a circuit consisting of a current generator, called the Norton generator (I_N), in parallel with the generator's internal resistance (R_N), across which the load (R_L) is connected. The current I_N of the generator is then divided between R_N and R_L. Fig. 27-1b shows the Norton generator and its load. This is the Norton equivalent of some complex circuit such as that shown in block form in Fig. 27-1a.

3. To determine the Norton current I_N, short-circuit the load and calculate the current through the short in the original circuit. This short-circuited current is I_N. To calculate I_N it may be necessary to use Ohm's and Kirchhoff's laws.
4. To determine the Norton resistance R_N which acts in parallel with the Norton generator, the same technique is applied as in finding the Thévenin resistance in the preceding experiment. Proceed as follows: Open the load at the two terminals in question in the original network. Replace all the voltage sources with their internal resistances. Then calculate R_N, the resistance at the open-load terminals, looking back into the circuit.
5. Norton's theorem applies to a circuit with one or more power sources.
6. Once the complex network has been replaced by the Norton equivalent circuit the current I_L through the load may be found by applying this formula (see Fig. 27-1b):

$$I_L = \frac{I_N \times R_N}{R_N + R_L}$$

8. In this experiment we will verify Norton's theorem experimentally for a dc network containing two voltage sources.

SELF-TEST

Check your understanding by answering these questions:
1. Consider the circuit of Fig. 27-2a. $V = 120$ V, $R_1 = 10\ \Omega$, $R_2 = 390\ \Omega$, $R_3 = 600\ \Omega$, and $R_L = 270\ \Omega$. Assume the internal resistance of the battery V is zero, in the equivalent Norton's circuit,
   (a) $I_N = $ _____ A
   (b) $R_N = $ _____ Ω
   (c) $I_L = $ _____ A
2. Consider the circuit of Fig. 27-5a. $V_1 = 45$ V, $V_2 = 75$ V, and assume the internal resistance of these batteries is zero. $R_1 = 450\ \Omega$, $R_2 = 1,500\ \Omega$, and R_3 (the load) $= 470\ \Omega$. In the Norton equivalent circuit,
   (a) $I_N = $ _____ A
   (b) $R_N = $ _____ Ω
   (c) $I_L = $ _____ A

Fig. 27-6. Experimental circuit for verifying Norton's theorem.

MATERIALS REQUIRED

- Equipment: Two voltage-regulated dc supplies; two multirange milliammeters or VOMs capable

of measuring up to 50 mA; EVM
- Resistors: ½-W 3,300-Ω, 5,100-Ω, 10,000-Ω, and others as required in step 8 (optional)
- Miscellaneous: 10,000-Ω 2-W potentiometer

PROCEDURE

1. Connect the circuit of Fig. 27-6a. S_1 is *open.* Turn regulated power supplies V_1 and V_2 **on** and set $V_1 = 40$ V and $V_2 = 20$ V.
2. S_2 is *open.* Close S_1. Be certain that V_1 and V_2 are maintained at 40 V and 20 V, respectively. With milliammeter M, measure the current I_L in R_3. Record in Table 27-1, in column "I_L Measured Original Circuit."
3. Close S_2, shorting R_3. Measure the current I_N and record under column labeled "I_N Measured."

NOTE: This is the short-circuit current of the Norton equivalent generator.

4. Remove the voltage supplies V_1 and V_2 from the circuit. Remove R_3 from the circuit. Short points D to F and A to B.

NOTE: It is assumed that the internal resistance of the voltage-regulated power supplies is zero.

S_1 is still closed, S_2 open. Measure the resistance between points F and G and record under "R_N Measured." This is the resistance in shunt with the Norton equivalent generator.

5. Connect the circuit of Fig. 27-6b. S_1 is open.

TABLE 27-1.

R_L	I_N (mA)		R_N (ohms)		I_L (mA)		
						Measured	
	Measured	Computed	Measured	Computed	Norton	Original Circuit	Computed
5,100							

NOTE: R_n is a 10,000-Ω 2-W adjustable resistor. Set R_n to the resistance measured in step 4.

R_3 is the 5,100-Ω load resistance from the previous circuit. M_1 and M_2 are milliammeters to measure the Norton current and the load current, respectively. Turn the voltage output of V down to *zero*.

6. Close S_1. Slowly turn up the output of the power supply until the *current* measured by M_1 equals I_N, the Norton generator current measured in step 3. Now, M_2 measures the current I_L, in R_3. Record the current measured by M_2 under "I_L Measured Norton."

7. **Power off.** In the circuit (Fig. 27-6a) compute the value I_N for the Norton equivalent generator and record in Table 27-1, under "I_N Computed." Show your computations.

   Compute also the value R_N, the Norton equivalent generator shunt resistance, and record under "R_N Computed." Show your computations.

   Now find the value I_L in R_3 using the Norton equivalent generator with the values computed in this step. Record this value of I_L under "I_L Computed." Show your computations.

Extra Credit

8. Verify that Norton's theorem is true for the circuit of Fig. 27-6a for other values of R_L by substituting for R_L, in turn, each of three different-valued resistors whose resistance lies between 560 and 12,000 Ω. Measure the current I_L in R_L in the circuit of Fig. 27-6a and also in the Norton equivalent circuit of Fig. 27-6b. Record in Table 27-1 the values of R_L used and all the other pertinent data. Explain in detail the method you used.

QUESTIONS

1. How does the value of I_L measured in the circuit of Fig. 27-6a compare with that measured in the Norton equivalent circuit of Fig. 27-6b? With that computed? Should they be the same? Explain.
2. Are the computed and measured values of I_N the same? Should they be? Explain.
3. Are the computed and measured values of R_N the same? Should they be? Explain.
4. Discuss the data in Table 27-1 and explain any unexpected results.
5. Does the experiment confirm the validity of Norton's theorem for the circuit of Fig. 27-6a? Explain.
6. Does the experiment "prove" Norton's theorem for any circuit? Explain.
7. If values of R_3 (Fig. 27-6a) other than 5,100 Ω are used in the experiment, what must the value of I_N in Fig. 27-6b be? Explain.

Answers to Self-Test

1. (a) 0.3; (b) 240; (c) 0.141
2. (a) 0.05; (b) 346; (c) 0.021

EXPERIMENT 28

Magnetic Fields about Bar and Horseshoe Magnets

OBJECTIVES

1. To determine experimentally the pattern of magnetic lines of force about a bar magnet; about two bar magnets with like poles near each other; with unlike poles near each other
2. To determine experimentally the pattern of magnetic lines of force about a horseshoe magnet

INTRODUCTORY INFORMATION

In previous experiments we were concerned with a qualitative (descriptive) and quantitative (mathematical) understanding of electricity, and with the measurement of electrical quantities. This concern was dictated by the consideration that an electronic technician deals primarily with the effects of electricity and must have a thorough understanding of electricity and electric circuits. There is another area, namely that of magnetism, which we must now stop to consider and comprehend, for electricity and magnetism are closely associated. Electricity is generated by a coil of wire rotating in a magnetic field. Moreover, electronics makes extensive use of magnetically operated devices such

as relays, meters, transformers, door openers, and alarm systems.

In the experiments on magnetism which follow, no attempt will be made to make measurements of magnetic quantities, for this is generally outside the domain of the electronics technician. Rather, our experiments will be descriptive, dealing with magnetic effects, in an attempt to gain some understanding of the nature of magnetism. For a quantitative discussion of magnetism the student is referred to a standard text on electricity and magnetism.

Some Physical Properties of Magnets

Certain materials, called magnets, exhibit unique properties. A magnet attracts iron and other ferrous materials. Between two bar magnets, properly oriented (Fig. 28-1a), a force of attraction exists, and one magnet is pulled to the other. If one of the bar magnets is rotated through 180° (Fig. 28-1b), the magnets repel each other.

If a small bar magnet, such as that in Fig. 28-2, is freely suspended at its center of gravity, one end of the magnet turns towards the north. If the magnet is spun and permitted to come to rest again, the same end points to the north. We call the north-seeking end of the magnet the north pole, designated N, and the other end the south pole, desig-

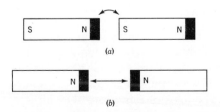

(a)

(b)

Fig. 28-1. (a) Two bar magnets properly oriented are attracted toward each other. (b) If one bar magnet is turned 180°, the magnets repel each other.

Fig. 28-2. A bar magnet free to turn about its center of gravity will point toward the north.

nated S. A magnetic compass, used for navigation, utilized this characteristic of a magnet as far back as the second century A.D. The earth itself is known to be a huge magnet with magnetic poles close to, but not coincident with, the geographic poles. The needle of a magnetic compass aligns itself with the magnetic poles of the earth.

If a bar magnet is cut in two (Fig. 28-3), each of the pieces exhibits the properties of a bar magnet. If we then cut each piece into many more pieces, each smaller piece continues to act like a small bar magnet with a north and a south pole. We cannot isolate just a single pole to any tiny piece.

A magnet does not attract some materials. Air, wood, paper, and glass, to mention but a few, will *not* be attracted by a magnet. These materials are nonmagnetic. Other materials such as iron and its compounds exhibit marked magnetic properties, and are designated ferromagnetic. These substances may be magnetized by placing them in contact with or in the magnetic field of a *natural magnet*. Moreover, some of these continue to act as magnets when the inducing magnet has been removed. Certain mixtures of aluminum, nickel, and cobalt are used to form very strong permanent magnets, called *alnico* magnets. These are used in radio loudspeakers and other magnetic devices. Hard steel also retains its magnetism. Soft iron is a temporary magnet, for when the inducing magnet is removed, the soft iron loses its magnetism.

Magnetic Field about a Bar Magnet

The force associated with a magnet is said to exist in the magnetic field about the magnet. This magnetic field is presumed to consist of lines of force whose concentration and configuration may be demonstrated experimentally.

Place a cardboard or a piece of glass over a bar magnet and sprinkle iron filings on the cardboard. Tap the cardboard or glass gently so that the iron filings are free to move. The iron filings arrange themselves in a unique pattern (Fig. 28-4). There is a noticeable concentration of the iron filings along curved lines, which we call "lines of force." The greatest concentration of these lines of force appears at the ends or poles of the magnet. As the lines of force move out from the magnet, the separation between them increases. That is, the magnetic field gets weaker the further we are from the magnet. We would therefore expect that the force of attraction of a magnet is greatest in the immediate vicinity of the magnetic poles where the lines of

magnetic force are most heavily concentrated, a fact which is readily verified experimentally.

All the magnetic lines in the field about a magnet constitute the magnetic flux. The stronger the magnetic field, the more lines of force there are in the flux.

A diagrammatic representation of the lines of force about a bar magnet is shown in Fig. 28-5. Note that each field line is a continuous closed loop, leaving the N pole, traveling around the magnet to the S pole and through the magnet back from the S pole to the N pole. The N to S direction of the field lines is purely arbitrary. Note that field lines do *not* cross each other, but push apart. Field lines are most heavily concentrated at the poles.

Fig. 28-5 is a two-dimensional view of the magnetic field about a bar magnet. A three-dimensional view would show that the magnetic field is indeed three-dimensional.

Unlike Poles Attract

If the N pole of a magnet is brought near the S pole of a second magnet, there is a force of attraction between the two magnets; the lines of force tend to shorten and pull the two magnets together. The field pattern about the two separated magnets now appears as in Fig. 28-6, and resembles the pattern about a single bar magnet, except for the visibility of the lines between adjacent poles of the two magnets, which are indeed extensions of the lines which complete their loops through the magnets.

Like Poles Repel

If like poles of two magnets are brought near each other, the two magnets are repelled and push apart. The field pattern about these two magnets would appear as in Fig. 28-7. The lines of force close to the adjacent like poles of the two magnets are distorted and push apart.

Magnetic Induction

Magnetic materials more readily admit magnetic lines of force than nonmagnetic materials. When a magnetic material, such as soft iron is placed in the magnetic field of a permanent magnet, this effect may be noted. The lines of force near the soft iron are attracted to the soft iron, as though it were a magnet, and pass through the soft iron rather than through the space around it (Fig. 28-8). In fact, the

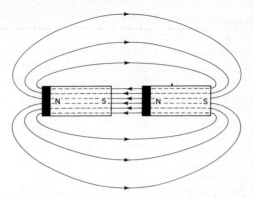

Fig. 28-6. Lines of force between unlike poles reinforce, and the magnets are attracted.

Fig. 28-3. If a bar magnet is cut into two pieces, each piece exhibits the properties of a bar magnet.

Fig. 28-4. Distribution of iron filings in lines of force about a bar magnet.

Fig. 28-7. Like poles repel. Field pattern about two magnets when like poles of the magnets are brought near each other.

Fig. 28-5. Field of force indicated by magnetic field lines.

Fig. 28-8. Action of soft iron on magnetic field about a permanent magnet.

Magnetic Fields about Bar and Horseshoe Magnets **133**

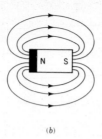

(a)   (b)

Fig. 28-9. (a) Magnetic field in the air gap of a horseshoe magnet is more concentrated than the magnetic field in (b) the air gap of a bar magnet whose poles are equally spaced.

soft iron, while in the field of the permanent magnet, acts like a magnet and will attract iron filings to itself. We say that the soft iron has been magnetized by *induction*. It will loose its magnetism when it is removed from the field of the permanent magnet. Soft iron, then, is a temporary magnet. Moreover, since soft iron provides an easier path for magnetic lines of force than does air, soft iron is said to be more permeable than air.

Magnetic Field about a Horseshoe Magnet

A horseshoe magnet may be considered a bar magnet bent in the form of a horseshoe (Fig. 28-9a). The magnetic poles remain at the ends of the magnet, but geometry of the magnet concentrates the lines of force in the air gap between the poles. Compare the lines of force in the air gap of a horseshoe magnet (Fig. 28-9a) with those of a bar magnet (Fig. 28-9b) whose poles are as far apart as those in the horseshoe magnet. It is apparent that the field in the air gap of the horseshoe magnet is more highly concentrated. Horseshoe magnets with semicircular soft-iron pole pieces are used in meter movements. The magnetic flux in their air gap is uniform and highly concentrated. These are two requirements for a sensitive, linear meter movement.

SUMMARY

1. The force associated with a magnet exists in the three-dimensional magnetic field surrounding the magnet.
2. Though magnetic lines of force cannot be seen, their effects can be observed by the concentration of iron filings along well-defined paths about the magnet.
3. A line of force is the path which an isolated

magnetic pole (if such a thing were possible) would travel in moving from one of the poles of the magnet to the other.
4. A magnet has two poles, north and south. If the magnet were suspended at its center of gravity and were free to turn, the *north* pole of the magnet would point in a northerly direction.
5. The field lines about a magnet are closed loops moving in the direction N to S (by conventional agreement).
6. Lines of magnetic force repel each other and tend to spread apart.
7. The lines of force about a magnet do not cross.
8. Like poles of magnets repel, unlike poles attract.
9. Lines of magnetic force meet less resistance in magnetic than in nonmagnetic materials. Therefore if a piece of soft iron (magnetic) is placed in the field of a magnet, it will distort the field by channeling in itself the lines of force in the vicinity.
10. Magnetic materials may be magnetized by induction, that is, in the presence of a magnetic field. Some will retain their magnetism after the inducing magnet is removed. Others, such as soft iron, will lose their magnetism when the inducing force is removed. Soft iron is therefore in the class of temporary magnets.
11. A knowledge of magnetism is essential to the electronics technician because electronics makes extensive use of magnetically operated devices.

SELF-TEST

Check your understanding by answering these questions:

1. Magnetic lines of force are visible to the naked eye, as is evident from the pattern of iron filings about a magnet. _____ (true/false)
2. The force of a magnet exists in its _____ _____.
3. Magnetic lines will cross each other close to the poles of the magnet. _____ (true/false)
4. The north pole of a bar magnet which is freely suspended at its center of gravity and is free to turn points to the _____ (north/south) magnetic pole of the earth, obeying the law that _____ (like/unlike) magnetic poles attract each other.
5. Some applications of magnetism in electronics are:
   (a) _____

(*b*) _____
(*c*) _____
(*d*) _____

6. Electricity may be generated by a coil of wire _____ in a _____ _____ .

7. The magnetic field in the air gap between the poles of a _____ _____ is more concentrated than the field in the air gap between the poles of a _____ _____ .

8. Some nonmagnetic materials are:
(*a*) _____ ; (*b*) _____ ; (*c*) _____ ;
(*d*) _____ .

9. Strong permanent magnets may be made from a mixture of _____ , _____ , and _____ . These are then called _____ magnets.

10. Magnetic lines of force meet less resistance in air than in soft iron. _____ (true/false)

MATERIALS REQUIRED

- Equipment: Polaroid camera and holder, if available
- Miscellaneous: Two bar magnets, magnetic compass, horseshoe magnet, iron filings and cardboard (8½ × 11 in.)

PROCEDURE

1. Examine the compass assigned to you. How can you tell which is the N pole? _____
_____ .
Identify the N pole of the compass.

2. The poles of the two bar magnets and the horseshoe magnet assigned to you are unmarked. How can you determine which is the N, which the S pole of each of these magnets? _____
_____ .
Identify the N and S poles of each of these magnets.

3. Place a cardboard over one of the bar magnets assigned to you. Sprinkle iron filings on the cardboard, around the magnet. Gently tap the cardboard until a recognizable pattern is seen. This is the pattern of a cross section of the lines of force about the bar magnet. Photograph or draw the pattern for future use. Identify the position of the poles of the magnet and indicate the direction of the lines of force.

4. Now place the two bar magnets, in line, so that the N pole of one is adjacent to and about *one* (1) in. away from the S pole of the other. Repeat step 3.

5. Turn one of the bar magnets (in step 4) 180°, so that the N pole of one is adjacent to and about *one* (1) in. away from the N pole of the other. The magnets are still in line. Repeat step 3.

6. Place a cardboard over a horseshoe magnet and repeat step 3.

QUESTIONS

NOTE: Submit drawings or photographs of all field patterns you obtained in this experiment.

1. Compare the magnetic field pattern about a bar magnet, which you obtained in this experiment, with that in Fig. 28-4.

2. From the distribution of iron filings about the bar magnet, was it evident that lines of magnetic force are continuous loops? If not, in which of the experimental patterns was it more evident?

3. How did the field pattern you obtained when you faced like poles of the two bar magnets compare with that in Fig. 28-7? Comment on the differences.

4. How did the field pattern you obtained when you faced unlike poles of two bar magnets compare with that in Fig. 28-6? Comment on the differences.

5. What evidence have you, if any, that magnetic lines of force repel each other?

6. How can you verify experimentally that soft iron is more permeable than air?

Answers to Self-Test

1. false
2. magnetic field
3. false
4. south; unlike
5. (*a*) relays; (*b*) transformers; (*c*) electrical generators; (*d*) speakers
6. moving; magnetic field
7. horseshoe magnet; bar magnet
8. (*a*) wood; (*b*) paper; (*c*) glass; (*d*) air
9. aluminum; nickel; cobalt; alnico
10. false

29

Magnetic Field Associated with Current in a Wire

OBJECTIVES

1. To verify experimentally the existence and direction of a magnetic field about a wire carrying current
2. To determine experimentally the pattern of magnetic lines of force about a solenoid

INTRODUCTORY INFORMATION

Current Producing a Magnetic Field

The physicist Hans Christian Oersted observed that a compass needle was deflected when placed in the vicinity of a wire carrying an electric current. Moreover, he observed that the direction in which the compass needle pointed depended on its position relative to the wire and on the direction of the current.

These observations may be illustrated with the help of the diagram in Fig. 29-1. *W*, the wire carrying current, is perpendicular to the plane of the table through which the wire is drawn. A compass placed flat on the table in position 1 deflects as shown. The direction of the compass needle changes when the compass is moved to positions 2, 3, and 4.

When current in the wire is shut off, the compass needle always points to the north, regardless of its position on the table relative to the wire. Moreover, when current in the wire is turned on again but reversed, the compass needle is again deflected by the force about the wire. However, this time the needle points in the opposite direction from its original orientation in positions 1 through 4 in Fig. 29-1.

These findings point to the conclusions that

1. A magnetic field is developed about a wire carrying current.
2. The direction of the magnetic field depends upon the direction of current in the wire.
3. The magnetic field appears to be circular about the wire.

Subsequent researchers showed that the magnetic field lies in a plane perpendicular to the current-carrying wire, that the magnetic field developed by the current *is* circular with the current-carrying wire at the center, and that the magnetic field intensity is greatest close to the wire.

One other fact may be noted. A magnetic field is developed about any moving electrical charge, whether that moving charge is a negative electron or is positive. Moreover, the motion of the electrical charge need not be restricted to a wire, for an electron beam, moving in the vacuum of a cathode ray tube, has associated with it a magnetic field which has the same characteristics as the field about a wire carrying current.

Experiments with the effects of the magnetic field about a wire lead to a rule which makes it

Fig. 29-1. Deflection of a magnetic compass in the vicinity of a wire carrying current.

Fig. 29-2. Left-hand rule showing direction of magnetic lines of force about a wire carrying electron current.

possible to predict the direction of the lines of force. That rule, based on conventional current flow (from the positive terminal of a battery, through the wire, to the negative terminal) is known as the right-hand rule. In this text, electron current, rather than conventional current, is used and so we will modify the rule for the *left hand*. The left-hand rule states: If the fingers of the left hand encircle the wire and the thumb of the left hand points in the direction of electron flow in the wire, the fingers point in the direction of the magnetic lines of force (Fig. 29-2). The right-hand rule is identical except that direction of conventional current is opposite from electron current, and the right hand is used.

Magnetic Field about a Solenoid

We observed in the previous experiment that when two magnetic fields were brought near each other they interacted and the lines of force were distorted. We can extend this observation to note that the lines of force of two magnetic fields *aid each other* when their lines of force are in the *same direction*, and *oppose (cancel) each other* when their lines of force are in opposite directions. Wire shaped in the form of a solenoid coil (Fig. 29-3) utilizes this characteristic to form an electromagnet, when current is sent through the solenoid. The circular lines of

magnetic force about the wire in the solenoid combine when they are in the same direction, and cancel when oppositely directed, to form a field pattern which is very much like that of a bar magnet. Note that the magnetic lines of force are most heavily concentrated inside the coil, just as they are in a bar magnet. Moreover the distribution of the field lines outside the coil is very similar to the field distribution about a bar magnet.

The location of the N and S poles of a solenoid may be determined by another left-hand rule. If you grasp the solenoid so that the fingers point in the direction of electron current, then the thumb points in the direction of the N pole (Fig. 29-4). Observe that the position of the poles may be reversed by changing the direction of current or changing the direction of the winding.

The form about which a solenoid is wound may consist of magnetic or nonmagnetic material. However, soft iron is generally used, because the iron core, having a higher permeability than nonmagnetic materials, increases the flux density inside the core. When current is turned on the iron core is magnetized and acts like a bar magnet. The magnet loses its "strength" when current is turned off. This is another reason for using a soft-iron core: to create a temporary magnet when it is needed and eliminate it when it is not needed.

The strength of a solenoid magnet may be increased by increasing the number of turns of wire in the coil or by increasing the current in the coil.

We can understand magnetic effects better by comparing these effects with those in an electric circuit. In an electric circuit the variables are V, electromotive force; I, the resulting current; and $\mathcal{R}$, the opposition to current in the circuit. In a magnetic circuit the equivalent variables are *mmf*, magnetomotive force; ϕ, the lines of magnetic force produced; and $\mathcal{R}$ (reluctance), the opposition of the

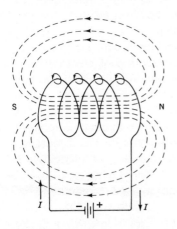

Fig. 29-3. Magnetic field about a solenoid.

Fig. 29-4. Left-hand rule for locating the N pole of a solenoid magnet.

circuit to producing lines of force. Algebraically,

$$mmf = At \qquad (29\text{-}1)$$

where A equals the amperes in the circuit (coil), and t equals the number of turns in the coil. The formula which relates the three variables in a magnetic circuit is

$$\phi = \frac{mmf}{\mathcal{R}} \qquad (29\text{-}2)$$

This is comparable to Ohm's law for electric circuits.

Reasoning from formulas (29-1) and (29-2) it becomes apparent that the strength (ϕ) of a solenoid magnet is directly proportional to the number of ampere-turns (At); that if the current (A) in the coil is changing, the strength of the magnet is changing in direct proportion to A. That is, an increase or decrease in current will result in a proportionate increase or decrease in the strength of the resulting magnetic field.

Industrially, solenoids in the form of relays are used as electrical switches to turn a process on and off. Solenoids are also used as valves to permit a liquid to circulate in a system or to turn the liquid off. Bells, buzzers, and many other devices are solenoid-operated.

SUMMARY

1. A magnetic field is developed by a moving electrical charge.
2. Circular magnetic lines of force appear around a wire carrying current. The magnetic field is at right angles to the wire and may be visualized as a cylindrical "sleeve" surrounding the wire.
3. The direction of the lines of force depends on the direction of current. If the wire is grasped in the left hand with the thumb pointing in the direction of electron current, the fingers point in the direction of the circular magnetic lines of force (Fig. 29-2).
4. If the wire is wound in the form of a coil and current is permitted in this solenoid coil, the solenoid exhibits a magnetic field pattern very much like the field about a bar magnet (Fig. 29-3).
5. The poles of this electromagnet may be located by grasping the coil with the left hand, the fingers pointing in the direction of current in the windings. The thumb then points to the N pole of the magnet (Fig. 29-4).
6. The position of the poles of a solenoid magnet may be reversed by (a) reversing the current or (b) reversing the direction of the winding.

7. Solenoids are frequently wound on soft-iron cores. When current is turned on, the iron core acts like a temporary bar magnet, but it loses its magnetism when current is turned off.
8. The strength of a solenoid magnet may be increased by increasing the current or the number of turns, or both.
9. A magnetic circuit may be compared to an electric circuit in which:
   (a) mmf (magnetomotive forces) is the equivalent of V (electromotive force).
   (b) ϕ the resulting lines of magnetic flux is the equivalent of I.
   (c) $\mathcal{R}$ the opposition to ϕ is the equivalent of R.
10. In a magnetic circuit

$$\phi = \frac{mmf}{\mathcal{R}}$$

11. In a magnetic circuit $mmf = At$.
12. In a solenoid magnet, the strength of the magnet (ϕ) is directly proportional to A, the current in the coil; that is, ϕ increases as A increases, and ϕ decreases as A decreases.
13. Solenoid magnets are used to actuate the switching mechanism in a relay, a bell, buzzer, and other electromagnetic devices.

SELF-TEST

Check your understanding by answering these questions:
1. An electric current creates a magnetic field. _____ (true/false)
2. The magnetic field at any one point about a wire carrying current lies in a plane _____ to the wire at that point.
3. If an observer looks along a wire carrying electron current, in the direction of electron flow, the direction of the magnetic field about the wire appears _____ (clockwise/counter clockwise) to the observer.
4. Soft iron is more permeable than air. Therefore a solenoid with a soft-iron core has a more intense field in the core than does a solenoid with an air core. _____ (true/false)
5. The left-hand rule can be used to determine the identity of the _____ in a solenoid.
6. The strength of an electromagnet can be increased by increasing (a) _____ (b) _____ .

Fig. 29-5. Experimental circuit to determine characteristics of the magnetic field about a solenoid.

MATERIALS REQUIRED

- Power supply: Variable dc source, low-voltage, high-current
- Equipment: 0–1 A ammeter; polaroid camera and holder, if available
- Resistors: 25-W 15-Ω
- Miscellaneous: SPST switch, magnetic compass, iron filings, cardboard (8½ × 11 in.), 16 ft of #18 varnish insulated copper wire, 2-in. cylindrical (½-in. diameter) hollow coil form (cardboard, plastic, or copper tubing); 2-in. round soft iron core for coil form

PROCEDURE

Solenoid without Core

1. Use #18 varnish insulated copper wire to construct a solenoid coil. On a hollow cylindrical form, cardboard, plastic or copper tubing, 2 in. long by ½ in. in diameter, wind clockwise in two or three layers, a 100-turn coil. Scrape the ends of the wire free of insulation and secure one end at conductive terminal *A* mounted at one end of the coil form. Similarly secure the other scraped end at conductive terminal *B* mounted at the other end of the coil form. Mark ends *A* and *B*.

2. Connect the coil constructed in step 1 in the circuit of Fig. 29-5. End *A* of the coil is connected to the positive terminal of the supply; end *B*, through a resistor *R* and ammeter *M* to the negative terminal of the supply. Power is **off.**

3. Set a magnetic compass down on the lab table so that it is not near any magnets or magnetic materials. Orient the compass case so that the compass pointer is set on north N, as in Fig. 29-6. Place end *A* of the coil about 2 in. away from the point marked *E* on the compass. Orient the solenoid coil so that its axis is in line with the line of the compass drawn from W to E, as in Fig. 29-6. The solenoid coil, with no current in it, should have no effect on the compass pointer.

4. Turn the power supply **on** and adjust its output so that there is ¾ A in the coil, as measured on *M*. Record in Table 29-1 the direction of the compass pointer. *Data for steps 4 through 14 should be recorded in Table 29-1.*

5. Maintaining the W–E orientation of the axis of the solenoid coil, slowly push the coil so that end *A* is ½ in. away from point E on the compass. The effect of the solenoid on the compass pointer should be more evident than in step 4. Record the direction of the compass pointer.

6. Reduce current in the solenoid to 100 mA. The solenoid is still ½ in. from point E on the compass, as it was in step 5. Record the direction of the compass pointer.

7. Reduce solenoid current to zero. Record the direction of the compass pointer. Power **off.**

Solenoid with Soft-Iron Core

8. Place a 2-in. long soft-iron core inside the solenoid coil. Move the solenoid so that end *A* is again 2 in. from the point on the compass

Fig. 29-6. Axis of the solenoid is in line with the line drawn from W to E. When there is no current in the solenoid, the compass points to the north N.

Magnetic Field Associated with Current in a Wire **139**

TABLE 29-1.

Step	Current in Solenoid, amperes	Solenoid Distance from E, inches	Current Polarity		Direction of Pointer
			A	B	
4	3/4	2	+	−	
5	3/4	1/2	+	−	
6	1/10	1/2	+	−	
7	0	1/2	+	−	
9	3/4	2	+	−	
10	3/4	1/2	+	−	
11	1/10	1/2	+	−	
12	0	1/2	+	−	
13	3/4	1/2	−	+	
14	0	1/2	−	+	

marked E. The solenoid and compass are oriented as in step 3.

9. **Power on.** Adjust the power supply output until there is 3/4 A current in the solenoid. Record the direction of the compass pointer.
10. Move the solenoid coil until it is 1/2 in. away from point E on the compass, maintaining the W–E orientation of its axis. Record the direction of the compass pointer.
11. Reduce solenoid current to 100 mA. Record the direction of the compass pointer.
12. Reduce solenoid current to zero. Record the direction of the compass pointer.
13. **Power off.** Reverse polarity of current meter leads. Reverse also polarity of power supply leads. **Power on.** Adjust the supply for 3/4 A current in the solenoid coil. The solenoid is still 1/2 in. away from the point marked E on the compass, as in step 10. Record the direction of the compass pointer.
14. **Power off.** Record the direction of the compass pointer.

Magnetic Field about a Solenoid

15. The soft-iron core is still inside the solenoid coil, as in steps 8 through 14. The solenoid is connected as in step 13. **Power on.** Set the current in the solenoid to 3/4 A.
16. Position a cardboard plane over the solenoid. Sprinkle iron filings on the cardboard and gently tap the cardboard until the filings settle into an identifiable pattern. Photograph or draw this pattern.

QUESTIONS

Confirm your answers to the following questions by referring specifically to your experimental data.

1. What verification have you of the fact that there is a magnetic field associated with current in a coil?
2. In step 10 which end of the electromagnet, A or B, is the N pole? Why?
3. In step 13 which end of the electromagnet is the N pole? Why?
4. Is the magnetic polarity of the electromagnet affected by the direction of current in the electromagnet? How?
5. Do the results of your experiment confirm the left-hand rule for determining the poles of an electromagnet? Explain.
6. Is the strength of the electromagnet affected by the amount of current in the solenoid? How?
7. What does the magnetic field pattern about a solenoid coil electromagnet resemble?

Answers to Self-Test

1. true
2. perpendicular
3. counterclockwise
4. true
5. poles
6. (a) number of turns; (b) amount of current

30 EXPERIMENT

Inducing Voltage in a Coil

OBJECTIVES

1. To verify experimentally that a voltage is induced in a coil when the lines of force of a magnet cut across its windings
2. To verify experimentally that the polarity of the induced voltage depends on the direction in which magnetic lines of force cut the coil windings

INTRODUCTORY INFORMATION

Electromagnetic Induction

In the preceding experiment we established that current, which is defined as the *movement* of electrical charges, generates a magnetic field. In this experiment we will demonstrate that a moving magnetic field which cuts across a conductor will generate a movement of electrical charges in the conductor; that is, it will produce a voltage in the conductor and if the conductor is part of a complete circuit, current will flow in the conductor.

If a zero-center galvanometer is connected to the ends of a metal rod, and the rod is placed in the air gap between the poles of a permanent magnet, there will be no movement of the galvanometer pointer when the rod has come to rest in the magnetic field. However, if the rod is moved up, *cutting* the lines of magnetic force, the galvanometer will deflect showing current, as in Fig. 30-1. If the rod is then moved down, the galvanometer will again deflect, but in the opposite direction. Once the rod has come to rest, the galvanometer will register zero current. Now, if the rod is permitted to rest and the *magnet* is moved up, away from the rod, there is again current flow in the rod. Then if the magnet is brought down toward the rod, there is current flow

in the opposite direction. Note that in each case the rod was perpendicular to the lines of force of the magnet and the effect of a relative motion, that is, a motion of the rod or the magnet, was to *cut* the magnetic lines of force. The result was to induce a voltage in the conductive rod, causing current, as registered by the galvanometer. If the rod is now moved parallel to the lines of force between the poles of the magnet, there is no deflection of the galvanometer. *The conclusion we reach is that a relative motion between the conductive rod and the magnetic lines of force will induce a voltage in the rod only when the lines of force are cut by the rod.*

Polarity of Induced Voltage

The fact that the galvanometer pointer deflected either to the left or to the right of zero, depending on the direction of cutting of the lines of force, is

Fig. 30-1. The galvanometer registers current when there is motion up or down, that is, when the lines of force are cut by the rod.

evidence that the polarity of the voltage induced in the conductive rod depends upon the *direction* of cutting of the lines of force. The polarity of the induced voltage can be established by Lenz's law.

Lenz's law states that the induced voltage must be of such a polarity that the direction of the resulting current in the conductor sets up a magnetic field about the conductor which will oppose the *motion* of the inducing field. An example will illustrate the meaning of Lenz's law.

Consider the solenoid wound about an air core (Fig. 30-2). The ends of the solenoid are connected to a galvanometer, as shown. If a bar magnet is pushed into the coil, so that the S pole of the magnet enters the left side of the coil, the current resulting in the solenoid should set up a magnetic field whose S pole will be at the left side of the solenoid. Since like magnetic poles repel, Lenz's law would be satisfied, for the induced current in the solenoid would have set up a magnetic field which opposes the inducing field. To verify that a S pole is in fact produced at the left end of the solenoid, observe the direction in which the galvanometer pointer deflects. Note that it moves to the right, indicating a voltage whose polarity is shown in Fig. 30-2. Applying the left-hand rule, grasp the solenoid so that the fingers point in the direction of electron current, as shown by the arrows on the solenoid core. The thumb then points to the N pole. It is apparent, then, that at the left end of the solenoid winding there is a S pole, at the right end a N pole, and we have confirmation of Lenz's law.

If the solenoid is next moved out of the core, the left end of the solenoid will become a N pole, thus attempting to restrain the bar magnet from moving out of the core. The voltage induced in the solenoid will be of the opposite polarity and the galvanometer will deflect to the left.

Counter EMF

The discussion to this point has dealt with the observation that a voltage is induced in a wire or coil, when there is relative motion between the wire or coil and the magnetic field of a permanent magnet, that is, when the wire or coil cuts the lines of force of the magnet. The facts are more general, however. The effect of induction in a coil or conductor is achieved when there is relative motion between the coil or conductor and the field of *any* magnet, including an electromagnet.

In a previous experiment it was demonstrated that a magnetic field exists about a coil carrying current. The coil carrying current is in fact an elec-

Fig. 30-2. Verifying Lenz's law. The polarity of the magnetic field created by the induced current in the solenoid opposes the polarity of the inducing field.

tromagnet and may be substituted for the permanent magnet in the discussion earlier in this experiment.

The term "relative motion" is not limited merely to the physical movement of a magnet near a conductor, or of a conductor in a magnetic field. Relative motion may exist without any physical movement whatsoever. Consider a coil through which an increasing or decreasing (not a steady state) current is flowing. In the case of an increasing current the magnetic field about the coil is increasing or expanding, that is, it is a *moving field*. When the current in the coil is decreasing, the magnetic field about the coil is decreasing or collapsing, again a moving magnetic field. A conductor held stationary in this moving magnetic field is in effect *cutting* the lines of magnetic force. Therefore a voltage will be induced in the conductor. Expanding and collapsing magnetic fields will cause voltages of opposite polarity to be induced in the conductor.

Consider again a coil through which an increasing or decreasing current is flowing. The expanding or collapsing (moving) lines of magnetic force about the coil *cut the windings of the coil itself*. Accordingly a voltage is induced in the coil, which by Lenz's Law will oppose the inducing force. That is, the voltage induced in the coil will have a polarity opposite to the voltage which initially caused current to flow in the coil. The voltage induced in the coil may therefore be termed a *counter emf*.

Magnitude of Induced Voltage

The amplitude of voltage induced in a solenoid depends directly on (*a*) the *number* of turns N of wire in the solenoid and (*b*) the *rate* at which the lines of flux are cut by the windings of the solenoid. Thus the more windings, the higher will be the induced voltage across the ends of the coil. Similarly, the faster the magnetic flux is cut by the windings, the higher will be the voltage produced across the solenoid. A mathematical formulation of this law was first given by the physicist Michael Faraday. It is suggested that the student refer to a standard text on electromagnetism for the exact formula, which is the mathematical statement of Faraday's law.

SUMMARY

1. A voltage is induced in a conductor when the conductor *cuts* the lines of force in a magnetic field. Cutting the lines of force can be achieved by either moving the conductor or moving the magnetic field.
2. The polarity of the voltage induced in the conductor is determined by the direction of cutting of the lines of force. Thus if a positive voltage is induced, say by a conductor cutting the lines of force in a *downward* direction, a negative voltage will be induced by an *upward* cutting of these same lines of force.
3. The polarity of the induced voltage can be predicted by Lenz's law, which states: The polarity of the voltage induced in a conductor must be such that the *magnetic field set up by the resulting current* in the conductor will oppose the motion of the inducing magnetic field.
4. The amplitude of voltage induced in a solenoid when its windings cut magnetic lines of force will depend directly on the number of turns in the winding and on the rate at which the magnetic flux lines are cut by the windings (Faraday's law).

SELF-TEST

Check your understanding by answering these questions:
1. A moving magnetic field will _____ a _____ in a copper wire, if the lines of magnetic force _____ the wire.
2. The polarity of voltage induced in a conductor will depend upon the _____ of cutting of the lines of force by the conductor.
3. A conductor moving parallel to the lines of force in a magnetic field _____ (will/will not) have a voltage induced in it.
4. The more turns there are in a solenoid whose windings cut the lines of force in a magnetic field, the greater will be the voltage induced in it, all other things being equal. _____ (true/false)
5. By means of _____ law, it is possible to predict the polarity of voltage induced in a coil.

MATERIALS REQUIRED

- Equipment: Galvanometer or very sensitive electronic microammeter which can be zero centered
- Miscellaneous: Solenoid—100 turns of #18 varnish insulated copper wire wound in three layers on a 3 × 1 in. (diam.) hollow cardboard or plastic cylindrical form

PROCEDURE

1. Connect a galvanometer or sensitive zero-centered dc microammeter, set on its lowest range, to the terminals of the air core solenoid as in Fig. 30-2.
2. Position a bar magnet near the solenoid, with the S pole closest to the left end of the solenoid, as in Fig. 30-2.
3. With the bar magnet stationary, does the meter show any induced voltage? Record your observation in Table 30-1.
4. Now push the bar magnet into the solenoid so that the south pole of the magnet enters the air core first. Does the meter show any induced voltage? What is the polarity of the voltage? What is the current amplitude? Record these observations in Table 30-1.

5. Now pull the bar magnet out of the solenoid. Does the meter show any induced voltage? What is the polarity of the voltage? Current amplitude? Record these observations in Table 30-1.
6. Now reverse the bar magnet so that the N pole is closest to the left end of the solenoid. Insert the bar magnet into the air core. Does the meter show any induced voltage? Polarity? Current amplitude? Record your observations in Table 30-1.
7. Repeat step 5.
8. Again plunge the magnet into the air core, but more rapidly than in step 6. Observe and record in Table 30-1 the amplitude of current in the circuit.

TABLE 30-1.

Step	Condition	Voltage Polarity	Current Amplitude (μA)
3	Magnet stationary		
4	S pole of magnet near left end of solenoid. Magnet is pushed into solenoid.		
5	Magnet pulled out of solenoid.		
6	N pole of magnet near left end of solenoid. Magnet is pushed into solenoid.		
7	Magnet pulled out of solenoid.		
8	N pole of magnet near left end of solenoid. Magnet is pushed into solenoid more rapidly than in step 6.	X	
9	Same as step 8 but more rapidly.	X	
11	Solenoid plunged down over magnet.		X
12	Solenoid pulled up, away from magnet.		X

9. Remove the magnet and again plunge it into the air core, but more rapidly than in step 8. Observe and record in Table 30-1 the amplitude of current in the circuit.

10. Now position the bar magnet vertically on the table, with the N pole on the table and the S pole in the air above the table. The magnet will remain stationary.

11. With the meter still connected to the solenoid plunge the solenoid down over the magnet. Observe and record in Table 30-1 polarity of voltage induced.

12. Pull the solenoid out of the magnet. Observe and record in Table 30-1 polarity of voltage induced.

QUESTIONS

1. What are the conditions for electromagnetic induction of a voltage in a solenoid?

2. What verification have you that a voltage is in fact induced in a solenoid when a bar magnet is plunged into it or pulled out of it? Refer specifically to your data to support your answer.

3. What verification have you that it is the *relative* motion involved in cutting of lines of force which induces a voltage in the solenoid? Refer specifically to your data to support your answer.

4. What is Lenz's law?

5. What verification have you that Lenz's law is true? Refer specifically to your data to support your answer.

6. What is Faraday's law?

7. Do you have any verification of all or part of Faraday's law? Explain.

Answers to Self-Test

1. induce; voltage; cut
2. direction
3. will not
4. true
5. Lenz's

EXPERIMENT 31

Characteristics of a dc Meter Movement

OBJECTIVES

1. To determine experimentally the sensitivity (I_m) of a meter movement
2. To determine experimentally the internal resistance (R_m) of a meter movement
3. To determine experimentally the linearity of the meter movement

INTRODUCTORY INFORMATION

The Moving-coil Meter Movement

The measurement of dc voltage and current in electric and electronic circuits is usually effected by meters employing a moving-coil meter movement. This type of movement is based on the principle of the galvanometer developed by the French physicist Arsene d'Arsonval. D'Arsonval's galvanometer was a basic meter which measured very small currents. Though it was substantially modified later by Edward Weston and others, the basic moving-coil meter movement in use today is still known as the d'Arsonval movement.

Construction

The moving-coil meter consists of a horseshoe magnet with semicircular pole pieces made of soft iron added at the ends. A circular soft-iron core is positioned in a uniform magnetic field provided by the magnet and the pole pieces (Fig. 31-1a). Very fine phosphor bronze nonmagnetic wire wound on a rectangular aluminum frame (Fig. 31-1b) constitutes the moving coil. The coil assembly is centered in the air gap between the soft-iron core and the magnet (Fig. 31-1c). Each end of the coil is con-

nected to a spring. The two springs are symmetrically mounted on the coil, one at the top, the other at the bottom. At the center of the springs are pivots (Fig. 31-1d) which fit into jeweled bearings to reduce friction, thus permitting the coil to move freely in the air gap. The free ends of the two springs are brought to the two meter terminals. These serve as the terminals of the coil. A pointer is attached to the coil assembly and moves with it. The pointer can be positioned so that its rest or zero position is at the center or at the left side. A zero adjust screw is used to compensate for minor changes in coil position. Stops are provided at the left and right side of the magnet to limit the degree of movement of the pointer.

Operation

When a potential difference is placed across the meter terminals from an external circuit, current flows through the meter coil, setting up a magnetic field around it. This magnetic field interacts with the field of the permanent magnet and the torque developed causes the coil to turn on its pivots. The number of degrees of rotation of the moving coil is determined by the strength of the magnetic field about the coil, which in turn is proportional to the amount of current in the coil. The greater the current, the greater the rotation, and therefore the greater is the deflection of the pointer attached to the moving-coil assembly. A scale, calibrated in suitable units, is used to read the measurement made in the circuit.

The direction of current through the coil determines the polarity of its magnetic field and the direction of rotation of the coil assembly. If current direction is reversed in a meter whose zero position is at the left, the pointer attempts to move

Fig. 31-1. Construction and arrangement of parts in a moving-coil meter movement (*reprinted from Army Technical Manual TM11-664*).

downward to the left, off the scale. It is restrained by the *stop* on the left. Since the system requires a clockwise movement of the pointer, improper direction of current makes measurement impossible. It is apparent, therefore, that the moving-coil meter is a polarized device, and polarity must be observed in measuring voltage or current with it.

The springs play an important role in the operation of the meter. First they act as a restraining force on the rotation of the coil. Hence, they must be precision-wound to permit accurate measurements. The springs also act as conductors to carry current to and from the coil. Their action brings the pointer back to zero rest after the measurement is completed, and there is no longer current in the coil. Finally, they are wound in opposite directions to compensate for temperature changes, because a temperature change affects both springs in the same manner. Since they are oppositely wound, the effect of an expansion or contraction of one is counter-

acted by an expansion or contraction of the other acting in the opposite direction. Hence, the net change is zero.

The aluminum frame on which the coil is wound prevents the tendency of the pointer to oscillate or "hunt" before it comes to rest. This is accomplished by the eddy currents induced in the frame as it moves in the magnetic field of the meter. These eddy currents set up a magnetic field about the frame which, by Lenz's law, opposes the action of the inducing field of the permanent magnet. This slows down or damps the action of the moving coil when its current is either increasing or decreasing, thus minimizing its tendency to hunt.

Sensitivity of the meter movement

The amount of current (I_m) required to produce full-scale deflection of the pointer is called the "sensitivity" of the meter movement. Meters have sensitivities ranging from a few microamperes to many milliamperes. The smaller the current required to produce full-scale deflection, the greater is the sensitivity of the movement.

The sensitivity of the movement depends on the strength of the permanent magnet and the number of ampere-turns of the coil. The more turns on the coil, the smaller is the coil current required for full-scale deflection. Of course, the more turns, the greater is the internal or dc resistance of the coil.

The characteristics which are therefore associated with a meter movement are its sensitivity and internal resistance. Another characteristic derived from these two is the ohms/volt rating. The significance of this characteristic will be more evident when the loading effects of voltmeters are considered.

Meter linearity

In the design of the meter movement great pains are taken to provide a uniform magnetic field in the air gap where the coil is located. The purpose is to assure linear changes in deflection with linear changes in current through the coil. In a linear meter the number of degrees of rotation is directly proportional to the current in the coil. Thus, full-scale deflection results when 100 percent of the rated current is present in the coil. Half-scale deflection occurs when there is 50 percent of the rated current, etc. A linear meter movement makes possible a linear scale.

Meter accuracy

Meter movements can be built with a high degree of precision and hence can be made accurate within a fraction of a percent. These are intended for very precise laboratory measurements. General-purpose meters, however, are usually rated at 2 percent accuracy.

The percentage of accuracy is based on the full-scale reading of the meter. Hence, for readings lower than full scale, the percentage of error is greater than the rated value. The significance of this fact is more apparent when the movement is converted into a voltmeter. Thus, a meter rated at 2 percent on the 100-V range is accurate within 2 V for any value measured on this range. The percent of error in a 10-V reading on this range could conceivably be as high as 20 percent. It is evident, therefore, that meter readings close to full-scale deflection are the most accurate, if circuit loading, or the effects of the meter on the circuit, are ignored.

Circuit loading will be discussed in detail later in the sections dealing with voltage and current measurement.

Determining the Characteristics of a Meter Movement

The instrument engineer must know the characteristics (internal resistance and sensitivity) of the meter movement in designing voltmeters, current meters, and ohmmeters based on this movement. These characteristics are specified by the manufacturer of the movement. They may also be measured by the technician in the laboratory.

Sensitivity

One method of measuring the sensitivity (I_m) of a meter movement is shown in the circuit in Fig. 31-2. The meter movement under test is placed in series with a rheostat R, a limiting resistor R_1, a "standard" current meter M, and a battery V. Rheostat R is adjusted until the meter-movement pointer of the meter under test reaches full-scale deflection. The current required to effect full-scale deflection is read from the "standard" current meter. This is the required *sensitivity* (I_m) of the meter movement under test.

The accuracy of the "standard" current meter limits the accuracy of the measurement.

Internal resistance

A calibrated rheostat R, or decade box, connected in series with a linear movement under test, receiving voltage from a voltage-regulated dc source V, as in Fig. 31-3, may be used to measure the internal resistance (R_m) of the meter movement. The technique is based on the assumption that deflection of the meter movement varies linearly with current in the movement. (Linearity of the move-

Fig. 31-2. Circuit for measuring sensitivity (I_m) of a meter movement.

Fig. 31-3. Circuit for determining the internal resistance (R_m) of a meter movement.

ment may be verified by using the circuit of Fig. 31-2.)

Measurement of R_m is achieved in the following manner:

First, R is adjusted to the value R_1 for full-scale deflection of the movement pointer. The value of R_1 is read from the calibrated dial(s) of the rheostat.

Next, R is adjusted to the value R_2 for half-scale deflection of the pointer. R_2 is read from the dial(s) of the rheostat. Then the internal resistance of the movement may be calculated from the relationship

$$R_m = R_2 - 2R_1 \qquad (31\text{-}1)$$

An analysis of the circuit of Fig. 31-3 will show why the method described yields the correct value of R_m. In measurement 1,

$$V = I_m(R_1 + R_m) \qquad (31\text{-}2)$$

In the second measurement,

$$V = \frac{I_m}{2}(R_2 + R_m) \qquad (31\text{-}3)$$

Eliminating V from Eqs. (31-1) and (31-2), we have

$$I_m \times (R_1 + R_m) = \frac{I_m}{2}(R_2 + R_m) \qquad (31\text{-}4)$$

Solution of Eq. (31-4) leads to

$$2R_1 + 2R_m = R_2 + R_m$$

and
$$R_m = R_2 - 2R_1 \qquad (31\text{-}5)$$

If a laboratory-type resistance decade box is used as the rheostat R, together with a low-voltage source, an accurate measurement of R_m may be made (that is, accurate within the tolerance of the decade box).

Equations (31-2) and (31-3) set one requirement for the dc source. The voltage V delivered for both measurements must be the same. Therefore, a well-regulated power source is required, or V must be closely monitored and maintained at the same level

for the two load conditions, that is, for I_m and $I_m/2$.

Another requirement is that R remain low in value. A concrete example will explain why this is necessary. Suppose V is a 5-V source and the meter a 0–1 mA, with an internal resistance of 50 Ω. For full-scale deflection

$$R + R_m = \frac{V}{I_m} = \frac{5}{0.001} = 5,000\ \Omega$$

Since $R_m = 50\ \Omega$, R must be set at $R_1 = 4,950\ \Omega$ for full-scale deflection. Similarly, at half-scale deflection, R must be set at $R_2 = 9,950\ \Omega$. At this resistance level it is very possible to make a measurement error. For if R were set at $R_2 = 9,900\ \Omega$ instead of 9,950, the current in the meter circuit would increase from 500 microamperes (μA) ($\frac{1}{2}$ mA) to 502 μA. On a 1,000-μA (1-mA) scale, a 2-μA change is hardly noticeable. It is therefore possible that the erroneous setting of R_2 could be selected. Now, substituting the values $R_1 = 4,950$ and $R_2 = 9,900$ in Eq. (31-5), we discover that $R_m = 9,900 - 2\ (4,950) = 0$, a totally meaningless conclusion.

To overcome this difficulty, the applied voltage V must be very low, so that the limiting resistance R can be low. To illustrate, consider again the 0–1 mA movement whose $R_m = 50\ \Omega$. If, in the circuit of Fig. 31-3, $R = R_1 = 25\ \Omega$ for full-scale deflection, then $V = I_m(R_1 + R_m) = 0.001 \times (75)$, and $V = 75 \times 10^{-3}$ V = 75 millivolts (mV). With the 75-mV source, a value of $R = R_2 = 100\ \Omega$ is required for half-scale deflection. In this case, a change of 75 Ω in R causes a change of 500 μA ($\frac{1}{2}$ mA). Thus, an error in setting R can readily be observed. And so, for effective measurement of R_m by the method of Fig. 31-3, a very low-voltage regulated dc voltage source is required.

A Preferred Procedure for Measuring R_m

The measurement of R_m for a low-sensitivity movement can be made with a good degree of accuracy even if a very low regulated dc source is not available, by the circuit of Fig. 31-4. The output V of a variable voltage regulated supply is set at 10 V. R_1, a 10,000-Ω potentiometer, is connected as a rheostat in series with R_3, a 5,600-Ω resistor. R_2 is a calibrated decade rheostat whose resistance may be varied from 0 to 10,000 Ω. (A low-value potentiometer will serve if a decade rheostat is not available.)

Measurement of R_m begins with switch S_2 open and S_1 closed. R_1 is adjusted for full-scale deflection

Fig. 31-4. Experimental circuit for measuring internal resistance (R_m) of a meter movement.

of the movement. Next, switch S_2 is closed, throwing R_2 in parallel with R_m. R_2 is now adjusted for half-scale deflection of the movement. The resistance of R_2, read from the calibrated scale of the rheostat, is the required R_m, the internal resistance of the movement.

NOTE: If a potentiometer is used for R_2, the resistance of R_2 can be measured with an ohmmeter, after half-scale adjustment is made. Open S_2 before measuring R_2.

This procedure introduces a small error because there is a slight increase in circuit current when R_2 is placed in parallel with R_m. This increase in circuit current actually appears in R_2, and therefore the current I_2 in R_2 is slightly higher than $I_m/2$, the current in the meter movement. However, if R_m is small, the increase in current is negligible and may be ignored.

Taut-band Meters

A recent development in meter design features *taut-band construction*, which is a change from the conventional pivot-and-jewel type just described. The moving element of the taut-band movement is supported by two thin metal ribbons, kept under tension by elastic members at each end of the assembly (see Fig. 31-5). This type of construction eliminates the friction associated with the moving parts in the pivot-and-jewel arrangement.

The main advantage of the taut-band movement is its sensitivity; it can operate with very limited power—less than 1 microwatt. A disadvantage of the taut-band movement is that it does not have the fast response of the pivot-and-jewel construction, because it does not develop as much torque. More-

over rocking movements can cause pointer vibration and the instrument is more subject to breakdown under shock than is the pivot-and-jewel type.

Taut-band meters are used where highly sensitive movements are required.

SUMMARY

1. Nondigital meters used to measure dc voltage and current usually employ a moving-coil meter movement called a d'Arsonval meter.
2. The actuating elements of a d'Arsonval movement consist of a permanent horseshoe magnet and a moving-coil assembly freely suspended on a frame in the air gap between the semicircular pole pieces of the permanent magnet and a circular soft-iron core.
3. A very fine nonmagnetic wire is used for winding the moving coil on a rectangular aluminum frame.
4. Direct current is brought to the coil from the external circuit which is to be measured. The current in the coil sets up a magnetic field which interacts with the permanent magnetic field of the horseshoe magnet and creates a torque, causing the freely suspended moving-

Fig. 31-5. Annular taut-band meter movement (Simpson Electric Company).

coil assembly to rotate. A pointer attached to the moving coil moves along a calibrated scale. Readings are made from the scale.

5. The amount of current required to cause full-scale deflection of the pointer is called the sensitivity of meter movement (I_m).

6. The sensitivity of the movement depends directly on the strength of the horseshoe magnet and on the number of ampere-turns of the coil.

7. The direction of rotation of the coil depends on the direction of current in the coil. The dc meter movement is therefore a polarized device.

8. The resistance of a movement (R_m) is the resistance of the coil winding. The more turns of winding the greater is the resistance of the movement. A very sensitive meter movement normally has a higher internal resistance than a less-sensitive movement.

9. The electrical characteristics of a meter movement of concern to the technician and engineer are sensitivity (I_m) and internal resistance (R_m).

10. In the design of a meter movement pains are taken to ensure a uniform magnetic field in the air gap where the coil is located. A uniform field ensures equal changes in deflection of the meter pointer for equal increments of current in the coil. This characteristic is called the *linearity* of the meter movement.

11. The percentage of accuracy of a d'Arsonval movement is based on the full-scale reading of the meter. Therefore, for readings lower than full-scale, the percentage of error is greater than the rated value.

12. It is possible to determine the sensitivity (I_m) of a meter movement experimentally by connecting the movement in series with a standard dc current meter, a rheostat or resistance-decade box, a current-limiting resistor, and a power source. The power source and/or decade box are adjusted for full-scale deflection of the meter movement. The current (I_m) required for full-scale deflection of the movement under test is then read from the standard meter (Fig. 31-2).

13. A preferred method for determining the internal resistance (R_m) of a meter movement is to connect the meter movement in series with a rheostat, a limiting resistor, and a power source (Fig. 31-4). The rheostat is then adjusted for full scale deflection of the meter movement. Now a resistance decade box is connected in parallel with the meter movement. The decade box is adjusted for half-scale deflection of the

meter pointer. The resistance read on the decade box at this point is the required R_m. Of course, it is essential that the output voltage of the source V remain the same during this process.

14. A method for determining the linearity of a meter movement is to connect the meter movement in series with a standard dc current meter, a rheostat (or resistance decade box), a current limiting resistor, and a voltage source (Fig. 31-2). The rheostat is adjusted to effect fixed percentage changes of full-scale deflection of the pointer of the meter movement under test. The increments of current required to effect these changes are then read from the standard-current meter. If equal deflections of the pointer (read along the linear scale of the meter movement) are effected by equal increments of current (read on the standard-current meter), the movement under test is linear.

SELF-TEST

Check your understanding by answering these questions:

1. A galvanometer is a _____.

2. The coil assembly of a d'Arsonval meter movement rotates on its pivots because of the _____ developed by the interaction of the magnetic field of the horseshoe magnet and the magnetic field set up in the coil when there is _____ in it.

3. The amount of current required to cause full-scale deflection of the pointer is called the _____ of the movement and is designated _____.

4. The smaller the current required to cause full-scale deflection of the pointer, the _____ _____ is the movement.

5. The sensitivity of the meter movement depends on the strength of the _____ _____ and on the number of _____ _____ of the coil.

6. If equal increments of current in the moving coil cause equal increments of angular deflection of the meter pointer, the movement is said to be _____.

7. The internal resistance (R_m) of a meter movement is the resistance of the _____.

8. Percentage of meter accuracy is based on _____-_____ _____ of the meter.

9. One method of measuring the sensitivity of a meter movement is to connect a series circuit consisting of the meter movement, a _____

_____ meter, and a _____ _____ _____. The current in the circuit is then adjusted for _____ _____ _____ of the movement, and the current required to do this is read from the _____-_____ _____.

10. The preferred method of measuring the internal resistance (R_m) of a meter movement requires the use of a _____ _____ _____ connected in _____ (series/parallel) with the meter movement.

MATERIALS REQUIRED

- Power supply: A variable regulated dc supply
- Equipment: "Standard" multirange milliammeter whose lowest range is 0–1 mA (approximately); rheostat (resistance decade box) 0–10 kΩ adjustable in 1-Ω steps; meter movement whose $I_m = 1$ mA
- Resistors: ½-W 3,300-, 5,600-, 15,000-Ω
- Miscellaneous: Two SPST switches; 10,000-Ω 2-W potentiometer; 10,000-Ω 2-W potentiometer (if decade box is not available)

PROCEDURE

Measuring Sensitivity of a Meter Movement

1. Connect the meter movement whose sensitivity you wish to measure in the circuit Fig. 31-2. S is open. V is a 4½-V battery or power supply set to 4.5 V. Rheostat R is set to 10,000 Ω, at the start.

CAUTION: Check to see that the "standard" current meter (an accurate commercial current meter may be used as the "standard") and the meter movement under test are connected with the proper polarity.

2. Close S and slowly reduce the resistance of R until full-scale deflection is indicated on the meter movement. Read the value I_m from the "standard" meter and record in Table 31-1.

Linearity Check

3. Increase the resistance R until the meter-movement (under test) pointer is deflected three-fourths of the way on its scale. Measure the current (using the standard meter) at this level, and record in Table 31-1 under column labeled "I (¾)."

   Measure also the current required to cause half-scale deflection and quarter-scale deflection of the meter-movement pointer, and record in Table 31-1, under columns "I (½)" and "I (¼)," respectively. For quarterscale deflection it will be necessary to replace R_1 with a 15,000-Ω resistor.

   Draw a graph of measured current (vertical axis) versus pointer deflection (horizontal axis). If the graph is a straight line, then deflection of the meter movement varies linearly with meter current.

Measuring R_m

4. Connect the circuit (Fig. 31-4). Switches S_1 and S_2 are open. Set V at 10 V, and R_1 at maximum resistance.
5. S_2 is still open. Close S_1. Gradually decrease resistance of R_1 until M reads full-scale deflection.
6. Close switch S_2. Adjust R_2 until M reads half-scale deflection. Read the value of R_2 from the calibrated dial and record in Table 31-1, under column "Measured Value R_m."

QUESTIONS

1. What determines the accuracy of your measurement of I_m?
2. Is the scale on the meter movement under test linear or nonlinear? What do your measurements indicate concerning meter-movement linearity?
3. What is the value of voltage across the meter movement required to cause full-scale deflection of the pointer?
4. What two characteristics describe a meter movement?
5. What error is introduced in measuring the R_m of a meter movement using the experimental circuit (Fig. 31-4)? Why was this error ignored in this experiment?

TABLE 31-1.

	Measured Values			Measured Value
I_m	I (¾)	I (½)	I (¼)	R_m

6. Why does the circuit of Fig. 31-3 require a constant source voltage V to measure R_m?

Extra Credit

7. Explain how you would measure the R_m of a high-sensitivity meter movement, as for example a movement whose $I_m = 50$ μA and whose $R_m = 2,500$ Ω. Show the circuit design you would use, the value of dc voltage required, and the values of all resistors and rheostats used. Discuss the factors relating to the accuracy of your procedure.

Answers to Self-Test

1. very sensitive moving-coil dc meter movement
2. torque; current
3. sensitivity; I_m
4. more sensitive
5. permanent magnet; ampere-turns
6. linear
7. coil
8. full-scale deflection
9. standard current; resistance decade box; regulated voltage supply; full-scale deflection; standard current meter
10. resistance decade box; parallel

32

Voltmeter Multipliers

OBJECTIVES

1. To develop by analysis a formula for determining the value of the multiplier required to convert a meter movement into a specified range voltmeter
2. To verify, experimentally, the value of the multiplier required to convert the meter movement into a voltmeter of a specified range
3. To calibrate the scale of the experimental voltmeter

INTRODUCTORY INFORMATION

Nondigital meters used in electronics consist of basic meter movements connected in suitable circuit arrangements. The basic movement has a moving element to which a pointer is attached. Current in the meter movement results in a force which acts on the moving element. The pointer is deflected, and moves in an arc along a scale suitably calibrated in the units being measured. These units may be volts, ohms, amperes, etc., as in a VOM which performs all these functions of measurement.

In Experiment 31 you learned that a meter movement has two characteristics associated with it:

Fig. 32-1. Constructing a voltmeter.

1. Sensitivity—the current required for full-scale deflection. For example, a 0-1 milliammeter movement requires 1 mA of current to give full-scale deflection of the pointer.
2. The resistance of the movement.

Voltmeter Multipliers

The circuit components of a *voltmeter* depend on these meter-movement characteristics. Thus suppose it is required to construct a 0-100 voltmeter from a basic 0-1 milliammeter movement whose resistance is 200 Ω.

Figure 32-1 shows the basic circuit arrangement for this voltmeter. *A* and *B* are the terminals of the meter where the test leads are inserted. *A* and *B* are coded (−) and (+) respectively from the polarity of the movement. When measuring voltage, *A* is placed on the negative side of the voltage source and *B* on the positive terminal. Otherwise, current through *M* is reversed and the movement may be damaged. *M* is the meter movement. It is shown as a 200-Ω resistance R_m, with a pointer, connected in series with a resistor R_s called the "multiplier." The purpose of R_s is to limit the current in this circuit to 1 mA when 100 V is applied across the meter test leads (*A* and *B*). This will cause full-scale deflection of the pointer, and the scale can be calibrated for 100 V at this point.

If a linear meter movement is used, deflection of the pointer is directly proportional to the current in the movement. The current (see Fig. 32-1) depends on the voltage across the terminals *AB* of the meter. The meter scale can therefore be calibrated in volts, though the movement itself measures (or responds to) current.

Ohm's law is used to find the value of the multiplier R_s. Consider the experimental voltmeter of Fig. 32-1. Since we wish to construct a 100-V meter, that is, a meter whose scale may be calibrated 100 V at full-scale deflection,

$$V_{\text{range}} = V_{AB} = 100 \text{ V}$$

and
$$I = I_m = 0.001 \text{ A}$$

Substituting these values and the value $R_m = 200$ in the equation

$$V_{\text{range}} = I_m(R_s + R_m) \qquad (32\text{-}1)$$

we get

$$100 = 0.001(R_s + 200) \qquad (32\text{-}2)$$

Solving Eq. (32-2), we find that

$$R_s = 100{,}000 - 200 = 99{,}800 \ \Omega \qquad (32\text{-}3)$$

The total resistance R_T of this 100-V meter is $100{,}000 \ \Omega$, since it equals $R_s + R_m$.

Formula for Determining Multiplier

A general formula for determining the value of the multiplier R_s required to convert a meter movement (I_m, R_m) into a voltmeter whose range is V_{range} can now be easily derived. Refer to Fig. 32-1. R_T is the total resistance of this meter and is equal to $R_s + R_m$. Thus:

$$R_T = R_s + R_m \qquad (32\text{-}4)$$

and
$$R_s = R_T - R_m \qquad (32\text{-}5)$$

But
$$R_T = \frac{V_{AB}}{I_m} = \frac{V_{\text{range}}}{I_m} \qquad (32\text{-}6)$$

Therefore, from Eq. (32-5)

$$R_s = \frac{V_{\text{range}}}{I_m} - R_m \qquad (32\text{-}7)$$

Simplifying Eq. (32-7) leads to

$$R_s = \frac{V_{\text{range}} - I_m \times R_m}{I_m} \qquad (32\text{-}8)$$

Either Eq. (32-7) or (32-8) may be considered a general formula for determining the value of a multiplier R_s, for a voltmeter of range, V_{range}. In these formulas R_s is the resistance of the multiplier in ohms; V_{range} is the range of the meter in volts; R_m is the resistance of the meter movement in ohms; and I_m is the sensitivity of the meter movement in amperes. Several examples will demonstrate how the formula may be applied.

Example 1

Convert a meter movement whose sensitivity $I_m = 0.001$ A, and whose internal resistance $R_m = 200 \ \Omega$, into a voltmeter of 100 V.

NOTE: This is the same movement as that in Fig. 32-1 and is the same problem as that solved in Eqs. (32-1) through (32-3).

Solution. Substituting the given values in Eq. (32-7) gives

$$R_s = \frac{100}{0.001} - 200 = 100{,}000 - 200 = 99{,}800 \ \Omega$$

Or substituting the given values in Eq. (32-8) gives

$$R_s = \frac{100 - 0.001(200)}{0.001} = \frac{99.8}{0.001} = 99{,}800 \ \Omega$$

The results are the same, whichever formula is used.

Example 2

Convert the meter movement in Fig. 32-1 into a voltmeter whose range is 10 V.

Solution. Substituting the given values ($I_m = 0.001$, $R_m = 200$, $V_{\text{range}} = 10$) in Eq. (32-7) gives

$$R_s = \frac{10}{0.001} - 200 = 10{,}000 - 200 = 9{,}800 \ \Omega$$

Similarly, it can be shown that the multipliers required to convert the meter movement of Fig. 32-1 into voltmeters of ranges 50 and 300 V, are respectively 49,800 Ω, and 299,800 Ω. These exact values are difficult to obtain. Hence 1 percent resistors are generally employed as multipliers in general-purpose voltmeters.

Calibrating the Voltmeter

Suppose Fig. 32-1 is the experimental 100-V meter. It is required to calibrate the scale of this voltmeter. If zero volts is applied ($V_{AB} = 0$) across the voltmeter, there is no current in the meter movement and hence no deflection of the pointer. This is the 0-V position of the pointer on the scale. If 50 V is applied, there will be $\frac{1}{2}$ mA of current in the movement, and therefore there will be half-scale deflection of the pointer. This will be the 50-V position on the scale. If 25 V is applied, there will be $\frac{1}{4}$ mA of current, and therefore there will be quarter-scale deflection of the pointer. This will be the 25-V position on the scale, and so on.

The entire 100-V scale is calibrated in this manner and appears as in Fig. 32-2. It is a linear scale because the meter movement used is linear.

SUMMARY

1. A basic dc meter movement may be converted into a voltmeter by adding a multiplier resistor (R_s) in series with the meter movement, as in Fig. 32-1.
2. The value of the multiplier R_s for a specified range V_{range} must be such that it will permit full-scale deflection of the pointer when the voltmeter terminals are placed across the voltage value V_{range}. This means that for full-scale deflection the total resistance of the voltmeter, $R_s + R_m$, must limit the current through the movement to I_m, the sensitivity of the movement.
3. Though the scale of a voltmeter is calibrated in volts, the meter itself is responding to current in the movement, that is, it is a current-actuated device.
4. A general formula for determining the value of multiplier R_s required to convert a meter movement whose sensitivity is I_m and internal resistance is R_m into a voltmeter of range V_{range} is

$$R_s = \frac{V_{range}}{I_m} - R_m$$

or

$$R_s = \frac{V_{range} - I_m \times R_m}{I_m}$$

where R_s and R_m are in ohms, V_{range} is in volts, and I_m is in amperes.
5. If a linear meter movement is used for a voltmeter, the scale of the voltmeter will also be linear.

SELF-TEST

Check your comprehension by answering these questions:

1. A dc voltmeter may be made by connecting a precision resistor called a _____ in

Fig. 32-2. Voltmeter scale.

_____ with a dc meter movement.
2. If the sensitivity of the meter movement used to build a dc voltmeter is 50 μA, then the current in a 50-V meter at full-scale deflection is _____.
3. The total resistance of a dc voltmeter consists of the _____ of the multiplier resistance and the resistance of the _____.
4. A 100-V meter is made from a meter movement whose sensitivity is 50 μA and whose internal resistance is 2,000 Ω.
   (a) The total resistance of the meter is _____ Ω.
   (b) The resistance of the multiplier is _____ Ω.
5. In the voltmeter of question 4 half-scale deflection of the pointer corresponds to _____ V; one quarter-scale deflection of the pointer corresponds to _____ V.

MATERIALS REQUIRED

- Power supply: Variable low-voltage dc source
- Equipment: Electronic voltmeter, meter movement (same as in Experiment 31), resistance decade box as in Experiment 31
- Resistors: ½-W as required
- Miscellaneous: One SPST switch, potentiometers as required in step 11 (optional)

PROCEDURE

(Show all your computations.)

Constructing a Voltmeter of Specified Range (30 V)

1. The meter movement you will use will be the same as that whose internal resistance R_m and sensitivity I_m were measured in Experiment 31. Record these in Table 32-1.
2. Compute and record in Table 32-1 the multiplier resistance R_s required to convert your meter movement into a 30-V meter. Choose a resistor (or combination of resistors) whose total resistance R_s is equal or close to the com-

TABLE 32-1.

Meter-movement Measured Values		Required Range V (volts)	R_s (multiplier) ohms		Measured Range V_1 (volts)	Resistance (R_{in}) of Experimental Voltmeter	
R_m	I_m		Computed	Measured		Computed	Measured
		30					
		10					

puted value. With an ohmmeter measure and record in Table 32-1 the resistance of the multiplier R_s.

NOTE: A resistance decade box may be used to select the proper value of R_s. When used, read the value of R_s from the calibrated dial.

Using the selected value of R_s, construct a voltmeter as in Fig. 32-1.

3. Connect the electronic voltmeter V_1 and the *experimental voltmeter* V_2 in parallel across the low-voltage supply V, and adjust the voltage V for 30-V output, as measured on the electronic voltmeter V_1 used as a standard (Fig. 32-3). Mark "30" the point on the *uncalibrated* scale of the *experimental* voltmeter (V_2) at which the pointer has stopped. If deflection is greater than full-scale, so state.

4. If exact full-scale deflection is not reached on experimental voltmeter V_2, adjust voltage supply V to obtain it and record under "measured range" the voltage V_1 required to obtain full-scale deflection of V_2, using the electronic voltmeter V_1 as the standard.

Calibrating the Experimental Voltmeter

5. Using V_1 as the standard of reference, calibrate the linear scale of experimental voltmeter V_2 in intervals of 5 V, from zero volts to full-scale deflection.

6. Compute the input resistance R_{in} of the experimental voltmeter V_2 and record in Table 32-1. Use the equation

$$R_{in} = R_T = \frac{V_{range}}{I_m}$$

7. Disconnect V_2 from the circuit of Fig. 32-3. Connect an ohmmeter across the terminals of experimental voltmeter V_2. Measure and record

in Table 32-1 the input resistance of V_2. Does the voltmeter pointer move? If it does, is it upscale or downscale?

8. Reverse ohmmeter leads and repeat step 7.

Experimental Voltmeter (10 V)

9. Repeat steps 2 through 4, constructing a 10-V range voltmeter. Record your data in Table 32-1.

10. Repeat steps 5 through 8, calibrating the 10-V meter and measuring its input resistance.

Design Problem (extra credit)

11. Design a three-range voltmeter, 5-V, 25-V, and 40-V, using the meter movement assigned to you in this experiment. Draw the circuit diagram. Include the design value for each multiplier R_s, and the values of R_m and I_m. Show all your computations. (You may not use a switch to change range.) Secure the necessary values of R_s and breadboard the three-range meter (you may use a resistance decade box or a potentiometer connected as a rheostat for R_s). Experimentally determine and record in Table 32-2 the voltage required for full-scale deflection on each of the ranges.

Fig. 32-3. Calibrating the experimental voltmeter.

TABLE 32-2.

Meter-movement Characteristics		Multiplier R (ohms)			Full-scale Voltage (volts)		
R_m, ohms	I_m, amperes	5 volts	25 volts	40 volts	5 volts	25 volts	40 volts

QUESTIONS

1. In Table 32-1 how do the required ranges V compare with the measured ranges V_1 for the experimental voltmeter on (a) 30-V range, (b) 10-V range? Explain any differences between V and V_1 for each specific range.
2. How do the computed and measured values of input resistance in Table 32-1 compare? If they are not the same, explain why.
3. Without using specially designed precision resistors, explain how you obtained a required value R_s for use as a multiplier in a single-range voltmeter.
4. In calibrating your experimental voltmeter, what factor(s) limit(s) the accuracy of calibration?
5. From your observations in steps 7 and 8, what conclusions, if any, can you draw about the effect of the ohmmeter on the experimental voltmeter pointer, when measuring the resistance of the experimental voltmeter? Explain.

Extra Credit

6. Explain in detail how you secured the proper values of R_s in the design of the three-range voltmeter. Explain how you determined full-scale deflection on each range and comment on the accuracy of your design.

Answers to Self-Test

1. multiplier; series
2. 50 μA
3. sum; meter movement
4. (a) 2,000,000; (b) 1,998,000
5. 50, 25

EXPERIMENT 33

Loading Effects of a Voltmeter

OBJECTIVES

1. To find by analysis the relationship between I_m and the ohms/volt rating of a voltmeter and to confirm this ohms per volt rating experimentally
2. To determine analytically and confirm experimentally the loading effects of a voltmeter
3. To consider analytically a method for eliminating the loading effects of a nonelectronic voltmeter

INTRODUCTORY INFORMATION

Ohms/volt Characteristic and Input Resistance of a Voltmeter

In Experiment 32 it was shown that the total input resistance that a voltmeter presents to the circuit under test, on any voltage range, is equal to $R_s + R_m$, where R_s is the multiplier for the specified range and R_m is the resistance of the movement. Thus, the input resistances of the experimental voltmeter in Fig. 33-1 on the 10-, 50-, 100-, and 300-V ranges are shown in Table 33-1.

Observe, in Table 33-1, that the ratio of input resistance to voltage on each range is 1,000. This constant ratio is called the ohms per volt (ohms/volt)

characteristic of the meter, and in the case of the meter under consideration is 1,000 Ω/V. This characteristic is also known as the *voltmeter sensitivity*.

Refer to the voltmeter in Fig. 33-1. If the range of this meter is V volts, that is, if there is full-scale deflection of the meter pointer when V volts is applied across the meter terminals, then

$$V = I_m(R_s + R_m) \qquad (33\text{-}1)$$

where I_m is the sensitivity of the meter movement and R_m is the internal resistance of the movement.

From Eq. (33-1) it follows that

$$\frac{R_s + R_m}{V} = \frac{1}{I_m} \qquad (33\text{-}2)$$

The ratio $(R_s + R_m)/V$ is the ohms/volt characteristic according to our definition. It is therefore apparent that the ohms/volt characteristic of a voltmeter is the reciprocal of the sensitivity $(1/I_m)$ of the meter movement.

$$\text{Ohms/volt} = \frac{1}{I_m} \qquad (33\text{-}3)$$

where I_m (in amperes) = sensitivity of meter movement.

Fig. 33-1. Equivalent circuit of a voltmeter.

TABLE 33-1. Input Resistance of Voltmeter Using a 0–1 mA Movement.

Voltage Range	Input Resistance	Ohms/volt: Input Resistance / Voltage
10	10,000	1,000
50	50,000	1,000
100	100,000	1,000
300	300,000	1,000

For example, using Eq. (33-3), we can show that the ohms/volt characteristic of a voltmeter using a 1-mA movement is 1,000, and of a voltmeter using a 50-μA movement is 20,000, etc.

The ohms/volt rating of a voltmeter multiplied by the set voltage range of the meter is the *input resistance* of the meter on that range; that is, it is the resistance which the circuit "sees" when the voltmeter is inserted in the circuit to measure voltage.

DC Voltmeter Circuit Loading

The ohms/volt rating is significant to the technician because it indicates the extent to which the voltmeter will "load" the circuit when used in voltage measurement. Consider the circuit of Fig. 33-2. Suppose it is desired to measure the voltage from A to B, using the 100-V range of a 1,000 Ω/V meter. Here the meter places 100,000 Ω resistance in parallel with the 4 megohms (MΩ) of the divider, and the total resistance R_{T_2} of the parallel circuit from A to B now becomes slightly less than 100,000 Ω. Suppose, however, that the figure 100,000 Ω is used. The equivalent circuit (Fig. 33-3) shows what happens when the meter is placed across the circuit.

The voltage across R_{T_2}, hence across the voltmeter, is $\frac{1}{11} \times 110 = 10$ V, and the meter would read 10 V. But this would not be the true voltage from A to B in the absence of the meter, since a computation shows that the voltage from A to B before the meter was connected would be $\frac{4}{5} \times 110 = 88$ V. The voltmeter has "loaded" the circuit. Obviously this is an undesirable situation because it can lead to serious errors. To minimize this loading, use a voltmeter with a much higher ohms/volt rating which is set on the highest range, so sufficient pointer deflection is obtained to make a voltage reading possible.

The effect of voltmeter "loading" in a series circuit may be generalized. Thus, in the circuit of Fig. 33-2, the computed voltage V_{AB} across R_2 is

Fig. 33-2. Voltmeter loading.

Fig. 33-3. Effect of voltmeter in the circuit.

$$V_{AB} = \frac{R_2}{R_1 + R_2} \times V \qquad (33\text{-}4)$$

When the leads of a voltmeter with input resistance R_{in} are placed across R_2, the voltage V'_{AB} measured between points A and B is

$$V'_{AB} = \frac{R_2 R_{in}}{R_1 R_2 + R_{in}(R_1 + R_2)} \times V \qquad (33\text{-}5)$$

The ratio V_{AB}/V'_{AB} is the ratio of the voltage across AB before the voltmeter is inserted in the circuit to the voltage after it is placed across AB. And it can be shown that

$$\frac{V_{AB}}{V'_{AB}} = \frac{R_1 R_2 + R_{in}(R_1 + R_2)}{R_{in}(R_1 + R_2)} = \frac{R_1 R_2}{R_{in}(R_1 + R_2)} + 1$$

$$(33\text{-}6)$$

From Eq. (33-6) it is apparent that the voltages V_{AB} and V'_{AB} are not the same. It can be seen further that V'_{AB} equals V_{AB} only when

$$\frac{R_1 R_2}{R_{in}(R_1 + R_2)} = 0 \qquad (33\text{-}7)$$

Since R_1 and R_2 are finite values, Eq. (33-7) is true only when R_{in} is infinitely large. It follows, then, that the input resistance of a voltmeter loads the circuit by changing the effective resistance across which the voltage is measured, but that the higher the value of R_{in}, the less is the loading of the circuit because as R_{in} increases, the value of

$$\frac{R_1 R_2}{R_{in}(R_1 + R_2)}$$

decreases.

Eliminating Loading Effects of a Nonelectronic Voltmeter

The voltmeter in Fig. 33-1 is not an electronic voltmeter, that is, it contains no amplifiers such as transistors or tubes. The input resistance of this nonelectronic voltmeter changes from range to range, as is evident from Table 33-1. It is theoretically possible to eliminate the loading effects of such a voltmeter by making two measurements with it, on different ranges of the meter.

Refer to Fig. 33-2. Let the voltage measured across AB on range 1 be V_1, and that measured across AB on range 2 be V_2, and let the ratio of voltage range 1 to voltage range 2 be a. (Assume voltage range 1 is a higher range than range 2. Therefore a is a number greater than 1.) It can be shown that the voltage V_{AB} which exists across AB before the voltmeter is inserted, that is, before the circuit is loaded, may be computed from the formula

$$V_{AB} = \frac{(a-1)V_1 V_2}{aV_2 - V_1} \qquad (33\text{-}8)$$

An example will show how Eq. (33-8) may be applied. Suppose range 1 is the 100-V range of a 1,000 Ω/V meter and $V_1 = 50$ V, and suppose range 2 is the 25-V range of the same meter and V_2 is 20 V. Then $a = (100/25) = 4$, and by substituting the values $a = 4$, $V_1 = 50$, $V_2 = 20$ in Eq. (33-8) we get

$$V_{AB} = \frac{(4-1) \times 50 \times 20}{4 \times 20 - 50} = \frac{3 \times 50 \times 20}{30} = 100 \text{ V}$$

That is, the unloaded voltage across AB is actually 100 V.

NOTES: (1) It is an interesting fact that this technique may be employed to find the true value of V_{AB} without knowing the values of $R_1 R_2$ or the voltage V applied to the circuit under test.
(2) This technique can be used only with a voltmeter whose input resistance differs on each of the ranges, for the ratio a previously defined as $a = (V_{\text{range 1}}/V_{\text{range 2}})$ is also the ratio $R_{\text{in}}(\text{range 1})/R_{\text{in}}(\text{range 2})$. That is

$$a = \frac{R_{\text{in}}(\text{range 1})}{R_{\text{in}}(\text{range 2})}$$

(3) The more accurate the measurements V_1 and V_2 are, the closer is the value of V_{AB} computed from Eq. (33-8) to the actual value of V_{AB} in the circuit.
(4) The input resistance of electronic voltmeters is usually the same for all ranges of the meter. Hence this technique cannot be used with such a meter. However, the input resistance of an electronic voltmeter is usually very high, so its loading effect in a circuit is relatively low.

SUMMARY

1. The input resistance R_{in} of a nonelectronic voltmeter such as that in Fig. 33-1 differs for every range of the meter.
2. If a meter movement of sensitivity I_m is used to construct a voltmeter of range V, then the input resistance of the meter on that range is

$$R_{\text{in}} = \frac{V}{I_m}$$

3. The ohms/volt characteristic of a voltmeter is defined as the ratio of the input resistance of the meter on a particular range (R_{in}) to the voltage of that range (V):

$$\text{Ohms/volt} = \frac{R_{\text{in}}}{V}$$

4. The ohms/volt characteristic of a nonelectronic voltmeter using a movement whose sensitivity is I_m is constant on each range. Moreover,

$$\text{Ohms/volt} = \frac{1}{I_m}$$

5. When a voltmeter is placed across a resistance R_{AB} connected between two points A and B in a circuit, it "loads" the circuit by changing the effective resistance between A and B. The "new" resistance which the circuit sees is the parallel combination of R_{AB} and R_{in}, the input resistance of the voltmeter on the specific range to which it is set.
6. As a result of voltmeter loading, the voltage measured may be very much lower than the non-loaded voltage which existed across the points A and B before the voltmeter was placed in the circuit.
7. Loading may be minimized by using high-sensitivity meter movements, that is, meter movements whose I_m is so low that the ohms/volt rating of the voltmeter, $1/I_m$, is very high. For electronic work, 20,000 Ω/V meters are usually used.

Fig. 33-4. Circuit to determine loading effects of experimental voltmeter.

8. Loading may also be minimized by using the higher rather than the lower ranges of a voltmeter, where a choice is possible.
9. The loading effects of a nonelectronic voltmeter may be eliminated by making two measurements with the same voltmeter, each on a different range. If the measurements on range 1 and range 2 are respectively V_1 and V_2, and if the ratio: Voltage range 1/Voltage range 2 = a, then the actual voltage may be found by substituting V_1, V_2, and a in the formula

$$V_{actual} = \frac{(a-1)V_1 \times V_2}{aV_2 - V_1}$$

SELF-TEST

Check your comprehension by answering these questions:
1. The input resistance of a 20,000 Ω/V meter on the 50-V range is _____ Ω; on the 300-V range its input resistance is _____ Ω.
2. A meter movement whose sensitivity $I_m = 10$ μA is used to construct a nonelectronic voltmeter. The ohms/volt characteristic of that meter is _____ Ω/V.
3. The input resistance of the voltmeter in question 2 on the 10-V range is _____ Ω.
4. In the circuit of Fig. 33-2, $R_1 = 1$ MΩ, $R_2 = 1$ MΩ, and $V = 100$ V. If a 1,000 Ω/V meter on the 100-V range is placed across R_2, the effective resistance R_{AB} between points A and B becomes _____ Ω.
5. The voltmeter in question 4 would read _____ volts instead of the unloaded value of _____ V.
6. If the meter in question 4 were set on the 50-V range, the voltage across R_2 would measure _____ volts.
7. Using the measurements in 5 and 6 to eliminate the loading effects of the voltmeter, the unloaded voltage $V_{AB} =$ _____ V.

MATERIALS REQUIRED

- Power supply: Variable low-voltage dc supply
- Equipment: Electronic voltmeter, meter movement (same as in Experiment 32), resistance decade box as in preceding experiment
- Resistors: 1/2-W, two 100,000-Ω; others as required
- Miscellaneous: One SPST switch, potentiometers as required

PROCEDURE

1. The voltmeters constructed in Experiment 32 will be used again in this experiment. Transfer from Table 32-1 to Table 33-2 the following *measured* values.
   a. Sensitivity I_m of the meter movement
   b. Voltage ranges of the experimental voltmeter, that is, V_1 on range 1 (required 30-V range) and V_2 on range 2 (required 10-V range)
   c. Multiplier R_s for range 1; for range 2
   d. Input resistance of the voltmeter on range 1; on range 2
2. Compute and record in Table 33-2 the ohms/volt rating of the voltmeter on range 1. Use the measured value V_1 and R_{in} for range 1 and substitute in the formula

$$\text{Ohms/volt} = \frac{V_1}{R_{in}}$$

3. Repeat step 2 for range 2.

TABLE 33-2.

Meter Movement Sensitivity (I_m) amperes	Measured Values				Computed Values	
	Voltmeter Range Volts	Multiplier (R_s) ohms	R_{in} ohms		Ohms/Volt	
					V/R_{in}	$\frac{1}{I_m}$
	V_1(range 1)					
	V_2(range 2)					

4. Using the measured value I_m, compute and record in Table 33-2 the ohms/volt rating of the voltmeter by substituting I_m in the formula

$$\text{Ohms/volt} = \frac{1}{I_m} \text{ (where } I_m \text{ is in amperes)}$$

Voltmeter Loading

5. Construct the experimental voltmeter in Fig. 33-1 range 1 (from Table 33-2). Construct also the circuit in Fig. 33-4. The electronic voltmeter (V_1) is used to measure the voltage V of the voltage supply. The *experimental voltmeter* (V_2) is used to measure voltage across R_1.
6. With both meters connected in the circuit, adjust V to read 40 V on V_1 (electronic voltmeter). Measure the voltage across R_1 as read on the experimental voltmeter V_2 and record in Table 33-3.
7. *Remove the experimental voltmeter V_2 from the circuit. Readjust the voltage source V, if necessary, to read 40 V on the electronic voltmeter V_1. Now connect the electronic voltmeter (V_1) across R_1 and measure and record the voltage across R_1 as read on V_1.*
8. Repeat the procedure in step 5 for range 2 of the experimental voltmeter.
9. Repeat steps 6 and 7 recording your measurements in Table 33-3.

TABLE 33-3. Voltage across R_1

Measured With		Computed
Experimental Voltmeter (V_2)	Electronic Voltmeter (V_1)	$V \times \dfrac{R_1}{R_1 + R_2}$
Range 1		
Range 2		

10. Compute and record in Table 33-3 the voltage which should appear across R_1, if no voltmeters were connected in the circuit, using the circuit values of R_1, R_2, and V and substituting in the formula:

$$V_{(\text{across } R_1)} = V \times \frac{R_1}{R_1 + R_2}$$

Extra Credit

NOTE: The readings with the experimental voltmeter are too crude (since the scale is not fully calibrated) to apply the formula in Eq. (33-8) to eliminate the loading effects of the experimental voltmeter. Hence it will not be attempted. It is possible to check the formula by repeating steps 5 through 9 using a commercial 1,000 Ω/V voltmeter instead of the experimental voltmeter.

11. Verify Eq. (33-8) using a commercial 1,000 Ω/V meter and the circuit of Fig. 33-4. Construct a special table for the data and substitute the values V_1, V_2, and a in Eq. (33-8). If a 20,000 Ω/V meter is used, let R_1 and R_2 be 1-MΩ resistors instead of 100-kΩ resistors.

QUESTIONS

1. Which meter loaded the circuit more, the electronic voltmeter or the experimental voltmeter? How do you know? Refer to the data in your tables to support your answer.
2. Which range of the experimental voltmeter loaded the circuit more and why? Refer specifically to the data in your tables to support your answer.
3. How did the two ohms/volt ratings (computed values) in Table 33-2 compare? If there was any difference, explain why.
4. What is the ohms/volt rating of a commercial VOM using
   (a) 10-mA movement?
   (b) 50-μA movement?
   (c) 1-mA movement?
   (d) 10-μA movement?
5. Of four voltmeters using, in turn, meter movements (a) through (d) (question 4), which one would cause the least amount of circuit loading in measuring voltage across R_1 in Fig. 33-4? Most amount of loading? Why?
6. In measuring an unknown dc voltage, which range of a multirange voltmeter should you use? Why?

Answers to Self-Test

1. 1,000,000; 6,000,000
2. 100,000
3. 1,000,000
4. 90,900 (approx.)
5. 8.3 (approx.); 50
6. 4.55 (approx.)
7. 48.2 (approx.)

EXPERIMENT 34 Current-meter Shunts

OBJECTIVES

1. To determine by analysis how a basic meter movement is converted into a specified higher-range current meter by the addition of a shunt
2. To verify experimentally the value of the shunt required to convert the meter movement into a current meter of a specified range

---

INTRODUCTORY INFORMATION

The basic meter movement is a current-reading meter. Thus, a 0–10 milliammeter movement requires 10 mA of current for full-scale deflection, 5 mA for half-scale deflection, etc. This movement could therefore be used without change as a 0–10 milliammeter to measure current in circuits where there is 10 mA or less.

Where it is necessary to measure more than 10 mA of current, it is possible to convert the meter into a higher-range current meter. Thus, if it is desired to measure the current in a circuit where there is 40 mA, a 0–50 milliammeter could be used.

To extend the range of the 0–10 milliammeter movement to read 50 mA of current at maximum deflection would require a "shunt." This is a resistor connected in parallel with the movement to provide another path for current. In Fig. 34-1, R_{sh} is the shunt.

Fig. 34-1. Milliammeter shunt.

Consider the movement which registers full-scale deflection with 10 mA of current. To convert it into a 0–50 milliammeter, it is necessary to send 40 mA through the shunt and 10 mA through the movement.

If the resistance R_m of the movement is known, it is a simple matter to determine the size of the shunt R_{sh}. Since the shunt and movement are connected in parallel, the voltage across each must be the same. Hence this voltage:

$$V_m = I_m \times R_m = I_{sh} \times R_{sh} \qquad (34\text{-}1)$$

where I_m is the current in amperes in the movement and I_{sh} is the current in amperes in the shunt. Therefore,

$$\frac{R_{sh}}{R_m} = \frac{I_m}{I_{sh}} \qquad (34\text{-}2)$$

and

$$R_{sh} = R_m \times \frac{I_m}{I_{sh}} \qquad (34\text{-}3)$$

If it is assumed that the resistance R_m of the movement is 200 Ω, and at full-scale deflection $I_m = 10$ mA and $I_{sh} = 40$ mA, the size of the required shunt is

$$R_{sh} = 200 \times \frac{10}{40} = 50 \ \Omega$$

As in the voltmeter application, R_{sh} must be a precision resistor.

To change the basic 0–10 mA movement into a 500-mA meter, a shunt R_{sh} would be used, whose size would be

$$R_{sh} = 200 \times \frac{10}{490} = 4.08 \ \Omega$$

A meter capable of measuring small currents accurately would require a more sensitive movement. Thus a 0–50 μA movement could be used as a current meter giving full-scale deflection at 50 μA. Shunts could then extend the range of the meter upward, as required.

Equation (34-3) is used to determine the value of the shunt resistor R_{sh} when the meter current I_m, the shunt current I_{sh}, and the meter-movement resistance R_m are known. If I_m represents the meter-movement sensitivity (i.e., current required for full-scale deflection), then $I_m + I_{sh} = I_t$ represents the maximum current which the meter can measure on that range. I_t, then, is the current range for the particular shunt R_{sh}. Equation (34-3) may now be written as

$$R_{sh} = R_m \times \frac{I_m}{I_t - I_m} \qquad (34\text{-}4)$$

In the case of the 10-mA meter movement which was converted into a 50-mA meter,

$$I_m = 10 \times 10^{-3} \text{ A, and } I_t = 50 \times 10^{-3} \text{ A}$$

It may sometimes be necessary to determine the current range I_t of a meter when the shunt resistor R_{sh}, meter-movement resistance R_m, and meter-movement sensitivity I_m are given. The following formula can be used to find I_t:

$$I_t = I_m \times \frac{R_m + R_{sh}}{R_{sh}} \qquad (34\text{-}5)$$

SUMMARY

1. A dc meter movement is a direct-current–reading device.
2. A dc meter movement whose sensitivity is I_m can be used as a direct-current meter which can measure currents equal to or less than I_m.
3. To measure direct currents higher than I_m, the meter movement must be converted into a higher range current meter by the addition of a *shunt*.
4. A shunt is a precision resistor (R_{sh}) connected in parallel with the meter movement.
5. At full-scale deflection, the current in the movement is I_m. If the current in the shunt is I_{sh}, then the range of the meter is I_t, where

$$I_t = I_m + I_{sh}$$

6. If R_m, I_m, and I_{sh} are known, then the resistance of the shunt may be computed from the formula

$$R_{sh} = R_m \times \frac{I_m}{I_{sh}}$$

7. If the range current I_t, I_m, and R_m are known,

the resistance of the shunt may be computed from the formula

$$R_{sh} = R_m \times \frac{I_m}{I_t - I_m}$$

8. If it is required to measure currents very much smaller than I_m, a meter movement of higher sensitivity than I_m must be used.

SELF-TEST

Check your understanding by answering these questions:

1. If a meter movement whose sensitivity is 1 mA is used as a current meter, it _____ (is/is not) possible to measure current greater than 1 mA with this movement alone.
2. It is required to convert a 1 mA (I_m) 50-Ω (R_m) dc meter movement into a 25-mA meter. The value of the shunt resistor (R_{sh}) must be _____ Ω.
3. At full-scale deflection a meter movement takes 50 μA of current. If the shunt in parallel with this movement takes 450 μA at full-scale deflection, the range of the meter is _____ μA or _____ mA.
4. If the resistance of the movement in question 3 is 2,500 ohms, the value of shunt resistance for the current meter above must be _____ Ω.

MATERIALS REQUIRED

- Power supply: Variable low-voltage direct current
- Equipment: EVM, 0–1-mA meter movement, the same as in Experiment 31, resistance decade box, same as in Experiment 33, 0–10 mA meter
- Resistors: ½-W 10,000-, 5,600-Ω
- Miscellaneous: Hand tools, 100-Ω 2-W potentiometer (if a decade box is not available); other components as required in step 16 (optional); one SPST switch

Fig. 34-2. Basic meter movement used as a 0–1 milliammeter.

Fig. 34-3. Using a shunt to extend range of a milliammeter.

Fig. 34-4. A standard current meter M_1 is used to che k the range of experimental current meter M.

PROCEDURE

Current Meter Shunt

1. Connect the circuit of Fig. 34-2, using a 0–1 milliammeter movement as the current meter M, and an EVM as the voltmeter V. *Observe meter polarity.*
2. Adjust the variable voltage source for full-scale deflection on M. Measure, and record in Table 34-1, the voltage V as measured on the EVM. The current I in the circuit is 1 mA.
3. **Power off.** Connect a rheostat R_{sh} in shunt with M as in Fig. 34-3. Set R_{sh} for maximum resistance. A resistance decade box R_{sh} set at 100 Ω is preferable to a 100-Ω rheostat.
4. **Power on.** Adjust R_{sh} until M reads half-scale deflection. Monitor the voltage source and hold the level at V, as in step 2. Ignoring the loading effect of the meter, has the current in the circuit of Fig. 34-3 changed value? M and R_{sh} now constitute a current meter whose range is _____ mA. Measure the voltage V and current I in the circuit and record in Table 34-1.
5. Increase the voltage V until there is full-scale deflection of M. Measure and record this voltage and the current I in R_1 as measured by the experimental current meter. **Power off.**

6. Remove shunt R_{sh} from the circuit. Do not change the setting of R_{sh}. Read the value of R_{sh} from the decade box dials or measure and record the resistance of R_{sh}.
7. How does the resistance R_m of the meter movement (see Experiment 31) compare with R_{sh} above?
8. Without changing the value of R_{sh}, place it in parallel with M as in Fig. 34-3. **Power is still off.**
9. Add a 1–10 milliammeter (M_1) to be used as a standard in the circuit, as in Fig. 34-4.
10. Close switch S and adjust the voltage V, if necessary, until there is full-scale deflection of M.
11. Record in Table 34-1 the current I read on M_1, the standard current meter. Power **off.**
12. Again remove R_{sh} from the circuit and measure its resistance. Record in Table 34-1.

NOTE: The value of R_{sh} in steps 6 and 11 should be exactly the same. If it is not, reset R_{sh} until its resistance is the same as in step 6 and repeat steps 10 and 11.

13. **Power off.** Analytically determine the value of R_{sh}, required to convert the meter movement M to a 0–5 mA current meter. Show your computations. Record this value in Table 34-2.
14. Remove R_{sh} from the circuit and set it to the

TABLE 34-1.

Step	V, volts	I, mA	R_{sh}, ohms
2		1 mA	X
4			X
5			X
6	X	X	
11	X		

TABLE 34-2.

Step	R_{sh}, ohms	I, mA
13		X
15	X	

TABLE 34-3.

Meter-movement Characteristics		Shunt Resistor R_{sh} (ohms) for			Current for Full-scale Deflection		
R_m, ohms	I_m, amperes	2 mA	5 mA	10 mA	2 mA	5 mA	10 mA
	0.001						

value determined in step 13. Place it in the circuit of Fig. 34-4. Replace the 10-kΩ limiting resistor R_1 with a 5.6-kΩ resistor.

15. **Power on.** Adjust V until meter movement M reads full-scale deflection. Record in Table 34-2 the current I measured by standard current meter M_1.

Design Problem (extra credit)

16. Design a milliammeter with the following ranges: 2 mA, 5 mA, 10 mA. Use the 0–1 mA meter movement assigned to you in this experiment. Draw the circuit diagram. In Table 34-3 include the design values for each shunt resistor R_{sh}. Show all your computations. You may use a switch to change range. Secure the necessary values of R_{sh} and breadboard the three-range current meter.

NOTE: A resistance decade box will serve for selecting the proper shunt value. Using a commercial milliammeter as standard, experimentally measure the amount of current required to cause full-scale deflection of the experimental current meter, on each range.

QUESTIONS

1. In step 5 how much current is there in the meter movement? In the shunt?
2. What is the range of the experimental current meter in step 5?
3. What is the computed (formula) value of R_{sh} required to convert the meter movement into a current meter of the same range as in step 5? Show your computations.
4. How does the computed value of R_{sh} compare with the measured value in step 6? Explain any difference.
5. How does the measured value of R_m (see Experiment 31) compare with the measured value of R_{sh} in procedural step 6? Should they be the same? Explain.
6. Explain in detail the experimental procedure you followed to test the accuracy of the experimental current meter. Use a circuit diagram.
7. In connection with the procedure in question 6, what factors limit the accuracy of this procedure?

Answers to Self-Test

Answers to slide-rule accuracy.

1. is not
2. 2.08
3. 500; 0.5
4. 278

35

OBJECTIVES

1. To determine analytically the effects of inserting a current meter in the circuit
2. To confirm experimentally the insertion (loading) effects of a current meter

INTRODUCTORY INFORMATION

Circuit Insertion Effects of a Current Meter

The effect of inserting a current meter in a series circuit is to add the resistance of the meter to the total resistance of the circuit. The circuit effect of inserting a current meter, then, depends on the resistance of the meter and the resistance of the circuit under test. For example, to measure current in the circuit of Fig. 35-1, the meter M would, as usual, be placed in series with the circuit, as in Fig. 35-2.

If it is assumed that the meter is a 0–10 milliammeter with 200-Ω resistance, then it would add 200 Ω to R_1 and R_2 and the battery would see a total resistance $330 + 470 + 200 = 1,000$ Ω. M would then read 10 mA in the circuit of Fig. 35-2.

However, computing the current in Fig. 35-1, we find

$$I = \frac{V}{R_T} = \frac{10}{800} = 12.5 \text{ mA}$$

There is actually 2.5-mA difference between the calculated value and the measured value, and the difference is obviously due to the resistance of the meter.

Of course, if $R_1 = 4,700$ Ω and $R_2 = 3,300$ Ω, the "insertion" effect of the milliammeter would be negligible.

In dc voltage measurement, minimum circuit loading requires the use of a *high-resistance voltmeter*. In current measurement, on the other hand, the *ammeter* should have *low resistance* to reduce circuit errors.

The error introduced by a current meter M connected in the series circuit (Fig. 35-2) may be readily determined for the general case. Let I_B be the current computed in the circuit before the meter is inserted, and I_A the current measured in the circuit, after the meter is connected. By Ohm's law

$$I_B = \frac{V}{R_T}$$

where $\qquad R_T = R_1 + R_2 \qquad$ (35-1)

and $\qquad I_A = \dfrac{V}{R_T + R_m} \qquad$ (35-2)

Fig. 35-1. Circuit where current is to be measured.

Fig. 35-2. Connecting milliammeter to measure current.

where R_m is the resistance of the meter.

The error I_{error} caused by meter insertion is

$$I_{error} = I_B - I_A \qquad (35\text{-}3)$$

By substituting in Eq. (35-3) the values of I_A and I_B from Eqs. (35-1) and (35-2), and solving, the value of I_{error} is found to be

$$I_{error} = I_A \times \frac{R_m}{R_T} \qquad (35\text{-}4)$$

From Eq. (35-4) it is evident that I_{error} is zero when $R_m = 0$. It is clear, therefore, that no error is introduced when the meter resistance is zero. Since the meter resistance is never zero, some error is introduced in measurement, but this error can be made relatively small by using an ammeter with low internal resistance.

SUMMARY

1. When a current meter is inserted in a circuit, the meter adds its resistance in series with the circuit resistance. The total resistance of the circuit is therefore higher with the current meter in than it was before the meter was added.
2. This increase in circuit resistance is called the loading effect of a current meter.
3. The total current in the circuit, with the current meter in, is less than the current in the circuit with the meter out.
4. To minimize the loading effect of a current meter, use a *low-resistance* meter.
5. The current error introduced in the circuit may be calculated from the formula

$$I_{error} = I_A \times \frac{R_m}{R_T}$$

where I_A is the current measured in the circuit,

R_m is the resistance of the current meter, and R_T is the total resistance in the circuit before the meter is added.

SELF-TEST

Check your understanding by answering these questions:
1. An ammeter is connected in _____ (series/parallel) with the circuit to measure current in the circuit.
2. The insertion effect of a current meter is to _____ resistance to the circuit.
3. The effect of connecting a current meter in a circuit then is to _____ the circuit.
4. The current measured by a current meter is actually _____ than the current in the circuit before the meter was added.
5. In the circuit of Fig. 35-1, $R_1 = 120\ \Omega$, $R_2 = 280\ \Omega$, and $V = 10$ V. The computed circuit current $I_B =$ _____ mA.
6. When a current meter M whose resistance (R_m) is 100 Ω is connected in the circuit of question 5, the meter reads _____ mA (I_A).
7. From Eq. (35-4), the error current $I_{error} =$ _____ mA.
8. The error current determined by subtracting I_A from I_B (that is, $I_B - I_A$) = _____ mA.

MATERIALS REQUIRED

- Power supply: Variable low-voltage dc source
- Equipment: Electronic voltmeter, 0–1 mA meter movement and resistance decade box (the same as in Experiments 31 through 34)
- Resistors: ½-W 47-Ω, 100-Ω
- Miscellaneous: 500-Ω 2-W potentiometer; one SPST switch

PROCEDURE

1. Connect the circuit of Fig. 35-3. **Power off.** V is an electronic voltmeter; M is the 0–1 mA meter movement acting as a 1-mA current meter. Switch S is *open*.
2. Set the arm of the 500-Ω potentiometer, acting as a circuit voltage control, at A. The voltmeter V will then measure the voltage V delivered by the power supply.
3. Turn power supply **on** (switch S is still *open*) and adjust V for 5 V as measured by the voltmeter V. Keep power supply output at this

Fig. 35-3. Experimental circuit to determine loading effect of a current meter.

TABLE 35-1.

Step	V_{BC}	I (amperes)			R_1, ohms	R_m, ohms	I_{error}, amperes
		I_A	I_B	$I_B - I_A$			
4 through 8		0.001					
10		0.001					

level. Turn the arm of the voltage-control potentiometer down to C until V reads *zero* volts. *Close switch S.*

4. Slowly turn the arm of the voltage control up until M shows full-scale deflection (1 mA). This is the current I_A in Table 35-1. Record in Table 35-1 the voltage V_{BC} as measured on the voltmeter. **Power off,** switch S open.

5. Remove R_1 from the circuit. Measure its resistance and record in Table 35-1. Record also the value of R_m taken from Experiment 31, Table 31-1.

6. Compute and record in Table 35-1, the current I_B in the circuit of Fig. 35-3, with the meter M removed, but with voltage V_{BC} the same as in step 4:

$$I_B = \frac{V_{BC}}{R_{1(measured)}}$$

7. Compute and record in Table 35-1, $I_B - I_A$.

8. Using Eq. (35-4), compute I_{error} and record in Table 35-1, using the values I_A, $R_m + R_{1(measured)} = R_T$.

9. Using a 47-Ω resistor as R_1 again, connect the circuit of Fig. 35-3. Switch S is open.

10. Repeat steps 2 through 8 recording your results in Table 35-1.

QUESTIONS

1. What effect on circuit current, if any, results from inserting a milliammeter to measure current in a circuit? What is this effect called?
2. For steps 4 through 8 in Table 35-1 compare the values I_{error} and $I_B - I_A$. Explain the relationship, if any, between these values.
3. Which value of $I_B - I_A$ (that is, which error current), if any, was greater, that in step 7, or that in step 10? Why?
4. Which current meter will load a circuit more, a high-resistance meter or a low-resistance meter? Why?

Answers to Self-Test

1. series
2. add
3. load
4. less
5. 25
6. 20
7. 5
8. 5

EXPERIMENT 36 The Series Ohmmeter

OBJECTIVES

1. To determine analytically and confirm experimentally how a basic meter movement may be converted into a series ohmmeter
2. To calibrate experimentally the scale of a series ohmmeter

INTRODUCTORY INFORMATION

The ohmmeter is an instrument used to measure the resistance of resistors and of other circuit components. In addition, it is used to locate open or shorted components and to determine circuit continuity. Resistance is read on a meter scale calibrated in ohms.

Like the nonelectronic voltmeter and ammeter, this basic test instrument utilizes a dc current-actuated meter movement. However, unlike the voltmeter and ammeter, the ohmmeter requires a self-contained source of voltage.

The circuit in Fig. 36-1 illustrates a series ohmmeter. M, a 0–1 mA movement with internal resistance R_m equal to 200 Ω, is connected in series with a current-limiting resistor R_1, a 500-Ω rheostat R_2, and a 3-Volt battery V. A and B are jacks where test leads are inserted. These leads are connected across resistor R_X, whose resistance we wish to measure. The circuit values in Fig. 36-1 were assigned to simplify the discussion of the operation of this instrument.

When the test leads at A and B are shorted together, the circuit is complete and there is current in M. R_2 can be adjusted so that the total resistance that the battery "sees," including R_m, is 3,000 Ω. It is apparent, by Ohm's law, with $R_T = 3,000 \ \Omega$ and

$V = 3$ V, that $I = 1$ mA is the current in this circuit. The meter will therefore read full-scale deflection. This is the zero position of the ohmmeter, at the right-hand side of the scale, and it occurs when A and B are shorted and R_2 is adjusted for full-scale deflection. R_2 is called the "zero-ohms adjust."

After the meter has been adjusted for zero ohms, the test leads are separated. The circuit is broken and the meter pointer returns to the open circuit position on the left side of the scale. When the leads are connected across a resistor, the circuit is again complete and there is current in the meter. However, the circuit current is now less than that required for full-scale deflection (1 mA), because the total series resistance R_T has increased. Thus, if a 1,000-Ω resistor is inserted between A and B (i.e., if a 1,000-Ω resistor is being measured with the experimental ohmmeter), $\frac{3}{4}$ mA will flow through M and the pointer will be deflected three-quarters of the way. This position on the ohmmeter scale (Fig. 36-2) can now be calibrated as 1,000 Ω.

If a 2,000-Ω resistor is measured, $\frac{3}{5}$ mA of current will flow through M. The pointer will deflect three-fifths of the scale, and this will be the 2,000-Ω point on the scale.

It can be shown similarly that in this ohmmeter $\frac{1}{2}$ deflection corresponds to 3,000 Ω, $\frac{3}{8}$ deflection

Fig. 36-1. The basic series ohmmeter circuit.

Fig. 36-2. Ohm scale.

corresponds to 5,000 Ω, $3/13$ deflection to 10,000 Ω, $3/53$ deflection to 50,000 Ω, etc. No deflection corresponds to an infinite resistance, that is, an open circuit between A and B. The symbol for infinity is ∞.

Several facts are apparent from the ohmmeter scale. It is evident that it is not linear, but crowds toward the left-hand side. The need for other ohmmeter ranges is also apparent. Thus, the scale of Fig. 36-2 would not be adequate for measuring a 10-Ω resistor. However, a range in which the center of the scale is, say, 30 Ω, could be used readily for this measurement. Another ohmmeter circuit arrangement would be required for a range giving a 30-Ω calibration at the center of the scale.

It should also be evident why an ohmmeter must not be used in a circuit to which power is applied. Consider the circuit of Fig. 36-3. If it is necessary to measure R, power must be removed before the ohmmeter leads are placed across it. Otherwise, 100 V would be applied to the ohmmeter and the meter would burn out.

The calibration of a series ohmmeter scale can be generalized. Any point on an ohmmeter scale can be found from the equation

$$R_X = R_T \frac{(I_m - I_X)}{I_X} \qquad (36\text{-}1)$$

where R_X = unknown resistance
R_T = total circuit resistance, with test leads shorted, giving full-scale deflection of the meter movement
I_m = meter-movement sensitivity, that is, current for full-scale deflection (ohmmeter current with test leads shorted)
I_X = ohmmeter current with test leads connected across R_X

The results of our previous discussion on scale calibration can now be verified by means of Eq.

(36-1). For example, in the circuit of Fig. 36-1, if a resistor R_X inserted across AB causes the meter current I_X to decrease to $3/4$ mA, we may find the value of R_X by substituting in Eq. (36-1) these values:

$$R_T = 3,000 \ \Omega$$
$$I_m = 1 \times 10^{-3} \ A$$
$$I_X = 3/4 \times 10^{-3} \ A$$

$$R_X = 3,000 \times \frac{(1 - 3/4) \times 10^{-3}}{3/4 \times 10^{-3}} = 1,000 \ \Omega$$

The center-scale value of a series ohmmeter, that is, the value of R_X which will give half-scale deflection of the pointer, may readily be found from Eq. (36-1), by letting $I_X = I_m/2$. It follows, then, that

$$R_X = R_T \frac{I_m - I_m/2}{I_m/2} = R_T \qquad (36\text{-}2)$$

Equation (36-2) states that the center-scale value of a series ohmmeter is equal to R_T, the total circuit resistance, with test leads shorted, required for full-scale deflection of the meter pointer.

The series ohmmeter in Fig. 36-1 cannot be used to measure accurately low-valued resistances because these resistances are crowded together on the low end of the scale, on the right. Consider the range of resistances on a tenth of the scale at the extreme right end. The meter current, corresponding to these ends points, is $I_X = I_m$, and $I_X = 9/10 \ I_m$.

The corresponding values of R_X, from Eq. (36-1), are

$$R_X = 0, \text{ and } R_X = 333\frac{1}{3} \ \Omega$$

It is evident, therefore, that in the last tenth of meter scale arc, the resistances measured are between 0 and $333\frac{1}{3}$ Ω. It would be impossible to distinguish an accurate measurement of 5 Ω or 10 Ω, etc., in this part of the arc. For the measurement of such low-valued resistors, an ohmmeter with a much lower center-scale reading must be used. A series-type ohmmeter would not be employed in such an application; a shunt or series shunt-type ohmmeter would.

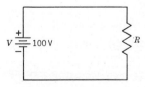

Fig. 36-3. Why must power be removed before using ohmmeter to measure R?

SUMMARY

1. The ohmmeter is used to measure resistance.
2. The series ohmmeter utilizes a basic meter movement in series with circuit resistors, including a zero-ohms adjust and a self-contained voltage source. This is a nonelectronic meter.
3. The resistance we wish to measure is connected in series with the circuit of the series ohmmeter. In Fig. 36-1 the points for connecting the resistor are shown as *AB*.
4. The meter is set on zero, by shorting the test leads at *A* and *B* and setting the zero adjust control for full-scale deflection. This position is zero ohms, on the right-hand side of the scale.
5. Infinite resistance is marked on the left-hand end of the scale, when there is *zero* current in the meter, that is, when the circuit of the meter is not complete. See Fig. 36-2 for the 0 and ∞ points on the scale.
6. If the total resistance required for full-scale deflection of the series ohmmeter is R_T, then half-scale deflection is the calibration point for resistance R_T. In Fig. 36-2, half-scale deflection is 3 kΩ.
7. The calibration of the scale of a series ohmmeter can be achieved by use of the formula

$$R_X = R_T \frac{I_m - I_X}{I_X}$$

where R_X is the resistance value in ohms on the scale corresponding to I_X, the ohmmeter current with test leads connected across R_X; R_T is the total circuit resistance, with test leads shorted, giving full-scale deflection of the meter movement; and I_m is the meter-movement sensitivity, that is, meter current required for full-scale deflection.
8. The ohmmeter scale is a nonlinear scale, with the high-value resistances crowded at the ∞ end of the scale, and the low-value resistances crowded together at the *zero* end of the scale.
9. The purpose of the zero-ohms control (rheostat) is to compensate for deterioration of the battery used to power the ohmmeter. Thus as the internal resistance of the battery *increases* with use, the resistance of the zero-ohms adjust must·be *decreased* to permit full-scale deflection of the pointer at zero ohms.
10. An ohmmeter must *never* be used in a circuit where power is applied to prevent the ohm-

meter from being damaged and to ensure a correct reading of resistance. Also, to conserve the battery, the ohmmeter test leads should remain open (not shorted) when the ohmmeter is not in use.

SELF-TEST

Check your understanding by answering these questions:

1. In a series nonelectronic ohmmeter the resistance to be measured is connected in _____ (series/parallel) with the meter movement.
2. Zero resistance on the series ohmmeter scale occurs when there is _____ (zero/full-scale) deflection of the meter pointer.
3. As the self-contained battery in an ohmmeter is used, its internal resistance _____ (increases/decreases).
4. The zero ohms control is used to _____ for changes in the _____ _____ of the ohmmeter battery.
5. A series ohmmeter utilizes a 1½-V battery, a 50-μA meter movement and the appropriate resistors, and zero ohms rheostat. Assuming the meter has been properly zeroed, half-scale deflection of the pointer occurs when a resistance of _____ Ω is being measured.
6. The 10-μA point on the scale of this meter would be calibrated _____ Ω.
7. The 48-μA point on this meter (question 5 and 6) would correspond to _____ Ω.
8. This meter _____ (would/would not) be useful for measuring a resistance of 30 Ω.

MATERIALS REQUIRED

- Power supply: 1.5-V battery
- Equipment: 0–1 milliammeter movement (same as in Experiment 34), EVM, resistance decade box

Fig. 36-4. Experimental ohmmeter.

- Resistors: ½-W 330-, 1,000-, 1,200-, 1,500-, 2,200-, 3,300-, and 56,000-Ω
- Miscellaneous: 500-Ω potentiometer; alligator-type test leads

Fig. 36-5. Calibrating ohmmeter scale.

PROCEDURE

1. Connect the circuit of Fig. 36-4. Alligator clip leads should be connected at *A* and *B* to act as ohmmeter test leads.
2. Short the test leads together and adjust R_2 (Zero Adj.) for full-scale deflection of the pointer. Calibrate the linear scale (Fig. 36-5) for zero resistance at this point. Do not disturb the setting of R_2.

   With the test leads shorted, measure V, the battery voltage (use an EVM), and record in Table 36-1. Open the test leads.

   Using the measured value of V and the known deflection sensitivity I_m of the meter movement (1 mA), compute, and record in Table 36-1, the total resistance R_T, which will give full-scale deflection of the pointer in the experimental ohmmeter (Fig. 36-4). Record also the known values I_m and R_m.
3. Open the test leads when the ohmmeter is not in use. To conserve the battery, the circuit should remain open.
4. Calibrate the scale ∞ at this point, that is, with test leads open.
5. **a.** Compute and record the size resistor R_X which, when measured, will give half-scale deflection of the meter. Show the computation.

   b. Verify this computation by measuring with the experimental ohmmeter a resistor equal in value to the computed value above. Set the resistance decade box to the proper value of R_X. Calibrate the scale (Fig. 36-5) for this value.

6. Compute, and record in Table 36-1, the value of R_X required to limit the meter current to the following values: $I_X = 0.9\ I_m; 0.8\ I_m; 0.7\ I_m; 0.6\ I_m; 0.5\ I_m; 0.4\ I_m; 0.3\ I_m; 0.2\ I_m; 0.1\ I_m; 0\ I_m$.
7. Calibrate the ohmmeter scale at these points.
8. Measure, and record in Table 36-1, the following resistors using the experimental ohmmeter; 330, 1,000, 1,500, 2,200, 3,300, and 56,000 Ω.

Design Problem (extra credit)

9. Design an ohmmeter whose center-scale value is 50 Ω, using the same meter movement and battery as in the experimental ohmmeter (Fig. 36-4). Draw the circuit diagram, showing all values. Show your computations. Explain how you would calibrate the scale of this meter.

HINT: Try a series shunt, or shunt-type ohmmeter.

QUESTIONS

1. On what does the center-scale reading of a series ohmmeter (connected as in Fig. 36-4) depend?
2. If the 1.5-V battery in the ohmmeter of Fig. 36-4 deteriorated and developed only 1-V output under load, would it be possible to zero the ohmmeter? Explain.
3. In applying Eq. (36-1) to calibrate a series ohmmeter, what is assumed about the relationship between pointer deflection and meter current?

TABLE 36-1.

V, volts	I_m, mA	R_m, ohms	R_T, ohms	\multicolumn{11}{c}{R_X (ohms), when I_X (mA) is:}
				1	0.9	0.8	0.7	0.6	0.5	0.4	0.3	0.2	0.1	0
Rated value R_X				330	1,000		1,500		2,200		3,300		56,000	
Measured value R_X														

4. Why must the test leads of an ohmmeter never be left shorted for any length of time?
5. If we wished to make an $R \times 10$ series ohmmeter using the same meter movement as in the experimental ohmmeter, what circuit changes would be required?
6. Why must power be removed from a circuit when using an ohmmeter to measure resistance in the circuit?
7. In your experiment, how accurately was the scale calibrated? Refer to the results of your measurements in Table 36-1.
8. A circuit contains two resistors connected in parallel, and their individual values must be measured with an ohmmeter. Explain the procedure.

9. Why is an ohmmeter scale nonlinear? What type of scale is it?

Answers to Self-Test

1. series
2. full-scale
3. increases
4. compensate; internal resistance
5. 30,000
6. 120,000
7. 1,250
8. would not

37
EXPERIMENT   Design of a Volt-Ohm Milliammeter

OBJECTIVES

1. To design a dc VOM which will meet specified range requirements
2. To calibrate the scales of the VOM

NOTE: This may be assigned as a research and development term project for the student to complete independently. It is not intended as a regular laboratory experiment.

INTRODUCTORY INFORMATION

In Experiments 31 through 36 you studied the characteristics of a meter movement and its application as a circuit element in the measurement of voltage, current, and resistance. You learned how to determine the characteristics of a meter movement; how to design multipliers and shunts for specified voltage and current ranges; how to calculate the value of the series resistance required to convert the movement into a series ohmmeter of specified range. This knowledge may now be applied to the design and construction of a multirange dc volt-ohm milliammeter (VOM).

Simple DC VOM

Several techniques may be employed in the design of a VOM. The simplest utilizes pin jacks for the selection of circuits and ranges, as in Fig. 37-1. Here a $1/2$-mA meter movement, together with the appropriate circuit, serves as a three-range milliammeter, a three-range voltmeter, and a two-range ohmmeter. For all measurements the black test lead is inserted into the COMMON jack. The red test lead is then plugged into one of the remaining jacks to select the proper function (current, voltage, or resistance) and range. For example, for current measurements of 1 mA or less, the red lead is inserted in the 1-mA jack. For currents between 1 mA and 10 mA, the 10-mA jack is used. For voltages less than 10 V, the 10-V jack is used, etc.

The resistance ranges are marked 100,000 Ω and 1,000,000 Ω. These markings represent maximum readable resistance on each respective range.

The milliammeter ranges employ a ring or Ayrton shunt, R_1, R_2, and R_3, whose operation will be ex-

Fig. 37-1. Simple dc VOM.

plained. The voltmeter multipliers are R_4, R_5, and R_6. The values of these multipliers are computed in the usual manner. It should be noted that in computing the values of the multipliers, both the current drawn by the meter movement M and the current in the series shunt arrangement, $R_1 + R_2 + R_3$, must be considered. Thus, M and its shunt act like a 1-mA movement. For example, if the resistance R_{AB} between points A and B is 50 Ω, then

$$1 \times 10^{-3}(R_4 + 50) = 10 \qquad (37\text{-}1)$$

and $\qquad R_4 = 10,000 - 50 = 9,950 \ \Omega$

(The current in R_4 for full-scale deflection is 1 mA.)

R_A and R_B are, respectively, ohm adjust rheostats for the 100,000-Ω and 1,000,000-Ω resistance ranges. On the 100,000-Ω range a 1.5-V battery is used, while on the 1,000,000-Ω range a 15-V battery is required. In computing the series range resistors, for example $R_A + R_7$, the meter movement and its shunt are again treated like a 1-mA movement. Therefore, when the meter is zeroed on this range, we can find $R_A + R_7$ by applying Ohm's law.

$$1 \times 10^{-3}(R_A + R_7 + 50) = 1.5 \qquad (37\text{-}2)$$

and $\qquad R_A + R_7 = 1,500 - 50 = 1,450 \ \Omega$

We can select a 1,200-Ω resistor for R_7 and a 500-Ω rheostat for R_A. Note that the center-scale reading on this range is 1,500 Ω.

Ayrton Shunt

The three-range current meter circuit, by itself, is shown in Fig. 37-2. M is a meter movement whose sensitivity $I_m = \frac{1}{2}$ mA $= 0.0005$ A, and whose internal resistance $R_m = 100$ Ω. We wish to determine the values of the ring shunt resistors R_1, R_2, and R_3 which will convert the meter movement into the current meter with 1-mA, 10-mA, and 100-mA ranges.

On the 1-mA range, the shunt consists of the sum of the three resistors. That is,

Fig. 37-3. Circuit on 10-mA range.

$$R_{sh} = R_1 + R_2 + R_3 \qquad (37\text{-}3)$$

At full-scale deflection there will be $\frac{1}{2}$ mA in the movement, and $\frac{1}{2}$ mA in the shunt. Therefore, the shunt and meter resistance are equal and

$$R_1 + R_2 + R_3 = 100 \ \Omega \qquad (37\text{-}4)$$

On the 10-mA range, the circuit will appear as in Fig. 37-3. The shunt now consists of $R_1 + R_2$. These are in parallel with M which is in series with R_3. Since $R_1 + R_2 + R_3 = 100$,

$$R_1 + R_2 = 100 - R_3 \qquad (37\text{-}5)$$

We can now solve for R_3. At full-scale deflection, the current in the meter movement is 0.0005 A, while the current in the shunt is 0.0095 A. Since voltage across $R_1 + R_2$ is equal to the voltage across $R_m + R_3$,

$$(0.0095)(100 - R_3) = 0.0005(100 + R_3) \qquad (37\text{-}6)$$

Solving Eq. (37-6), we find that

$$R_3 = 90 \ \Omega \qquad (37\text{-}7)$$

Substituting $R_3 = 90$ in Eq. (37-5), it is evident that

$$R_1 + R_2 = 10 \qquad (37\text{-}8)$$

Now consider the 100-mA range, Fig. 37-4. R_1 is the shunt, in parallel with the series combination of the meter movement, R_3 and R_2. Since

$$R_1 + R_2 = 10$$
$$R_1 = 10 - R_2 \qquad (37\text{-}9)$$

Fig. 37-2. Three-range milliammeter using a ring shunt.

Fig. 37-4. Circuit on 100-mA range.

Fig. 37-5. Switch-type dc VOM.

We can now solve for R_2. The current in the meter movement at full-scale deflection is again 0.0005 A, while that in the shunt is 0.0995 A. Therefore

$$0.0005(100 + 90 + R_2) = 0.0995(10 - R_2) \quad (37\text{-}10)$$

Solving Eq. (37-10) we find

$$R_2 = 9 \ \Omega \quad (37\text{-}11)$$

and

$$R_1 = 1 \ \Omega \quad (37\text{-}12)$$

This method may be extended to find the values of any number of resistors connected in a ring shunt.

Switch-type DC VOM

The dc VOM (Fig. 37-1) may be designed using only two jacks for the common (−) and hot (+) meter leads, employing a somewhat complex switching arrangement (Fig. 37-5). The test leads are plugged into the (−) and (+) jacks. Two switches are used, S_1 and S_2. Each is a three-position switch. S_2 is the function switch. In the position shown, the meter is set to measure current. In the center position of S_2, the meter will measure volts, and in the third position, ohms.

S_1 acts as the range selector. S_1 really consists of three independent switches, S_{1A}, S_{1B}, and S_{1C}.

These switches are on separate wafers, ganged together, so that they are actuated simultaneously. S_2 is shown in the current (ma) position. Note the COMMON arm of S_2 is connected to the + terminal of the meter. The + terminal of the meter in Fig. 37-5 is connected to the common arm of S_{1A}, which is shown in the 1-mA range. These, then, are the settings of S_1 and S_2 for measuring current on the 1-mA range. By keeping S_2 in the mA position, and rotating S_1, we may also select the 10-mA or 100-mA range.

When S_2 is in the volts position, the common arm of S_{1B} is connected to the + terminal of the meter. The 10-, 100-, or 1,000-V range may then be selected by S_{1B}. A similar arrangement is used for measuring resistance. However, since we assumed only two resistance ranges, switch S_{1C} has one open terminal, the position shown in Fig. 37-5.

Figures 37-1 and 37-5 illustrate two design arrangements for a simple dc VOM. For other designs the student is referred to the technical literature on VOMs.

SUMMARY

1. A nonelectronic multirange volt-ohm milliammeter (VOM) provides, in one instrument, facilities for measuring voltage, resistance, and current. The number of ranges included for

each function depends on the switch and/or circuit arrangement.

2. A dc meter movement is the common element employed for each function and each range of a VOM.

3. A simple VOM may eliminate a switching arrangement and provide jacks where the test leads are inserted for selecting function (voltage, resistance, or current) and range (Fig. 37-1).

4. A switch type of VOM eliminates the need for moving the positive test lead from jack to jack. In this type of instrument (Fig. 37-5) selection of function and range is accomplished by a rather complex switching arrangement.

5. In Fig. 37-5 the function switch S_2 is a single-pole three-position switch. Its terminal positions are marked mA, Volts, and Ohms.

6. In Fig. 37-5 the range switch S_1 is a three-pole triple-throw switch. The common arms of the switch S_{1A}, S_{1B}, and S_{1C} are ganged, and they work together for the selection of the proper range for current (S_{1A}), voltage (S_{1B}), and resistance (S_{1C}).

7. The current-meter shunts are frequently arranged in a ring with the meter movement—an arrangement called an Ayrton shunt.

8. The lowest range of the current meter is that which utilizes all the series-connected ring resistors as a shunt for the meter movement (Fig. 37-2).

9. In succeeding switch positions of the Ayrton shunt (Figs. 37-3, 37-4), some resistors in the ring are thrown in series with the meter movement and give the effect of a meter movement with increased internal resistance. For example, in Fig. 37-3 the meter movement acts like a movement whose resistance equals $R_m +$ R_3. In Fig. 37-4 the meter movement acts like a movement whose resistance is equal to $R_m + R_3 + R_2$.

10. Computation of the resistors in the Ayrton shunt then follows the system employed in computing a simple shunt resistor and a meter movement (Experiment 34).

SELF-TEST

Check your understanding by answering these questions:

1. The measurement of _____, _____, and _____ may be made with a VOM.

2. The circuit arrangement of a VOM uses the same _____ _____ for each function and range of the meter.

3. One or more _____ are used to select function and range of the meter.

4. A ring shunt, normally used in the design of the current shunts of a VOM, is called an _____ shunt.

5. In the circuit of Fig. 37-5, if M is a 50-μA, 2,000-Ω movement, and the lowest current range of the meter is 60 μA, then $R_1 + R_2 + R_3 =$ _____ Ω.

6. In the meter of question 5, if the next range of the meter is 0.5 mA, then $R_3 =$_____ Ω.

7. In the meter of question 5, the value of R_5 on the 100-Ω range is _____ Ω. Assume the Ayrton shunt is switched out of the circuit on volts.

8. Center-scale deflection on the highest resistance range of the VOM in question 5 is_____ Ω.

MATERIALS REQUIRED

As required in the design of the VOM (see Procedure)

PROCEDURE

1. The student will experimentally determine the characteristics of a meter movement (I_m and R_m) assigned to him by his instructor. He will check the linearity of the movement at intervals of $I_m/10$. In the comprehensive report, which is required, all circuits and the details of the measurement process should be clearly explained.

2. The student will design a VOM using this movement with the following specifications, or as specified by your instructor:

a. Three voltage ranges: 5-, 20-, 40-V
b. Three current ranges: 2-, 10-, 50-mA
c. One resistance range: 1,500-Ω center scale
In the accompanying report, show all computations and the circuit values determined for each of the components.

3. Submit the circuit design (showing all values, including voltage and tolerance) to your instructor.

4. After he has approved your design, secure the

proper components and breadboard the circuit. (For the multipliers, shunts, etc., use a resistance decade box.) Test it on each of the operating ranges. Include in your report the procedure employed in each check, and the results of your checks.

5. (Optional) Design a case for the instrument. Show all material, construction, and mounting details in the accompanying report.

6. Construct the case and mount the components. Again test your meter on each range. In the accompanying report, list and describe all electrical and mechanical components.

7. Calibrate the meter scales for volts, current, and resistance on each range. Draw the meter scales on the accompanying report.

8. Conclude your report by listing these specifications for your instrument:

a. Voltage ranges
b. Current ranges
c. Center-scale resistance
d. Ohms/volt rating of the meter
e. Accuracy

Explain how you determined instrument accuracy.

Answers to Self-Test

1. voltage; resistance; current
2. meter movement
3. switches
4. Ayrton
5. 10,000
6. 8,800
7. 1,998,000
8. 250 kΩ

38

Use and Care of the VOM

OBJECTIVES

1. To identify the controls and scales of a VOM
2. To operate this multimeter for measuring dc and ac voltage, direct current, and resistance

INTRODUCTORY INFORMATION

The commercial VOM combines within one instrument the functions of dc and ac voltage, direct current, and resistance measurement. Additional functions such as decibel, or dB, measurement (with which you are not yet familiar) are sometimes included.

The design of commercial VOMs follows the techniques discussed in previous experiments. Circuit arrangements, however, are more sophisticated, including ring shunts for current ranges and series and series-parallel circuits for resistance ranges. Complex switching circuits are employed for range and function selection.

The general-purpose commercial VOM popularly uses a 50 μA meter movement resulting in a 20,000-Ω/V dc voltmeter. The ohms/volt characteristic on ac volts is lower.

AC Voltmeter Function

A variety of meters and circuit arrangements are used for measuring ac volts. A common arrangement is one in which a "rectifier" in the meter changes alternating into direct current. The meter then measures this direct current. However, the scale of the meter is calibrated in terms of ac volts. The calibrations are generally in rms or "effective" values, which are 0.707 of peak value.

Figure 38-1 is a simple ac voltmeter circuit. Input ac voltage is applied across the meter terminals. The voltage is rectified by X_1, a solid-state device with which you will become more familiar in your study of basic electronics. The resulting direct current actuates the meter movement. The rectified current is limited by R_1, which therefore acts as a voltage-range multiplier. With the exception of X_1, this circuit is similar to a dc voltmeter circuit.

The circuit in Fig. 38-1 was simplified to explain the theory of operation. A more practical arrangement for a multirange ac voltmeter is shown in Fig. 38-2. A switch permits selection of five different ranges, 2.5 to 1,000 V, for measurement between the terminals labeled Neg. and Pos. Note that polarity need not be observed in ac measurements. The polarity designations at the meter terminal (which are used for both direct and alternating current) apply to dc measurements.

A separate 5,000-V ac jack may be used with the Neg. jack for measuring up to 5,000 V of alternating current.

The rectifier circuit in Fig. 38-2 is more complex than that shown in Fig. 38-1. Moreover, a shunt resistor R_{24} is placed across the meter movement. The purpose of this arrangement is to improve the linearity of the ac voltage scale. An improvement in linearity is achieved at the expense of meter

Fig. 38-1. Simplified ac voltmeter.

Fig. 38-2. Multirange ac voltmeter circuit (*Simpson Electric Co., Div. of American Gage & Machine Co.*).

sensitivity. The shunting effect of R_{24} reduces the ac ohms/volt characteristic of the voltmeter. Common ratings for ac voltmeters using a 50-μA movement vary between 1,000 and 10,000 Ω/V. Therefore, ac voltmeters using the same movement load the circuit more than dc voltmeters do. As in dc voltmeters, the higher the ohms/volt rating, the smaller the loading effect of the meter on the circuit. Obviously, also, the higher voltage ranges of an ac voltmeter cause less loading than the lower ranges.

The ac voltmeter is used in the same manner as the dc voltmeter except that measurements are made in ac circuits and polarity need not be observed. In electronics it is frequently desired to measure the ac voltage at a point where dc voltage is also present. To accomplish this the technician must insert a blocking capacitor, usually 0.1-μF, in series with the ac voltmeter leads. This will prevent the meter from reading the dc voltage but will permit ac voltage readings.

This blocking capacitor is frequently included in the design of the meter in the form of an additional meter jack labeled "output." One of the ac voltmeter leads is inserted into the output jack. The other is inserted into "common." Figure 38-3 shows this output function. Observe that with the exception of the 0.1-μF capacitor which was added, the circuit is identical with that of Fig. 38-2. The voltage-measuring circuit is affected by the 0.1-μF capacitor. Hence, readings are relative, not absolute.

Fig. 38-3. "Output" meter circuit (*Simpson Electric Co., Div. of American Gage & Machine Co.*).

Function and Range Selection

The methods of function and range selection vary with the type of meter and design. Figure 38-4 shows a commercial VOM. A switch selects the functions, labeled AC, +DC, and −DC. The AC position is for measuring ac volts and decibels. The switch is set to the DC position when it is desired to measure dc volts, direct current, or resistance. The output jack is used for measuring ac volts in a circuit where dc voltage is also present.

The range switch is used in conjunction with the function switch. The ranges inscribed on the panel are available at the + and − (common) terminal jacks of the meter. Observe that additional ranges are provided by jacks on the meter.

Zero Ohms is a manual control whose purpose is to zero the ohmmeter. Zero Ohms has the same function here as in the experimental ohmmeter, namely, to compensate for changes in the internal

Fig. 38-5. VOM scale.

resistance of the instrument-contained ohmmeter battery.

Meter Scales

Figure 38-5 shows five scales on the meter pictured in Fig. 38-4. The top scale is for ohms. Zero ohms is on the right-hand side. Center-scale reading is approximately 12.5 Ω. The scale is nonlinear and crowds in the region of high resistance. When the range switch is set on $R \times 1$, the readings are made directly from the ohm scale. On the $R \times 100$ or $R \times 10,000$ range, scale values are multiplied by 100 and 10,000 respectively.

The dc scale is found immediately below the ohm scale. On this scale, read dc volts when the range switch is set on any of the voltage ranges. When the range switch is set on any of the current ranges, the dc scale serves also for current readings. The range designation 10 V, 100 mA, etc., corresponds to full-scale deflection of the pointer.

There are two separate ac scales on this meter immediately below the dc scales. The upper scale is for ac voltages greater than 2.5 V. The lower ac scale is for voltages up to 2.5 V. Both scales are somewhat nonlinear, crowding at the zero end.

Decibels are read on the fifth and last scale of this instrument. This is a nonlinear, logarithmic scale.

Accuracy

The accuracy of the general-purpose VOM is usually rated at full-scale deflection and is 2 to 3 percent on dc function, and 5 percent on ac function. Because these ratings are at full-scale deflection, the percentage of accuracy of the meter may be lower on the lower levels of the range. For example,

Fig. 38-4. Commercial VOM—Simpson 260.

2 percent on the 250-V range is 5 V. This means that a 5-V inaccuracy can occur anywhere on the scale. At the 50-V mark on the 250-V range, the voltage may therefore be 50 ± 5 V. This represents a possible error of 10 percent at this level.

Care in the handling and use of a VOM

The technician must use common sense and care in using a multitester. Though ruggedly constructed and capable of giving many years of satisfactory performance if properly handled, a VOM may be easily damaged if it is used carelessly. Read the operating instructions before using the meter.

Following these simple rules will help secure trouble-free operation of the multitester. Before inserting the meter into the circuit be certain that:

1. The meter switch or switches have been set to the proper measurement function. If voltage is to be measured, the meter must be set for reading voltage. If current is to be measured, the meter switch or switches must be set for reading current, etc.
2. The meter switch or switches have been first set to the highest voltage or current range in measuring an unknown voltage or current. This will reduce the possibility of meter overload and damage.
3. The meter test leads have been plugged into the proper test jacks.
4. Polarity is being observed in measuring voltage and current.
5. Power is disconnected before resistance is measured in a circuit.
6. In measuring current the circuit has been broken so that the meter may be inserted in series with the circuit.
7. The voltage or current to be measured does not exceed the range or capabilities of the meter.
8. When work with the VOM is completed it should be left on the highest ac voltage range or on "transit."

CAUTION—DANGER: Improper handling of the multimeter or any instrument may result in personal injury. Therefore, in measuring voltage it is good practice to use just one hand. Keep the other hand behind your back. Hold the insulated part of the test prod and not the metal tip. In measuring current shut **power off** before breaking the circuit. Insert the milliammeter before turning **power on.** Always inform the instructor concerning any personal injury, damage to equipment, or any possible hazards.

SUMMARY

1. With a 20,000 Ω/V (dc) commercial VOM the following measurements can be made: dc and ac voltage, direct current, and resistance. Special decibel (dB) scales may be included.
2. A 20,000 Ω/V (dc) meter uses a 50-μA meter movement. The ac ohms/volt characteristic is less than 20,000 and may be between 1,000 to 10,000 Ω/V.
3. Polarity need not be observed in measuring ac voltages.
4. The "output" jack on a VOM is used to measure ac voltages where a dc voltage may also be present. In series with the output jack is a 0.1 μF blocking capacitor which blocks the dc but permits the ac voltage to be measured.
5. One or two switches are usually found on the meter panel to select function and range.
6. Usually one manual adjust, Zero Ohms, is found on the meter panel. This is used to zero the meter on resistance function, when the test leads are shorted together.
7. The number of scales on a VOM depends on the type and model. On all VOMs there is a separate ohms scale, usually the top scale. Zero ohms is on the right (for full scale deflection of the meter). The ohms scale is nonlinear and resistance values crowd on the left, the high-resistance end.
8. Ohms ranges are marked $R \times 1$, $R \times 100$, etc. On the $R \times 1$ range readings are made directly from the ohms scale. On the $R \times 100$ range, direct readings are multiplied by 100. On the $R \times 10,000$ range direct readings are multiplied by 10,000, and so on.
9. The dc scale is linear. Full-scale deflection corresponds to the range voltage or current, whichever function is in use. The numerical markings on the dc scale may be direct reading numbers, or it may be necessary to multiply the readings by a range multiplier. For example, if there are separate 10-V and 100-V ranges on the meter, but the scale is calibrated numerically to a maximum of 10, then on the 100-V range all direct readings must be multiplied by 10.
10. The ac voltage scale is somewhat nonlinear at the low end. There may be one or more sepa-

rate ac scales, which are calibrated in rms or effective values.

11. The accuracy of a VOM is rated at full-scale deflection and is usually 2 to 3 percent on dc function, and 5 percent on ac function. At the lower end of the range the amount of error may be appreciably higher than the full-scale rating. For example, a 2-V error at 100 V on the 100-V range is only 2 percent, but the same 2-V error at the 10-V marker (still on the 100-V range) constitutes a 20 percent error.

12. Observe the following rules in use and care of your VOM:
   (a) Read the manufacturer's instructions before use.
   (b) Be certain that the meter is set on the proper function.
   (c) If measuring an unknown voltage or current, set the meter on the highest range of that function.
   (d) Observe polarity in measuring dc voltage and current.
   (e) Never measure resistance in a circuit where there is power applied.
   (f) Do not exceed the range capabilities of the meter.
   (g) Leave meter on transit or highest ac range when not in use.
   (h) Use one hand in measuring high voltage. Always hold the insulated *probe* of the test lead, *not* the metal tip.

SELF-TEST

Check your understanding by answering these questions:

1. The dc and ac ohms/volt ratings of a commercial VOM are usually the same. _____ (true/false).
2. A 20,000 Ω/V (dc) meter uses a _____-_____ ampere meter movement.
3. The dc volts and ohms scale of a VOM are the same. _____ (true/false)
4. _____ shunts are normally used in a commercial 20,000 Ω/V VOM.
5. Polarity need not be observed in measuring ac volts. _____ (true/false)
6. The dc current scale of a VOM is usually the same as the _____ _____ scale.
7. AC scales are calibrated in _____ values.
8. An advantage of using the output jack on a VOM is that dc will be _____, and only _____ will be measured.
9. A disadvantage of using the output jack is that the scale calibrations are only _____, not _____.
10. Before using a VOM with which you are not familiar, be certain to read the _____ _____.
11. Never exceed the range _____ of a VOM.

MATERIALS REQUIRED

- Power supply: Regulated variable dc voltage; variable autotransformer
- Equipment: 20,000-Ω/V VOM; isolation transformer
- Resistors: ½-W 2,200-Ω; 5,100-Ω; 15 kΩ; 56-kΩ; 1-MΩ; 2.2-MΩ
- Miscellaneous: SPST switch

PROCEDURE

1. Familiarize yourself with the controls and scales of the VOM assigned to you. Read the instruction manual of the instrument before proceeding further.
2. Tabulate in Table 38-1 the controls, dc voltage ranges, dc current ranges, resistance ranges, center-scale reading on the $R \times 1$ range, ac voltage ranges, ohms/volt characteristic on dc volts and on ac volts, and rated meter accuracy on direct and alternating current.
3. Connect the circuit of Fig. 38-6. Use a variable auto transformer and an isolation transformer

to deliver the 40 V to the circuit. With the VOM set to read ac volts on the 50-V or next higher range, measure the voltage V_{FD} applied to the circuit and record in Table 38-2.

Measure also and record in Table 38-2 the ac voltages in the divider network. Indicate the range of the voltmeter used and the input resistance R_{in} (R_{in} = voltage range $\times$ ohms/volt) of the range used. (It may be necessary to reduce range when measuring across the 56,000-Ω resistor.) Compute the voltage, which is to be found across each resistor, if measurement were

Fig. 38-6. Ac voltage divider to illustrate voltmeter loading in high-resistance circuit.

Fig. 38-7. A low-resistance ac voltage divider: Loading effects.

TABLE 38-1. VOM Characteristics

Controls	
dc voltage ranges	
dc current ranges	
Resistance ranges	
Center scale reading on $R \times 1$	
ac voltage ranges	
Ohms/volt direct current	
Ohms/volt alternating current	
Accuracy (direct current)	
Accuracy (alternating current)	

made with an infinite impedance meter (that is, if there were no circuit loading), and record in Table 38-2. Show all computations.

NOTE: In a series-resistive voltage divider, Ohm's and Kirchhoff's laws apply just as they apply in a dc voltage divider.

4. Remove the ac voltage source and the isolation transformer. Repeat step 2, using a regulated dc source set to equal the ac voltage previously used. Where possible, use the same ranges on the dc voltage function as on the ac. Record your

measurements and computations in Table 38-2. (If the ac and dc voltage levels are the same, will it be necessary to recompute the dc voltage across each resistor, once you have computed the ac voltage?)

5. Connect the circuit of Fig. 38-7, again using the variable autotransformer and isolation transformer. Measure, and record in Table 38-3, the ac input voltage and the voltage across each resistor. Indicate the range of the voltmeter used and the input resistance R_{in} of the range used. Compute and record the unloaded voltage which

TABLE 38-2. High-resistance Voltage Divider

Points of Measurement	ac voltage		Meter		dc voltage		Meter	
	Measured	Computed	Range	R_{in}	Measured	Computed	Range	R_{in}
FD (input)		X				X		
AD								
AB								
BC								

TABLE 38-3. Low-resistance Voltage Divider

Points of Measurement	ac voltage		Meter		dc voltage		Meter	
	Measured	Computed	Range	R_{in}	Measured	Computed	Range	R_{in}
FD		X				X		
AD								
AB								
BC								

should appear across each resistor in the circuit. Show your computations.

6. Remove the ac voltage source and the isolation transformer. Repeat step 4, using a regulated dc source set to equal the ac voltage previously used. Wherever possible, use the same ranges on the dc voltage function as on the ac. Record your measurements and computations in Table 38-3.

7. **Power off.** Properly connect the VOM for reading circuit current. Set range switch to highest current range.

8. **Power on.** Measure and record the current in the circuit: _____ mA. Compute and record the circuit current: _____ mA.

9. Disconnect the circuit. Measure the resistance value of each resistor used and record in Table 38-4.

QUESTIONS

1. Explain the difference, if any, between the measured ac voltages and the corresponding computed values in Table 38-2.
2. Across which resistor in Fig. 38-6 was the loading effect of the ac meter greatest? Why?
3. Should the sum of the measured ac voltage (Table 38-2) across the resistors in Fig. 38-6 equal the applied voltage? Does it? If the sum does not equal the applied voltage, explain why.
4. In the circuit of Fig. 38-6 was the loading effect of the dc voltmeter greater, less than, or the same as the

TABLE 38-4.

	Resistance Value, ohms	
	Nominal	Measured
1		
2		
3		
4		
5		
6		

loading effect of the ac voltmeter? Prove your answer by referring to your data in Table 38-2.

5. Were the loading effects of the ac and dc voltmeters in the circuit of Fig. 38-7 as great as in the preceding circuit? Prove your answer by referring specifically to the data in Table 38-3.

6. If there was a difference in loading effects in the two circuits, explain why.

7. Do the measured values of the resistors fall within the rated tolerance range of the units? If not, why not?

8. Does the current meter load the circuit of Fig. 38-7? Prove your answer.

9. Is the ohms/volt characteristic of your meter higher on ac or on dc volts? Is this fact consistent with the results of your measurements? Explain.

10. List the possible sources of error in your measurements.

Answers to Self-Test

1. false
2. 50 micro
3. false
4. ring (or Ayrton)
5. true
6. dc voltage
7. effective (or rms)
8. blocked; ac
9. relative; absolute
10. operating instructions
11. capabilities

EXPERIMENT 39

Oscilloscope Operation—Triggered Scope

OBJECTIVES

1. To identify the operating controls of a triggered oscilloscope
2. To set up the oscilloscope and adjust the controls properly to observe an ac voltage waveform

INTRODUCTORY INFORMATION

The cathode-ray oscilloscope (CRO) or "scope," as it is familiarly known, is the most versatile instrument in electronics. The technician must therefore be able to operate this instrument and understand how and where it is used.

For purposes of this book, oscilloscopes will be typed as triggered or nontriggered. Triggered oscilloscopes are the more sophisticated of the two, can do more, and generally are used in industrial laboratories and plants, in engineering and technical school laboratories, and in any application requiring the study of low- and high-frequency waveforms, precise measurement of time, and timing relationships. At this point this very brief statement of the applications of triggered oscilloscopes must suffice. As the student progresses in the study of electronics, additional uses of this laboratory instrument will become evident.

The nontriggered scope is employed for viewing ac waveforms at test points in a circuit, where a knowledge of the existence or nonexistence of a waveform is usually all that is required. A nontriggered oscilloscope can also be calibrated for measuring the amplitude of ac waveforms.

NOTE: In this experiment we will be concerned with triggered oscilloscopes. If the oscilloscope you have in the laboratory is nontriggered, it is sug-

gested that you study the Introductory Information for background, and then go on to Experiment 40 which deals with nontriggered scopes.

What an Oscilloscope Does

An oscilloscope automatically graphs a time-varying voltage; that is, it displays the instantaneous amplitude of an ac voltage waveform versus time. Most triggered scopes also measure dc voltages. The indicator in an oscilloscope is a cathode-ray tube (CRT). Inside the cathode-ray tube are an electron gun assembly, vertical and horizontal deflection plates, and a fluorescent screen.

The electron gun emits a high-velocity low-inertia beam of electrons, which strikes the fluorescent screen and causes the screen to emit light. The intensity of the light given off by the screen is determined by the voltage relationships between the elements in the electron gun assembly. The manual control of brightness is effected by a control located on the oscilloscope panel.

The motion of the beam over the CRT screen is controlled by a deflection system which includes deflection voltages generated in electronic circuits outside of the CRT, and the deflection plates inside the CRT to which the deflection voltages are applied.

Figure 39-1 is an elementary block diagram of an oscilloscope. The CRT serves as the indicator on which electrical waveforms are viewed. These "signal" waveforms are applied to the vertical input on the oscilloscope and are processed by "vertical" amplifiers in circuitry external to the CRT. Since the oscilloscope must handle a wide range of signal voltage amplitudes, a vertical attenuator, a variable voltage divider, acts to set up the proper signal level for viewing. The signal voltage applied to the *vertical* deflection plates causes the electron

Fig. 39-1. Elementary block diagram of an oscilloscope.

beam of the CRT to be deflected vertically. The resulting up-and-down trace is significant in that *the extent of vertical deflection is directly proportional to the amplitude of signal voltage applied to the V input.*

To make it possible for the oscilloscope to graph a time-varying voltage, a linearly changing (time-base) deflection voltage is applied to the horizontal deflection plates. This voltage is developed, in electronic circuits external to the CRT, by a time-base or sweep generator. It is this sweep generator which is either triggered or nontriggered. We have classified an oscilloscope with a triggered time-base generator as a "triggered" scope, and one whose time-base generator is free-running (nontriggered) as a "nontriggered" scope.

Dual-trace Oscilloscopes

Triggered oscilloscopes with *two* traces are in common use. By means of an electronic switching arrangement two traces are developed on the screen of the scope. Dual-trace oscilloscopes make it possible to observe simultaneously two time-related waveforms at different points in an electronic circuit. Familiarity with the operation of dual-trace oscilloscopes will be helpful to the student in the study of electricity and electronics.

Triggered Oscilloscope—Elementary Considerations

Some triggered scopes utilize a single-frame construction; that is, the electronic circuits external to the CRT are mounted on a chassis. This chassis, together with the CRT mount and the front panel, constitute a single assembly. The sweep generator and the vertical amplifier and all the other elec-

tronic circuitry which goes to make up the oscilloscope are self-contained in this single unit. There are other triggered oscilloscopes which utilize a multiframe construction. These have separate assemblies for the sweep generator and the vertical amplifier which plug in to the main frame. The main frame holds the CRT, its associated circuits, and the remaining oscilloscope circuits.

Manual Operating Controls

Intensity. This control sets the level of brightness or intensity of the light trace on the CRT. Rotation in a clockwise (CW) direction increases the brightness. Intensity should not be set too high to prevent damage to the CRT screen.

Focus. This control is adjusted in conjunction with the "intensity" control for the sharpest trace on the screen. There is interaction between these two controls, so adjustment of one may require readjustment of the other.

Astigmatism. This is another beam-focusing control found on some oscilloscopes which operates in conjunction with the focus control for the sharpest trace. The astigmatism control is sometimes a screwdriver rather than a manual control.

Horizontal and Vertical Positioning or Centering. These are trace-positioning controls. They are adjusted so that the trace is positioned or centered both vertically and horizontally on the screen. In front of the CRT screen is an etched faceplate called the graticule. The etchings appear in the form of horizontal and vertical graph lines. Calibration markings are usually placed on the center vertical and horizontal lines on this faceplate.

Volts/Div. (also called Volts/cm). There are two concentric controls which act as attenuators of the vertical input signal waveform (which is to be viewed on the screen). The center control marked *Variable* is continuously variable for setting the height (vertical amplitude) of the signal on the screen. Its completely clockwise position is *calibrated* for making peak-to-peak voltage measurements of the vertical input signal. Volts/Div. is the outer of the two concentric vertical attenuators. It is a switched control. A dot on the control can be thrown to the calibrated voltage markings on the panel around the control. Thus when the variable control is set to its calibrated position, the setting of the Volts/Div. control determines the voltage which is equivalent to every division of vertical signal deflection on the screen. The manner in which voltage measurements are made will be discussed in detail in Experiment 41.

Time/Div. (also called Time/cm). There are two concentric controls which affect the timing of the sweep or time-base generator. The inner control is marked *Variable* and is continuously adjustable over each range of the Time/Div. control. The complete clockwise (CW) position of the variable control is calibrated for making time measurements of waveforms displayed on the screen. Time/Div. is the outer of the two concentric time-base generator controls. It is a switched control. A dot on the control can be thrown to the calibrated time markings on the panel around the control. When the variable control is in its calibrated position, the settings of the Time/Div. control determine the time it takes the trace to move horizontally across one division of the graticule. The manner in which the calibrated time base is used to make timing measurements will be treated in Experiment 41.

Triggering Controls. A simple calibrated time base usually has four triggering controls associated with it. Thus one oscilloscope has a:

1. *Level control.* There is a switch position, associated with this control, labelled *Auto*-(matic). In this position the trigger circuit is free-running and on each cycle triggers the sweep generator. Hence a trace always appears on the screen. The oscilloscope is frequently used in this mode of operation. When not in the automatic mode, triggering depends on some external or internal signal, and the setting of the level control determines the stability or synchronization of the sweep. In the nonautomatic mode there will be *no trace* on the screen in the absence of a triggering signal.
2. *Slope.* This switch is marked + and −, and its setting determines whether triggering of the

sweep is effected by the positive or negative portion of the triggering signal.
3. *Coupling.* This selects the manner in which trigger coupling is achieved. The particular oscilloscope we are describing has three coupling modes: AC slow, AC fast, and DC.
4. *Source.* The trigger signal may be *Ext*(ernal), *Int*(ernal), or *Line*. In this experiment this switched control will be set on Int.

NOTES:
(1) The controls described may have other names, depending on the manufacturer and oscilloscope model. However, once the operation of a triggered oscilloscope is understood, it will be relatively easy to operate other triggered oscilloscopes.
(2) A triggered oscilloscope usually has facilities for switching off the internal sweep generator. A horizontal input jack will then receive some external sweep voltage, apply it to the horizontal processing circuits, and thus cause a horizontal trace. Of course the calibrated Time/Div. controls of the oscilloscope do not operate for this external sweep and the time base is uncalibrated.
(3) There is a vertical signal input jack on the panel which receives the input signal, via a shielded coaxial cable terminated in a probe. There are various types of oscilloscope probes, for example, direct probes and low capacitance probes. We will be using a *direct* probe in this experiment.
(4) Additional features and controls are found on many oscilloscopes. However, knowledge of their operation is not needed at this point.

SUMMARY

1. A triggered oscilloscope can be used for the measurement of dc, as well as low- and high-frequency ac waveforms and time.
2. A nontriggered scope is normally used to observe low-frequency waveforms but cannot measure time directly.
3. An oscilloscope displays a graph of the amplitude of an ac waveform versus time.
4. A cathode-ray tube is the indicator or screen of an oscilloscope.
5. The purpose of the electron gun in a CRT is to emit an electron beam which strikes the screen and causes it to give off light.

6. The signal voltages applied to the vertical deflection plates of a CRT cause the beam to be deflected up and down.
7. The horizontal deflection plates receive the linearly changing deflection voltage which generates the time base.
8. The intensity control is used for setting the brightness of the trace.
9. The focus control is used to narrow the beam into the sharpest trace. There may be an auxiliary astigmatism control for focusing.
10. The horizontal and vertical centering or positioning controls are used to position the trace on the CRT screen.
11. The etched faceplate in front of the CRT face which appears as vertical and horizontal graph lines is called the graticule. Linear calibration markers (height and width) are frequently etched on the graticule.
12. The Volts/Div. or Volts/cm control is calibrated for the measurement of the amplitude of signal waveforms along the vertical axis.
13. The Time/Div. or Time/cm control is calibrated for the measurement of time along the horizontal axis.
14. The triggering controls determine the manner in which a trigger pulse is initiated to start the sweep generator.
15. The trigger can be run automatically, the mode which is frequently used.
16. The trigger circuit can be actuated on Int. by the signal waveform from within the oscilloscope circuits.
17. The trigger circuit can also be actuated on Ext. by an external signal voltage applied to an input jack labelled Ext. trigger.
18. Also, the trigger circuit can be actuated on Line position of the trigger switch by a line-derived voltage from within the oscilloscope circuits.

SELF-TEST

Check your understanding by answering these questions:
1. Both triggered and nontriggered scopes are used for the observation of ac waveforms. _____ (true/false)
2. Precise time measurements can be made both with triggered and nontriggered scopes. _____ (true/false)
3. The waveform seen on the screen of a CRT is a _____ of amplitude versus _____.
4. The _____ deflection plates of a CRT are the signal plates; the _____ deflection plates are for the time-base voltage.
5. Of the controls on an oscilloscope, those that affect the height of the signal are called _____ .
6. Those controls that affect the sharpness of the trace are called _____ and _____.
7. The etched faceplate in front of the face of the CRT is called the _____ .
8. A frequently used triggering mode of a trigger oscilloscope is _____ .
9. The controls that affect the up-and-down movement of the trace are called _____ .
10. The height of the waveform displayed on the oscilloscope screen is directly proportional to the _____ of the waveform.

MATERIALS REQUIRED

- Power supply: Source of 120 V ac
- Equipment: Triggered-type oscilloscope with calibrated time base, with calibrated vertical amplifier, with internal voltage calibrator, and with direct probe

PROCEDURE

NOTE: Before attempting the experiment, the student should read and become thoroughly familiar with the operating instructions of the oscilloscope.

CAUTION: Do not operate the scope with trace intensity too high.

Operating the Controls Which Affect the Trace

1. List each manual control and switch on your oscilloscope and state its function in Table 39-1. Include also the input jacks.

2. Turn the oscilloscope on. The scope normally

TABLE 39-1. Manual Controls and Switches and Their Functions

Control or Switch	Function

contains a protective time-delay relay. Wait until you hear the relay click in. This will occur after the oscilloscope is warmed up and ready for operation (about 1 to 2 min). Set the Time/Div. control to 1 millisec.

3. If a trace does not appear on the screen, check to see that the triggering switch is on Auto-(matic). If it is not, set in on Auto.

4. If there is still no trace, turn the Intensity control completely clockwise.

5. If there is still no trace, try the Positioning (centering) controls until a trace does appear.

6. Adjust the focus, astigmatism, and brightness controls for a clear, sharp trace. Center the beam both vertically and horizontally. Set the triggering *Slope* to +, triggering *Coupling* to AC or AC Fast, triggering *Source* to Int. The oscilloscope is now ready for viewing ac waveforms.

7. Throw off the controls you adjusted in steps 2 through 6 and repeat the entire procedure for setting up a trace. When you are satisfied that you can operate the controls properly, notify your instructor.

Viewing a Waveform

8. Connect the vertical input leads of the oscilloscope to the output of the voltage calibrator on the scope. Set the calibrator to 2 V output (approx.). The calibration waveform will be seen when the vertical attenuators (Volts/Div.) and Time/Div. controls are properly set.

9. Set the variable vertical attenuator on *Calibrated* and vary the Volts/Div. control for about three divisions of signal height.

10. Now set the Variable Time/Div. control to Calibrated and vary the Time/Div. switch until three cycles (approx.) of the calibration waveform appear on the screen.

11. Now, leaving the Height/Div. controls as set, change the setting of the calibrator output to $\frac{1}{2}$ V (approx.). What happens to the height of the waveform on the screen?

12. Reset the calibrator output to 5 V. What happens to the height of the waveform?

13. Reset calibrator output to 2 V, as in step 8. The waveform should now have a height of three divisions as in step 9.

14. *Increase* the sweep speed (Time/Div.) one setting on the Time/Div. switch (for example, if the switch was set to 0.5 ms/div., set it to the next lower time calibration, say, 0.2 ms/div.). What happens to the number of cycles on the screen?

15. *Decrease* the sweep speed one setting past its position in step 10. What happens to the number of cycles on the screen?

16. When you are satisfied that you understand the operation of all the manual controls on the scope, have your instructor throw them off. Readjust the operating controls until you can see the calibrator waveform, properly centered, on the screen, as in step 10.

17. Repeat step 16 until you are completely satisfied that you can operate the oscilloscope controls properly.

QUESTIONS

1. After the Volts/Div. controls were set to view the calibrator waveform, what happened to the height of the waveform when the calibrator voltage output was increased? Why?

2. After the Time/Div. controls were set to view three cycles of the calibration waveform, what happened to the number of cycles seen when the sweep speed was increased? decreased? Why?

3. Refer to Procedural steps 14 and 15. Was the frequency of the calibration waveform affected when the sweep speed was increased or decreased? What was changed?

4. What is the relationship, if any, between the number

of cycles of waveform and the setting of the Time/Div. control?

5. List the controls on your scope which affect the (a) height of the waveform; (b) brightness of the trace; (c) sharpness of the trace; (d) position of the trace; (e) triggering of the sweep generator.

Answers to Self-Test

1. true
2. false
3. graph; time
4. vertical; horizontal
5. Volts/Div. or Volts/cm
6. focus; astigmatism
7. graticule
8. automatic
9. positioning or centering
10. amplitude (voltage)

40

Oscilloscope Operation—Nontriggered Scope

OBJECTIVES

1. To identify the operating controls of a non-triggered oscilloscope
2. To set up the oscilloscope and adjust the controls properly to observe an ac voltage waveform

INTRODUCTORY INFORMATION

The triggered oscilloscope studied in Experiment 39 is used in industrial and research laboratories, in school laboratories, for general industrial applications as well as for radio and TV servicing. In some small radio and TV shops, and in some technical school laboratories, the nontriggered oscilloscope may still be found. This experiment is intended for those schools which use nontriggered oscilloscopes.

The nontriggered service-type oscilloscope is less expensive than the laboratory standard triggered oscilloscope, does not have a calibrated time base (sweep), and does not have calibrated vertical amplifiers. It is generally less sophisticated than the triggered oscilloscope. The controls have different names, though they are similar in function and are relatively simple to adjust and operate. The nontriggered oscilloscope has a free-running time-base generator.

Operating Controls

At this point the student is not expected to understand the circuitry or theory of operation of the non-triggered cathode-ray oscilloscope. However, he is expected to learn how to use it properly. For this reason he must understand the function and operation of each of the controls. These are:

Intensity. This control sets the level of brightness or intensity of the light trace on the cathode-ray tube. Rotation in a clockwise (CW) direction increases the brightness. Usually associated with this control is the **on-off** switch, which applies power to the scope.

Focus. This control is adjusted in conjunction with the "intensity" control for sharpest trace on the cathode-ray tube. There is interaction between these two controls. Hence adjustment of one may require readjustment of the other.

Horizontal and Vertical Centering. These are positioning controls for positioning or centering the trace. They affect the horizontal and vertical position of the trace.

Sweep Selector. This control in conjunction with the fine-frequency control makes it possible to set the scope for viewing waveforms of different frequencies. They are adjusted to the position which "stops" the waveforms on the screen.

Sweep Vernier. It operates in conjunction with the coarse frequency as described.

Sync. The sweep selector and vernier controls cannot usually "stop" a waveform completely. At their closest setting the waveforms may still drift (move). The sync control is adjusted to stop this drift. It is preferable to set the sync control at approximately 10 to 20 percent of its range. This control synchronizes the scope sweep frequency with the ac signal-voltage frequency.

Sync Selector. This switch is usually set to Int. for most applications.

Vertical-gain Selector. This control operates in conjunction with the vertical-gain vernier to adjust the height of the waveform by changing the overall vertical scope gain. The "V" selector is a coarse adjust. It is usually a calibrated attenuator.

Vertical-gain Vernier. This continuous-gain control sets the height of the waveform over each range of the vertical-gain selector. Thus the "V" selector is set for the approximate height of the waveform required. The "V" vernier is then used to set the waveform to the exact height required.

Horizontal Selector. A switch used for selecting the horizontal sweep or for setting the width of the trace in conjunction with the horizontal-gain vernier. This switch may not be found on all scopes.

Horizontal-gain Vernier. This control provides a continuous-width adjustment. It is a fine adjustment. The horizontal trace is usually adjusted to occupy ⁴/₅ the width of the screen.

Additional features and controls are found on many oscilloscopes. However, knowledge of their operation is not needed at this point. It should be noted here that similar controls on different oscilloscopes may have other names than those given. For this reason a chart is included here which lists some of the names by which each of the controls is known.

The oscilloscope has certain input jacks with which the technician should be acquainted. At this time he will be concerned only with the vertical input and ground jacks. A shielded cable is usually attached to these two jacks. The waveform to be viewed is brought into the oscilloscope by this shielded cable.

SUMMARY

1. Nontriggered oscilloscopes are sometimes used by the radio and TV service technician for the repair and adjustment of radio and TV receivers.

2. The controls which affect the trace on the cathode-ray tube are: *Intensity, Focus, Horizontal Centering,* and *Vertical Centering.* The name of the control is descriptive of its function.

3. The controls which affect the gain of the vertical or *signal* amplifiers are: *vertical-gain selector* and *vertical-gain vernier.* These are used to adjust the height of the signal waveform on the oscilloscope screen. The *selector* is a *coarse* attenuator; the *vernier* is a continuous or fine attenuator.

4. The controls which make it possible to view waveforms of different frequencies are: *sweep selector* and *sweep vernier.* The selector is a *coarse* control; the vernier is a continuously variable *fine* control. They operate in conjunction and are adjusted to the position which sets up the required number of cycles of the waveform to be viewed.

5. The *sync control* is adjusted to *stop* the waveform from drifting across the screen. The synchronizing source is normally the signal voltage which is being viewed on the oscilloscope when the *sync selector* is switched to Int.

6. The *horizontal selector* (not found on all scopes) and *horizontal-gain vernier* are used to adjust

Control Names	Alternate Names
Intensity—**off**	Intensity; intensity line **off**; intensity ac **off**; beam **off**
Focus	
Horizontal centering	H centering; horizontal position
Vertical centering	V centering; vertical position
Sweep selector	Sweep; sweep range; sweep frequency; coarse frequency; saw-tooth sweep; frequency range; range
Sweep vernier	Range frequency; vernier; frequency vernier; frequency
Sync	Sync adj.; sync lock; sync signal
Sync selector	Sync; function; synchronizing input control
Vertical-gain selector	V range; vert. attenuator; V sensitivity; vert. input control
Vertical-gain vernier	V vernier; vertical gain
Horizontal selector	H sel; horiz. sens; horizontal input control; H gain selector
Horizontal-gain vernier	H gain; horizontal gain

the width of the trace. The selector, when used, is a coarse control; the *vernier* a fine control.

SELF-TEST

Check your understanding by answering these questions:
1. An oscilloscope with a free-running time base is called a _____ _____ oscilloscope.
2. This type of oscilloscope _____ (has/does not have) a calibrated time base.
3. The time base generates the _____ on the screen.
4. The control which affects the brightness of the trace is the _____ control.
5. The _____ control is adjusted for the sharpest (clearest) trace.
6. The vertical and horizontal _____ controls are used to position the trace.
7. The controls which make the viewing of signals of a wide range of frequencies possible are: _____ selector and _____ vernier.
8. To prevent a signal waveform from drifting across the screen, the _____ control is used.
9. To set the amplitude or height of the waveform on the screen, we used the _____ _____ selector and _____ _____ vernier.
10. To set the width of the trace we use the _____ selector and _____ _____ vernier.

MATERIALS REQUIRED

- Power supply: Source of 120 V ac
- Equipment: Nontriggered oscilloscope with direct probe

PROCEDURE

NOTE: The student should read and become thoroughly familiar with the operating instructions of the oscilloscope which he is using before starting this experiment.

CAUTION: To prevent screen burnout, do *not* leave a stationary spot on the screen.

Getting Acquainted with Operating Controls

1. Set the sweep selector and sweep venier in the middle of their range.
2. Set the vertical and horizontal centering controls in the middle of their range.
3. Set sync selector to INT and sync completely off.
4. Turn scope ON and advance intensity control about ¾ of the way up. Set vertical-gain vernier completely counterclockwise.
5. Set horizontal gain about ¾ of the way up (clockwise).
6. Let scope warm up for a few minutes and then adjust vertical and horizontal centering controls and note effect on trace. Center the trace properly.
7. Adjust intensity and focus controls for sharpest trace and satisfactory brightness.
8. Vary the horizontal-gain vernier and note effect on trace width. Adjust this control for satisfactory trace width. Thus on a 5-in. screen, adjust for a trace 4 in. wide. Center the trace or beam.
9. Throw all the controls off and go through the entire procedure until a satisfactory trace has been obtained on the scope. When you are satisfied that you can operate the controls, notify your instructor.
10. The instructor will throw all the controls off and the student will readjust them properly. The student may not proceed until the instructor has approved this work.

Viewing a Waveform

11. Connect the vertical input cable of the oscilloscope to an ac signal source. There may be a test-signal or a calibrating-signal source on the oscilloscope. This ac signal is satisfactory.
12. Adjust the vertical-gain selector and vertical-gain vernier until the signal fills 80 percent of the screen in height. Recenter the beam if necessary.
13. Adjust the sweep selector and sweep vernier until there are just two cycles of signal voltage on the screen.
14. Turn the sync control up until the signal is sta-

tionary on the screen. This should be no higher than 20 percent of its range. Readjust the sweep vernier if necessary to stop the signal. There is interaction between the sweep vernier and sync control.

15. After the signal has been "stopped," turn the sync control all the way up and note the effect. Return sync to its previous setting (step 14).

16. Adjust vertical and horizontal centering and note the effect. Recenter the waveform properly.

17. Adjust focus and brightness controls and note the effect on waveform. Reset these controls properly.

18. Vary the vertical-gain vernier and note the effect on height and synchronization (stability). Reset the height as in step 12.

19. When you are reasonably certain that you know how to operate these controls, throw them all off and go through the entire procedure again. You should end up with two cycles on the screen.

20. Readjust the sweep vernier (and sweep selector if necessary) until you have stationary (a) one cycle, (b) three cycles, (c) four cycles.

21. Obtain your instructor's signature for your test setup.

QUESTIONS

1. After the waveform has been synchronized, what is the effect on synchronization, if any, of varying the vertical-gain vernier on either side of its setting?

2. Is the frequency of the test waveform changed as the sweep vernier is varied?

3. List the controls on your scope which affect the: (a) height of the waveform; (b) width of the trace; (c) brightness of the trace; (d) sharpness of the trace; (e) position of the trace; (f) steadiness of the waveform.

4. Why should you never leave just a bright spot on the screen?

5. List the sets of controls which interact with one another and indicate the effects of the interaction.

Answers to Self-Test

1. nontriggered
2. does not
3. trace
4. intensity
5. focus
6. centering
7. sweep; sweep
8. sync
9. vertical-gain; vertical-gain
10. horizontal; horizontal-gain

41

EXPERIMENT **41** Operation of an Audio-frequency Generator

OBJECTIVES

1. To identify the operating controls of an audio-frequency (AF) generator
2. To observe on an oscilloscope the signal delivered by an AF generator

INTRODUCTORY INFORMATION

Audio Oscillator (AF Signal Generator)

A dc power supply provides dc voltages for dc circuits. Similarly, the "signal generator" supplies ac voltages for ac circuits. These generators develop signals whose frequency is variable over a specified range.

AC frequencies cover a very wide spectrum, from a fraction of a cycle per second to thousands of millions of cycles per second. No one instrument has been designed to cover this extensive range. Many generators are available commercially to provide the various frequency needs in electronics.

One of the characteristics by which signal generators are identified is the frequency range which the generator covers. The audio-frequency (AF) generator supplies frequencies from several hertz up to 20,000 Hz approximately. It is called "audio" (which means sound) because this is the range of frequencies (approximately) to which the ear responds. This is not to say that the ear can hear the direct electrical signal delivered by the generator. But the electrical signal can be converted to a sound signal (a vibration in the air) by a suitable device such as a loudspeaker.

An AF generator often covers a range much wider than the audio frequencies. Thus, one "audio" generator supplies a signal whose frequency can be varied from several hertz up to 600 kHz. However, this coverage is achieved in several ranges.

Another characteristic which identifies a signal generator is the shape of the waveform it develops. Thus, there is a sine-wave generator whose output is sinusoidal (Fig. 41-1a), a square-wave generator whose output is the square waveform (Fig. 41-1b), and other waveform generators with which the student need not be concerned at the moment.

The usual controls and switches found on the front panel of a signal generator are:

1. On-off switch for applying power to the generator.
2. *Range.* A coarse control to select a specified frequency coverage (range); for example, 10 to 1,000 Hz or 1,000 to 100,000 Hz, etc.
3. *Frequency.* This is a continuously variable control to select a specified frequency within any range. Associated with this control is a calibrated frequency scale.
4. *Level (output) control.* This is used to set the voltage of the output signal. There may be two types of level controls on the panel, a coarse

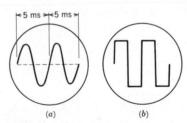

Fig. 41-1. AF generator develops (a) sine wave; (b) square wave.

Fig. 41-2. Sine wave seen on scope screen.

control and a continuously variable output control. Both types are voltage-divider networks. The coarse control is usually a decade attenuator which reduces the signal level in multiples of 10.

The output of a general purpose sine-wave generator is not metered. Its level must be measured by an external ac voltmeter. Some laboratory-standard signal generators include a metering circuit which measures the signal delivered at the output terminals. General-purpose AF generators can deliver a signal from a fraction of a volt to many volts. Thus, the approximate output range of one instrument is from a millivolt to 20 V.

A shielded cable is used to deliver the generator signal to the external circuit. This cable is detachable from the generator.

Characteristics of an ac Signal Voltage

Alternating-current voltages are identified by certain characteristics. These are:

The Waveform or Shape of the Voltage

This characteristic pertains to the manner in which the voltage varies between maximum and minimum. Thus, if the voltage is varying in a sinusoidal fashion, the scope will display the voltage as a sine wave (Fig. 41-2).

The Amplitude of the Voltage

This characteristic describes the difference between the positive and negative peaks of the voltage and is expressed in peak-to-peak volts (Fig. 41-3).

The Frequency of the Voltage

An ac periodic wave completes a number of cycles every second. The time interval for each of these cycles is called the period t. Thus the period or time

in seconds of one complete cycle of an ac voltage is found from the relationship

$$t = \frac{1}{F}$$

where F is the frequency of the voltage in hertz. Thus, if the voltage of Fig. 41-3 has a frequency of 60 Hz, the period is $1/60$ s. Hence, the time required for the scope to "graph" one cycle of the waveform is $1/60$ s.

Measuring the Period of an AC Voltage with an Oscilloscope

Since the period t and frequency F of an ac signal waveform are inversely related; that is, since

$$t = \frac{1}{F}$$

and

$$F = \frac{1}{t}$$

it is possible to calculate the frequency of a waveform if its period is known. If its frequency is known, its period may be computed.

The triggered oscilloscope has facilities for direct measurement of the period of an ac signal voltage. The characteristic which makes this possible is the calibrated sweep or time base of this type of oscilloscope.

In Experiment 39 you noted that to display a signal waveform on the screen, it was necessary to set the Time/cm control properly. It is this control which determines the rate at which the cathode-ray tube beam is moving across the screen to generate the trace. An example will show how the screen can be used to measure the period of the waveform displayed on the CRT.

Suppose the Time/cm control is set to 1 ms and suppose the calibrated face of the oscilloscope, called the graticule, is 10 cm wide. Then it takes 10 ms for the electron beam to move across the 10 cm screen. Now, if the width of a waveform is exactly 10 cm, as in Fig. 41-2, then the period of the waveform is 10 ms. That is, $t = 10 \times 10^{-3}$ s. The fre-

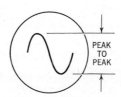

Fig. 41-3. Measuring peak-to-peak voltage.

quency of the displayed waveform can now be calculated, for

$$F = \frac{1}{t} = \frac{1}{10 \times 10^{-3}} = 100 \text{ Hz}$$

If two waveforms had been displayed on the 10 ms trace (Fig. 41-1), then the period of each cycle is 5 ms and the frequency is

$$F = \frac{1}{5 \times 10^{-3}} = 200 \text{ Hz}$$

NOTE: The nontriggered oscilloscope does not have a calibrated time base, so the trace can not be used directly for measurement of time. This should be evident from the fact that the nontriggered oscilloscope has an adjustable trace width. As the width is varied, the scanning rate (time/cm) of the trace changes. The trace width of a *triggered sweep*, however, is fixed, as is the scanning rate.

SUMMARY

1. A signal generator supplies ac voltages for ac circuits.
2. An AF signal generator provides ac signals in the audio range of frequencies, that is, 10 to 20,000 Hz (approximately).
3. There are other generators which develop signal voltages over other frequency ranges.
4. An AF generator may provide a sinusoidal voltage waveform, a square wave, or other waveshape, depending on the type and purpose of the generator.
5. The controls normally found on a general purpose AF generator are: *on-off* (for power), *range* (coarse frequency), *frequency* (continuously variable within each range), *level* (output).
6. The output of a general-purpose sine-wave generator is normally not metered, so it must be measured by an external voltmeter.
7. Laboratory standard AF generators do have metered outputs.
8. An ac voltage waveform may be identified by
   (*a*) Its waveform—is it a sine wave, square wave, triangular wave, etc.?

(*b*) The amplitude of the wave—usually measured in peak-to-peak volts.
(*c*) Its frequency (*F*) in hertz.
(*d*) The period or time (*t*) in seconds of one cycle of the waveform.
9. Frequency and period are related by the formulas

$$F = \frac{1}{t}$$

$$t = \frac{1}{F}$$

10. A triggered oscilloscope has a calibrated time base which may be used for measuring the period *t* of a cycle of a periodic waveform. Having measured *t*, *F* can be calculated.

SELF-TEST

Check your understanding by answering these questions.
1. The instrument which supplies a test ac signal is called a _____ _____.
2. The audio frequencies are in the range _____ Hz to _____ Hz (approximately).
3. The general-purpose AF signal generator has a metered output. _____ (true/false)
4. The characteristics which identify a periodic ac signal are:
   (*a*) _____
   (*b*) _____
   (*c*) _____
   (*d*) _____
5. If the period of an ac waveform is known, the frequency of that waveform may be calculated from the formula: _____ = _____.
6. A triggered oscilloscope has a _____ time base, which can be used to measure the _____ of an ac waveform.

MATERIALS REQUIRED

- Power supply: ac voltage source
- Equipment: AF signal generator; oscilloscope with direct probe

PROCEDURE

Familiarization with AF Signal Generator

1. Familiarize yourself with the signal generator issued to you. Study the instruction manual and learn the function and use of the operating controls on the front panel.
2. **Power on.** Connect the output leads of the sig-

nal generator to the vertical input of the oscilloscope. Set the output controls of the generator
in the middle of their range. Adjust the frequency control to 100 Hz.

NOTE: If you have a laboratory-type triggered
scope, do steps 3 through 10. If you have a nontriggered scope, do steps 11 through 16.

For Triggered Oscilloscope

3. Set the oscilloscope volts/div. selector until the
   waveform is deflected 3 divisions vertically
   (approximately). Set scope on automatic triggering and sync on Int. +. Adjust the oscilloscope Time/div. control for a display of two
   sine waves (approximately).
4. Reduce the output level of the generator signal
   and observe the effect on the height of the signal
   displayed on the oscilloscope. Is the stability of
   the presentation affected as the signal is reduced below a certain level?
5. Increase the level of the generator signal to
   maximum. Readjust, if necessary, the volts/div.
   control until the entire waveform is displayed
   on the screen. Is the volts/div. factor now
   higher or lower than in step 3?
6. Reset the frequency of the signal generator to
   200 Hz. There are now _____ cycles on
   the screen.
7. Reset the frequency of the signal generator to
   300 Hz. There are now _____ cycles on
   the screen.
8. Reset the oscilloscope Time/div. control until
   one or two cycles are displayed on the screen.
   Measure the width of one sine wave. _____
   div.
9. Multiply the Time/div. setting for this display
   by the width (number of div.) of the waveform.
   This is the period of the sine wave _____ s.
10. Using the formula $F = 1/t$, compute the fre-
    quency of the sine wave. (NOTE: f is in Hz, when
    t is in seconds.) $f =$ _____ Hz.

For Nontriggered Oscilloscope

11. Adjust the oscilloscope vertical-gain controls
    for a vertical deflection of 3 in. (approximately)
    on a 5-in. screen. Adjust the oscilloscope sweep
    and sync controls for a stable display of two
    sine waves. Sync should be set no higher than
    20 percent of its range.

12. Reduce the output level of the generator signal
    and observe the effect on the height of the sig-
    nal displayed on the oscilloscope. Is the stabil-
    ity of the presentation affected as the signal is
    reduced below a certain level?
13. Increase the level of the generator signal to
    maximum. Readjust, if necessary, the oscillo-
    scope vertical-gain controls so that the entire
    signal is displayed on the screen.
14. Do not vary the sweep or sync controls from
    their original setting when two cycles were dis-
    played on the screen.
15. Vary the frequency of the signal generator until
    there are four cycles on the screen. The fre-
    quency dial of the AF generator now reads
    _____ Hz.
16. Do not readjust sweep or sync controls of the
    oscilloscope. Vary the frequency of the signal
    generator until there are six cycles on the
    screen. The frequency dial of the AF generator
    now reads _____ Hz.

QUESTIONS

1. What is the range of frequencies which your AF generator can deliver?
2. Does there seem to be any relationship between the
   number of cycles displayed on the screen and the frequency setting of the generator, assuming there is no
   change in Time/div. (or sweep) and sync settings of
   the oscilloscope? If so, what is the relationship?
3. For the triggered oscilloscope, how did the signal frequency, computed in step 10, compare with the frequency setting of the generator? If they were not the
   same, explain why.
4. In comparing steps 3 through 10 of this procedure for
   triggered oscilloscope, with steps 11 through 16 for
   nontriggered scopes, what advantage can you see in
   the use of a triggered over a nontriggered scope?
5. What is the purpose of each of the following controls
   on the signal generator?
   (a) Range
   (b) Frequency
   (c) Output or level

Answers to Self-Test

1. signal generator
2. 10; 20,000
3. false
4. (a) waveform; (b) amplitude; (c) frequency; (d) period
5. $F = 1/t$
6. calibrated; period

42

EXPERIMENT **42**   Oscilloscope Voltage Measurements

OBJECTIVES

1. To check the calibration accuracy or to calibrate the oscilloscope for use as an ac voltmeter
2. To make ac voltage measurements with the scope
3. To make dc voltage measurements with the scope

INTRODUCTORY INFORMATION

An oscilloscope is used to view and measure ac waveforms. It has many other functions, but our concern in this experiment will be with its voltage measuring capabilities.

Measuring Voltages with the Scope

The height of the voltage waveform displayed on the oscilloscope screen is directly proportional to the peak-to-peak amplitude of the voltage. Thus, for the same settings of Volts/div., or vertical-gain controls, a 100-V signal will have twice the height of a 50-V signal. A 25-V signal will have one-fourth the height of a 100-V signal. Moreover, if we suppose for this same fixed setting of the vertical controls, every division of vertical deflection corresponds to 25 V of input, then a 25-V signal will have 1 division of height, a 50-V signal will have 2 divisions of height, a 75-V signal will have 3 divisions of height, and so on. Two and one half (2½) divisions of vertical deflection (height) will correspond to 62½ V, etc. This characteristic of the signal circuits makes it possible to use the oscilloscope for measuring ac voltages.

To measure dc, the method described above is also used. The ac-dc switch is placed in the dc position and the probe is connected to the point in the circuit where the dc voltage is. The ground lead of the scope is connected to the ground of the circuit. The number of divisions the trace rises above or falls below the zero base setting is a measure of the + or − dc voltage.

NOTE: In this experiment we shall refer to the number of centimeters (cm) or the number of inches (in.) of height of the voltage waveform, because these two markings are found on the graticules of oscilloscopes. The triggered oscilloscope frequently uses a graticule whose division markings are in centimeters. The nontriggered oscilloscope, in general, uses inch markings for the major divisions on its graticule. There is another difference between the graticules of these two types of oscilloscopes—the triggered-scope graticule is usually rectangular and appears as in Fig. 42-1a. It is shown here as 8 × 6 cm, though not all graticules have these dimensions. The graticule of the nontriggered scope is usually round and appears as in Fig. 42-1b. The round graticule covers the entire cathode-ray tube screen, which here is shown as 5 in. in diameter.

In measuring voltages with the oscilloscope, the procedure is to apply the signal voltage to the vertical input of the oscilloscope and to measure on the etched faceplate or graticule the number of divisions of height of the waveform. This number is then multiplied by the calibration factor of the oscilloscope, giving the peak-to-peak voltage of the waveform. For example, if the oscilloscope is set for 20 V/div., 3 div. of deflection (height) measures a 60-V signal. Similarly, if the scope is calibrated for 20 V/div., 3 div. of height also measures a 60-V signal.

It should be noted that there are no voltage scales on the graticule. There are simply linear markings whose value is determined by the calibration factor to which the volts/div. vertical-gain controls of the

202 *Basic Electricity*

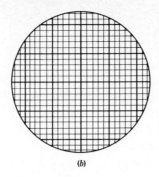

Fig. 42-1. (a) Graticule on a triggered-type oscilloscope which measures 8 × 6 cm. (b) Graticule on nontriggered type oscilloscope measures 5 in. in diameter.

oscilloscope are calibrated. Thus the volts/div. or vertical-gain controls correspond to the range controls of a voltmeter; the graticule markings correspond to the scale of the voltmeter.

Oscilloscope Calibration

Triggered Oscilloscope. The triggered oscilloscope has a calibrated vertical amplifier. There are two concentric input controls which affect, by attenuation, the height of the waveform displayed on the screen. The inner control is continuously variable. The switch-type volts/div. control changes the signal attenuation by a precalibrated factor. The inner control must be turned to its "calibrated" position, which is completely clockwise. Then, the marking to which the switch-type volts/div. control is set is the calibration factor of the signal-measuring vertical amplifier of the oscilloscope. Thus with the inner control set to its calibrated position, and the volts/div. control set to 0.5 V, every 1 div. of height corresponds to 0.5 V.

NOTE: These rules apply to a scope with a direct probe. If a low-capacitance probe is used, the reading must be multiplied by the attenuation factor of the probe.

The volts/div. control is frequently a decade-type attenuator; that is, it switches the voltage-range calibration by a factor of 10. For this type of attenuator the calibration marker settings might read: 0.01 V, 0.1 V, 1.0 V, 10 V, and so on.

Many triggered oscilloscopes contain a warning light which goes on when the inner control is not set to its calibrated position; that is, when the scope is uncalibrated. Measurements cannot be accurately made when this light is on. Before measurement, the inner control must be turned completely clockwise to its calibrated position. The warning light will then go out and the oscilloscope is again calibrated.

Triggered oscilloscopes have a self-contained voltage calibrator for checking the calibration of the volts/cm selector. The switch-type calibrator delivers a measured voltage at its output terminals. The measured outputs usually correspond to the positions of the volts/cm selector. Thus for the volts/cm selector described above, the calibrator would deliver 0.01 V, 0.1 V, 1.0 V, and so on, depending on the switch setting of the calibrator.

To check the calibration accuracy of the volts/cm selector, the technician would apply the oscilloscope probe to the calibrator output jack and note the height of the waveform displayed. Thus, if the calibrator were set for 1.0 V output, the calibrator signal should be deflected 1 cm in height, when the volts/cm selector is set to 1 V/cm.

Nontriggered Oscilloscopes. This type of scope can be easily calibrated for use as an ac voltmeter by proper setting of the two vertical controls, namely the *vertical-gain selector* (or simply vertical selector), and the vertical-gain vernier (frequently labelled vertical gain). It is possible to adjust these controls for any desired volts/in. setting within the capabilities of the oscilloscope.

The vertical-gain vernier is a continuously variable control, which in conjunction with the switch-type vertical selector can be set to a desired vertical sensitivity. For example, suppose a 1-V calibration signal is delivered to the input leads of an oscilloscope, and the vertical switch selector is set to the position labelled ×1. The vertical vernier can now be varied until the height of the signal on the screen is, say, 1 in. For these *specific settings* of the two input attenuator controls, the deflection sensitivity of the oscilloscope is now 1 V/in.

It may also be possible to find another position of the two controls for which the deflection sensitivity of the oscilloscope is, say, 3 V/in.

The vertical-gain selector, where found, is usually a calibrated divider or attenuator. The range "factor" is frequently 10. Thus, if the vertical-gain selector and vertical-gain vernier have been set for a vertical sensitivity of 10 V/in., switching to the next "range" lower (i.e., the range where less signal voltage is required to give a similar deflection) will result in a deflection sensitivity of 1 V/in. Switching back to the original range and then going one range higher will result in a deflection sensitivity of 100 V/in.

This characteristic of the vertical-gain selector makes it fairly simple to calibrate the scope for ac voltage measurements. As a matter of fact, it is possible to make almost all ac circuit voltage measurements by calibrating the scope at just two points, a deflection setting of 1 V/in. and a deflection setting of 2 V/in.

To eliminate the need for using a tape measure to determine the height of the waveform, scope manufacturers use a graticule, an etched faceplate placed over the screen of the cathode-ray tube. The etching is in the form of graph markings (Fig. 42-1b). The heavy horizontal and vertical lines are ½ or 1 in. apart. The lighter graph lines divide the space between heavy lines into equidistant markings. Some etched faceplates use cm markings (Fig. 42-1a). In that case, calibrations are read in terms of volts/cm rather than volts/in.

This graphed faceplate is useful for ac measurements. If the vertical-gain controls have been set for a deflection range of 10 V/in., and if there are five spacings in the inch, each spacing counts 2 V.

The procedure for using the nontriggered scope for ac measurements then requires that it be calibrated at several points, using a known voltage source. The setting of the vertical-gain vernier and vertical-gain selector is noted for each of these calibrated points. The vertical-gain vernier is left at the determined setting and is not changed. However, the vertical selector may be switched from range to range. When it is switched, its divider factor must be considered in determining the volts/in. sensitivity. The unknown ac voltage to be measured is applied to the vertical input of the oscilloscope and

Fig. 42-2. Sine wave on scope screen.

Fig. 42-3. Collapsing scope trace on nontriggered scope, to measure vertical waveform.

the height of the waveform deflection is measured. This measurement in inches or fraction of an inch is then converted into volts by multiplying inches by the volts/in. factor for which the scope has been calibrated.

For ease in measuring the height of the waveform, the horizontal trace may be narrowed or collapsed completely by setting the horizontal-gain control to zero. The horizontal and vertical centering controls may also be varied to orient the waveform. This facilitates vertical height measurements.

Figure 42-2 shows a sine wave whose peak-to-peak amplitude must be measured. Only the heavy horizontal lines, spaced 1 in. apart, are shown. The scope has been calibrated for 10 V/in. The horizontal trace is collapsed so that only a vertical line is seen. The ends of this line correspond to the peaks of the waveform A and B. The line is then centered as in Fig. 42-3. It is evident now that it is 2 in. high. Therefore, the voltage being measured is 20 V peak-to-peak.

NOTE: Scope calibration and measurement is usually peak-to-peak. The following is a sample calibration procedure. Assume a known voltage source of 25 V peak-to-peak for calibration and a range factor of 10 for the vertical-gain selector. The calibration voltage is fed into the scope binding posts labeled "vertical input." Since it is desired to obtain a convenient volts/in. relationship, the vertical vernier and vertical-gain selector are set up for 2.5 in. of vertical waveform deflection. The vertical-vernier (continuously variable) control is set at least 20 percent up from its minimum position when possible. The positions of the two controls are noted (see Fig. 42-4) and recorded for future use. See line 1, Table 42-1.

It is now a simple matter to complete the table for various settings of the vertical-gain selector. The vertical-gain selector is shifted one position and the effect upon the height of the waveform is noted. If it becomes larger, it indicates greater overall scope sensitivity. Thus, less signal voltage is required for the same 1 in. of deflection. If this

TABLE 42-1.

Known Calibrating Voltage	Deflection, inches	Vertical Setting		Volts/inch, A/B
		Vernier	Selector	
25	2.5	3.4	10	10
		3.4	1	1
		3.4	100	100

is the case the new selector position and new sensitivity position are noted as in line 2 of the table. If the opposite effect is noted upon the height of the waveform, the table is filled in as in line 3. On all ranges, however, the vernier position remains unchanged, for this particular calibration.

Voltage Calibration Sources for Nontriggered Scopes

Internal. Most oscilloscopes provide a self-contained measured ac voltage source for calibration purposes. This is usually a sampling of the line voltage and may be either fixed or adjustable. Some oscilloscopes have facilities for automatically feeding a measured voltage to the vertical input terminals by means of a function switch set to a "calibrate" position. On other oscilloscopes, the voltage is available at a test jack on the front panel. If a variable voltage is provided, either it is metered or the value may be read from the calibrated dials associated with the "calibration" level control. Whatever the calibration voltage, its peak-to-peak value will be indicated on the oscilloscope panel, or in the oscilloscope instruction manual.

External. For nontriggered oscilloscopes which do not have self-contained facilities, external voltage calibrators are available. These are line-derived adjustable voltage sources. The peak-to-peak output voltage level may be read on a self-contained meter or from calibrated dials. The output of the voltage calibrator is fed to the vertical input of the oscilloscope for calibration.

If an external voltage calibrator is not available, a sine-wave generator and an accurately calibrated

Fig. 42-4. Vertical attentuator (gain) controls on nontriggered scope.

ac voltmeter can be used. The sine-wave generator frequency control is set at 1,000 Hz, and the output control is set for a specified rms voltage as measured by the ac voltmeter used here as a standard. The rms value is then converted to peak-to-peak by multiplying by 2.82. This measured signal is fed to the input leads of the nontriggered scope. Scope calibration is then accomplished as previously described.

SUMMARY

1. Either a triggered or a nontriggered oscilloscope can be used for ac voltage measurements.
2. Triggered oscilloscopes contain calibrated signal amplifiers. These are used directly in the measurement of ac voltage waveforms and dc voltages.
3. Before a triggered oscilloscope is used to measure voltage, its calibration factors, volts/div. setting, should be checked by applying the measured calibration voltages available on the oscilloscope to the input terminals of the scope. If they check, it is ready to make measurements.
4. Nontriggered oscilloscopes are not calibrated and must be calibrated if they are to be used as voltmeters.
5. An accurate, known voltage source, whether provided by the oscilloscope calibrator or by an external calibrator, is used as an input to the nontriggered scope.
6. The vertical-gain selector of the nontriggered scope is switched to a desired setting, and the gain vernier is then adjusted to the desired height for the required volts/in. factor. The position of the vernier is marked on the oscilloscope. It must not be moved from this calibrated setting. However, the vertical-gain selector can be switched to its various positions, and it acts as a range multiplier. The scope is now calibrated for signal-voltage measurements.
7. The signal voltage is then applied to the vertical

input of the calibrated triggered or nontriggered oscilloscope. The vertical height of the signal is measured and multiplied by the volts/cm or volts/in. factor to which the oscilloscope is set. The result is the required voltage of the signal waveform being measured.

8. Oscilloscope signal amplifiers are normally calibrated to make peak-to-peak measurements.
9. There are instruments called voltage calibrators which may be used as external calibration sources. They deliver a measured signal voltage for calibration purposes.
10. For the nontriggered scope, if neither an internal nor an external calibrator is available, an ordinary sine-wave generator may be used, in conjunction with an accurate ac voltmeter. A measured 1,000 Hz signal can be used as the calibration source.

SELF-TEST

Check your understanding by answering these questions:

1. AC voltages _____ (can/cannot) be measured with an oscilloscope.
2. For peak-to-peak waveform measurements, the oscilloscope _____ amplifiers must be _____.
3. Triggered-type oscilloscopes _____ (do/

do not) have calibrated vertical amplifiers.
4. The volts/cm selector setting on a triggered scope determines the _____ factor by which the signal _____ of the waveform must be multiplied to get the voltage of the waveform.
5. A nontriggered scope has been calibrated for 10 V/in. on the ×10 range of the vertical-gain selector. At this setting the height of a waveform is 2.6 in. The voltage of the waveform is _____ V.
6. For the oscilloscope in question 5, the position of the vernier control is kept at the same point as it was when calibrated, but the vertical-gain selector is switched to the ×100 position. The height of a waveform on the screen is 1.5 in. The waveform voltage is _____ V.
7. Calibration voltages are available on most oscilloscopes, whether they are triggered or nontriggered. _____ (true/false)
8. If an oscilloscope does not have self-contained facilities for voltage calibration, an _____ _____ may be used.

MATERIALS REQUIRED

■ Power supply: Known ac voltage source of 18 V peak-to-peak; fused line cord
■ Equipment: Oscilloscope with direct probe, AF signal generator, electronic voltmeter or VOM
■ Resistors: ½-W 5,100-, 10,000-, and 15,000-Ω

PROCEDURE

NOTE: If you have an oscilloscope with calibrated vertical amplifiers, follow the procedural steps relating to that instrument. If your scope does not have calibrated vertical amplifiers, follow the steps relating to that instrument.

Checking Calibration Accuracy of Oscilloscope with Calibrated Vertical Amplifiers

1. Locate the calibration voltage jack and the calibration-voltage selector switch on your oscilloscope. Set the vertical amplifier-gain control on "calibrated."
2. Connect the direct probe of the oscilloscope to the calibration voltage jack and check the accuracy of the volts/cm selector on each of its ranges. Record the results of your check in Table

42-2, identifying the volts/cm range, the calibration voltage used to check that range, and the height of the calibration waveform on the screen. If your oscilloscope is not properly calibrated, notify your instructor. If calibration is proper, proceed to step 6.

Calibrating Oscilloscope with Noncalibrated Vertical Amplifiers

Calibration Using Internal Calibrator

3. Adjust oscilloscope controls for viewing the calibration voltage.
4. Apply the calibration voltage to the input terminals and adjust the controls for a display of

TABLE 42-2.

Volts/cm Setting	Calibration Voltage Volts, peak-to-peak	Height of Calibration Voltage on Screen

Fig. 42-5. AC voltage across divider network.

two cycles. Set the calibration level controls (if any) and the vertical-gain selector and vertical-gain vernier so that the deflection sensitivity of the oscilloscope is 1 V/in. on the ×1 vertical range. In Table 42-3 record the level of the calibration voltage and the settings of the vertical-gain selector and vernier for 1 V/in. If the vernier does not have a numbered scale, mark its position with a pencil.

Next, adjust the calibration level controls and vertical-gain selector and vernier for a deflection sensitivity of 2 V/in. In Table 42-3 record the level of the calibration voltage and the settings of the vertical-gain selector and vernier for 2 V/in. If the vernier does not have a numbered scale, mark its position.

Calibration Using External Calibrator, or AF Sine-Wave Generator and AC Voltmeter

NOTE: If an external voltage calibrator is used, set its output controls first for 1 V, then for 2 V as required in Table 42-3. If an AF sine-wave generator is used, set its output controls so that the 1,000 Hz

sine wave as measured on an ac voltmeter is first 0.354 V rms (corresponding to 1 V peak-to-peak), then 0.708 V rms (corresponding to 2 V peak-to-peak).

5. Repeat steps 3 and 4 recording your results in Table 42-3. Then proceed to step 6.

Measuring AC Voltages with Oscilloscope

6. Remove the vertical input cables from the calibrating voltage source and apply 18 V peak-to-peak (as measured with an ac voltmeter) across the circuit of Fig. 42-5. Using the oscilloscope calibrated at 2 V/div., measure, and record in Table 42-4, the peak-to-peak voltages from A to G, B to G, and C to G. Compute, and record in Table 42-4, the peak-to-peak voltages appearing at the measured points. Show your computations.

7. Reset the vertical controls for a deflection sensitivity of 1 V/div. Again measure and record in Table 42-4 the voltages at the indicated test points.

8. With the oscilloscope calibrated at 1 V/div., measure, and record in Table 42-4, the maximum signal voltage delivered by the audio-signal generator set at 1,000 Hz. If possible, also measure and record the minimum output of the AF generator at 1,000 Hz.

TABLE 42-3. Oscilloscope Calibration

Calibration Source	Calibration Voltage Level	Volts/in.	Settings of Vertical	
			Gain Selector	Gain Vernier
Internal		1		
		2		
External		1		
		2		

TABLE 42-4. Voltage Measurements

Test Points	Computed Volts peak-to-peak	Measured Volts peak-to-peak	
		Scope Calibrated 1 V/div.	Scope Calibrated 2 V/div.
A to G			
B to G			
C to G			

Audio Oscillator Output	Maximum Volts peak-to-peak	Minimum Volts peak-to-peak

Measuring DC Voltages with Oscilloscope

9. If you are using an ac-dc oscilloscope, switch to dc. Remove the 18 V ac source from Fig. 42-5, and connect a dc voltage source across the points *AG* of Fig. 42-5. Set the output of the dc supply at 30 V, as measured with an EVM. You now have a dc voltage divider. Set the calibrated vertical controls on your scope at 5 V/div.

10. Center the trace vertically on the screen. This is the zero reference line. With the oscilloscope measure and record in Table 42-5 the dc voltage across each set of test points listed. Now measure the same voltages with an EVM, and record your measurements in the appropriate column of Table 42-5.

QUESTIONS

1. (*a*) Does your oscilloscope have calibrated or noncalibrated vertical amplifiers? (*b*) Is it a triggered or nontriggered scope? (*c*) Is it a wideband scope?
2. What are the markings (settings or calibrations) of the vertical-gain selector found on your scope? Which of these is the most sensitive, that is, which requires least voltage for 1 division (cm or inch) of vertical deflection?
3. How do the computed and measured values in Table 42-4 compare for:
   (*a*) Scope calibration of 1 V/division?
   (*b*) Scope calibration of 2 V/division?
4. List the advantages, if any, in using the scope as an ac voltmeter.
5. What are the disadvantages of using the scope as an ac voltmeter?
6. If you were unable to get a measurement of the minimum-level signal voltage delivered by the audio-signal generator, explain why. Can you suggest a method for measuring the minimum-level voltage?
7. How would you calibrate the oscilloscope for rms measurements?
8. Can you convert the peak-to-peak measurements in Table 42-4 to rms measurements? How?
9. For what type of waveforms do the formulas relating peak, rms, and average voltage apply?
10. Which provided a more accurate dc voltage measurement, the scope or the voltmeter? Justify your answer by referring to your readings in Table 42-5 and to the computed voltages.

Answers to Self-Test

1. can
2. vertical or signal; calibrated
3. do
4. calibration; height
5. 26
6. 150
7. true
8. external calibrator

TABLE 42-5. DC Voltage Measurements

Test Points	Voltage Measured, V	
	Oscilloscope	Voltmeter
A to G		
B to G		
C to G		

43

OBJECTIVES

1. To measure the frequency of an ac sine wave by means of Lissajous patterns
2. To check the accuracy of the frequency calibration of an AF generator

INTRODUCTORY INFORMATION

Frequency Measurements

Oscilloscopes are used to measure ac voltages and to observe ac waveforms. They may also be employed, if they have a calibrated time base, to measure the period of a waveform. Once the period t is known, the frequency F of the waveform may be calculated by means of the formula

$$F = \frac{1}{t}$$

However, whether an oscilloscope does or does not have a calibrated time base, it may be used to measure the frequency of ac signals by the method of Lissajous patterns.

With this technique, the internal time base or "sweep" is switched *out*. The oscilloscope is set to Ext. sweep. A signal from an accurately calibrated generator, the frequency "standard," is applied to the horizontal input of the oscilloscope. The signal whose frequency F_V we wish to measure is applied to the vertical input. The frequency F_H of the standard signal generator is then varied manually until a stable pattern appears on the screen. This is called a Lissajous pattern. The ratio of the two signal frequencies may be readily determined from their characteristic pattern. Once the

ratio is found, the unknown frequency can be calculated.

Frequency ratios are given in terms of F_V/F_H. Assume that

$$\frac{F_V}{F_H} = k \qquad (43\text{-}1)$$

Then
$$F_V = kF_H \qquad (43\text{-}2)$$

If the two signals are sine waves, the Lissajous patterns are similar to those in Fig. 43-1. If the sine waves have the same frequency, that is, if $F_V = F_H$,

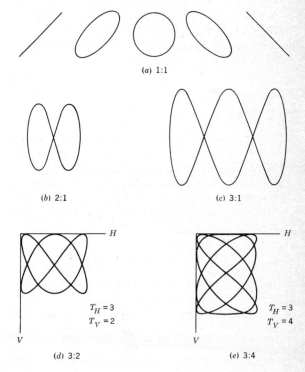

(a) 1:1

(b) 2:1

(c) 3:1

(d) 3:2 $T_H = 3$ $T_V = 2$

(e) 3:4 $T_H = 3$ $T_V = 4$

Fig. 43-1. Characteristic Lissajous patterns.

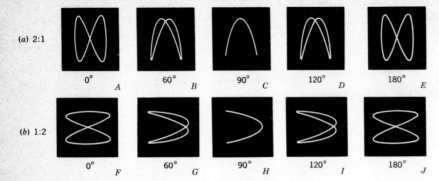

(a) 2:1

0°   A   60°   B   90°   C   120°   D   180°   E

(b) 1:2

0°   F   60°   G   90°   H   120°   I   180°   J

Fig. 43-2. Lissajous patterns for two signals varying in phase.

the resulting pattern shown in Fig. 43-1a will be either a straight line, an ellipse, or a circle, depending on the phase relationship of the two signals. In practice the pattern on the screen may vary slowly, assuming in turn each of these shapes. The reason for this variation is that the frequency of the signal generator drifts slightly, resulting in a phase change.

The pattern in Fig. 43-1b is characteristic of a 2:1 ratio, that is, $F_V/F_H = 2$, and $F_V = 2F_H$. Here the unknown signal frequency is twice that of the standard frequency. Other patterns are shown in Fig. 43-1c, d, and e.

The ratio of the frequencies may be determined by drawing a horizontal (H) and a vertical (V) line tangent to the loops in the closed pattern, as in Fig. 43-1d and e, and counting the number of points of tangency (T_H) to the horizontal line, and the number of points of tangency (T_V) to the vertical line.

The ratio of the two frequencies is

$$\frac{F_V}{F_H} = \frac{T_H}{T_V} \qquad (43\text{-}3)$$

and

$$F_V = F_H \frac{T_H}{T_V} \qquad (43\text{-}4)$$

In Fig. 43-1d, $T_H = 3$, $T_V = 2$, and $F_V = F_H \times {}^3\!/_2$. In Fig. 43-1e, $T_H = 3$, $T_V = 4$, and $F_V = F_H \times {}^3\!/_4$.

As in the case of Fig. 43-1a, signals whose frequency and phase relative to each other are varying slightly will produce Lissajous patterns which will vary. In Fig. 43-2a, $F_V = 2F_H$, and in Fig. 43-2b, $F_V = {}^1\!/_2 F_H$. In each case we see that the pattern depends on the phase relationship between F_V and F_H.

Frequency Calibration of an AF Generator

The accuracy of the frequency calibration of AF sine-wave generators may be checked against stan-

dard audio signals broadcast by the Bureau of Standards, against secondary standard-frequency sources which may be available in the laboratory, and against the power-line frequency. Regardless of the signal source, an oscilloscope can serve as the indicator if the Lissajous-pattern technique is employed. In this experiment you will check the calibration accuracy of an AF sine-wave generator, using the power-line frequency as a standard for comparison.

SUMMARY

1. The frequency of periodic sine waves may be measured with an oscilloscope by means of Lissajous patterns.
2. To employ this technique, the time base is switched out. A signal F_H from an accurately calibrated sine-wave generator, the frequency standard, is applied to the *horizontal* input of the oscilloscope. The signal whose frequency we wish to measure, F_V, is applied to the vertical input of the oscilloscope.
3. The frequency F_H of the standard generator is varied until a stable pattern appears on the screen. This is called a Lissajous pattern (Figs. 43-1b, c, and Fig. 43-2).
4. The ratio of the frequencies F_H and F_V may be determined by drawing horizontal and vertical line tangents to the loops in the closed pattern and counting the number of points of tangency (T_H) to the horizontal line, and the number of points of tangency (T_V) to the vertical line (Figs. 43-1d, e).
5. The ratio of the two frequencies is

$$\frac{F_V}{F_H} = \frac{T_H}{T_V}$$

6. We read the standard frequency F_H from the

standard signal generator, and substitute F_H, T_H, and T_V in the formula

$$F_V = \frac{T_H}{T_V} \times F_H$$

7. F_V is the required frequency.
8. If $F_H = F_V$, then a pattern changing continuously from a straight line to an ellipse to a circle will be seen (Fig. 43-1a).
9. The Lissajous-pattern method may be used for checking the calibration accuracy of an AF sine-wave generator.

SELF-TEST

Check your understanding by answering these questions:
1. The Lissajous-pattern method of checking the frequency of a sine wave requires the use of a calibrated time base. _____ (true/false)
2. To use the method of Lissajous patterns requires the use of an accurately _____ sine-wave generator which serves as the _____ standard.
3. The standard generator is connected to the _____ input terminals of the oscilloscope; the sine wave whose frequency is unknown is applied to the _____ input terminals.
4. If the Lissajous pattern is a circle, straight line, or ellipse, then the frequency of the standard generator and the unknown signal are _____.
5. If the number of points of horizontal tangency $T_H = 5$, the number of points of vertical tangency $T_V = 2$, and the standard frequency is 100 Hz, then the unknown frequency $T_V =$ _____ Hz.

MATERIALS REQUIRED

■ Equipment: Oscilloscope; AF sine-wave generator; filament transformer with 6.3-V secondary (source on regulated power supply may be used); fused (1 A) line cord

PROCEDURE

1. Connect the circuit of Fig. 43-3. T is a 6.3-V filament transformer whose primary winding receives 120 V at 60 Hz (or 50 Hz) line voltage. Apply the 6.3 V of the secondary to the "horizontal" input of the oscilloscope.

NOTE 1: The line-isolated 6.3-V filament source of a power supply may be used instead of a filament transformer.

NOTE 2: Some oscilloscopes contain facilities for "line" sweep. In the "line" position of the sweep-selector switch, a line-derived sinusoidal voltage is automatically applied to the horizontal input of the oscilloscope. For this type of oscilloscope an external 6.3-V sine wave is not required.

2. Adjust the oscilloscope controls for proper viewing. Set the oscilloscope-sweep speed control for Ext. sweep. Adjust the horizontal-gain control for 4 units of deflection.
3. Apply the output of an AF sine-wave generator to the vertical input of the oscilloscope. Set the range switch of generator so that the power-line frequency falls within the set generator range. Adjust the output controls of the AF generator and the vertical-gain control of the oscilloscope for 4 units of vertical deflection.
4. Vary the frequency of the generator until a single circular pattern appears. This pattern may vary from a circle to an ellipse to a straight line. Draw this pattern in Table 43-1. If the AF sine-wave generator has a zero control for correcting generator frequency calibration, zero the generator at the power-line frequency. Record the frequency F_V as shown on the dial of the generator. Record the number of points of horizontal tangency T_H and vertical tangency T_V.

Fig. 43-3. Circuit for checking frequency calibration.

TABLE 43-1. Lissajous-pattern Frequency Measurements

Step	Lissajous Pattern	No. of Points of Tangency T_H	No. of Points of Tangency T_V	Generator Frequency F_V Dial Setting	Generator Frequency F_V Computed
4					
5		1	2		
		3	2		
		5	3		
		2	1		
		3	1		
		4	1		

Using Eq. (43-4), compute and record the frequency F_V of the signal generator.

5. Vary the frequency of the generator until a pattern appears showing one point of horizontal tangency $T_H = 1$ and two of vertical tangency $T_V = 2$. Record this pattern and the frequency calibration shown on the dial of the generator. Compute the frequency F_V and record in Table 43-1. Repeat for each of the T_H/T_V ratios shown in the table.

QUESTIONS

1. What determines the accuracy of frequency measurements when using Lissajous patterns?

2. From the data in Table 43-1, comment on the calibration accuracy of the AF sine-wave generator.
3. Draw the Lissajous pattern if the frequencies applied to the oscilloscope are in the ratio $F_V/F_H = 4/3$.

Answers to Self-Test

1. false
2. calibrated; frequency
3. horizontal; vertical
4. equal
5. 250

44

Characteristics of an Inductance

OBJECTIVES

1. To observe experimentally the effect of an inductance on current in a dc and ac circuit
2. To verify experimentally the formula for inductive reactance

$$X_L = 2\pi F L$$

INTRODUCTORY INFORMATION

Inductance and Reactance of a Coil

In dc circuits, resistors limit the amount (amplitude) of current. Resistors also oppose current in ac circuits. In addition to resistors, reactive components, namely, inductors and capacitors, impede currents in ac circuits.

In our experiments with magnetism we observed that a magnetic field exists about a coil carrying current. We observed also that a voltage is induced in a conductor when the conductor is *cut* by a magnetic field. By Lenz's Law the polarity of the induced voltage in the conductor is such that any current resulting from it will produce its own magnetic field which will oppose the effect of the original magnetic field.

In the circuit of Fig. 44-1 a sine wave of voltage V causes current through the coil to increase and decrease sinusoidally. This changing current creates a moving magnetic field about the coil which therefore *cuts* the windings of the coil, inducing a voltage V' in these windings. By Lenz's Law this voltage is in opposition to the source voltage V, and can therefore be termed a counter emf. When V is positive, V' is negative but smaller than V. The resultant emf which powers the circuit is $V -$

V'. It is evident therefore that because of the counter emf the ac current in the coil is less than it would have been if V were a dc source and a steady-state current were flowing. This property of the coil which opposes a *changing* current is termed the *inductance* of the coil.

The unit of inductance is the henry (H), named in honor of the scientist Joseph Henry. The number of henrys in an inductor can be measured by means of an inductance bridge, an instrument with which you are not yet familiar. Inductance (L) is that characteristic of a coil which *opposes* a change in current. The *amount* of opposition offered by an inductor is called inductive reactance (X_L). Inductive reactance is measured in ohms.

The inductive reactance of a coil is not constant but is a variable quantity. X_L depends on the inductance L and on the frequency F of an alternating current. X_L may be computed from the formula

$$X_L = 2\pi F L \qquad (44\text{-}1)$$

where π is the constant 3.14, F is the frequency in hertz of the alternating current in the inductance, and L is the inductance in henrys.

Some important facts may be deduced from Eq. (44-1). The first is that X_L varies directly with frequency, that is it is directly proportional to F. This

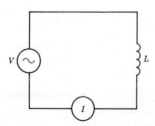

Fig. 44-1. Effect of an inductance on current in an ac circuit.

Fig. 44-2. Variation of X_L with F for a 1.59-H inductor.

variation may be shown graphically by a specific example.

Problem

Draw a graph of X_L versus F for an inductor $L = 1.59$ H.

Solution. Find X_L at each of a series of frequencies. Thus, when

$$F = 0$$
$$X_L = 2\pi FL = 2(3.14)(0)(1.59) = 0$$

when $F = 100$ Hz,

$$X_L = 6.28(100)(1.59) = 1,000 \ \Omega$$

Similarly, it can be shown that when

$$F = 200 \text{ Hz}, X_L = 2,000 \ \Omega$$
$$F = 300 \text{ Hz}, X_L = 3,000 \ \Omega$$
$$F = 400 \text{ Hz}, X_L = 4,000 \ \Omega$$
$$F = 500 \text{ Hz}, X_L = 5,000 \ \Omega, \text{ etc.}$$

Figure 44-2 shows the variation of X_L versus F for $L = 1.59$ H. It is a straight-line graph. Hence we say that the relationship between X_L and F is a *linear one.*

A second fact which may be deduced from Eq. (44-1) is that for direct current, that is, for $F = 0$, $X_L = 0$. This fact is consistent with the definition of inductance, as that characteristic of a coil which opposes a *change* in current. A steady-state direct current involves no change. Hence an inductance has no effect on a dc current which is constant in amplitude, that is, $X_L = 0$ for such a direct current.

Equation 44-1 also defines the relationship between X_L and L. If F is kept constant, X_L increases or decreases as L is increased or decreased, respec-

tively. Again, if X_L were plotted as a function of L, a linear graph would result.

Resistance of a Coil

An inductor consists of a number of turns of insulated wire wound around a core. There may be only several turns, or there may be thousands of turns. The more turns there are on a particular core, the greater is the inductance of the coil. The diameter of the wire used in winding a coil depends on the maximum current which that coil will be required to pass. The larger the diameter of the wire, the greater the current capabilities of the coil. A smaller-diameter wire will sustain less current. If more current is permitted to flow in a coil than that for which it is rated, the coil will overheat and may burn away the insulation around the windings, thus shorting windings together. Or an overheated coil may burn out, that is, the wire may open. In either event the coil will become defective.

In electronics we generally deal with relatively small currents. Hence, very fine diameter wire is normally used in the construction of coils. Since the resistance of a wire is directly proportional to its length, the resistance of inductors consisting of many turns of fine wire will be relatively high. Thus, the characteristics of inductance *and resistance* are associated with an inductor. Though the resistance of a coil cannot be separated from the inductance, a schematic representation of a coil with inductance L and resistance R (Fig. 44-3) shows these two characteristics as though they were lumped constants, acting in series. The resistance can be measured with an ohmmeter. A coil may therefore be represented by its series equivalent form (Fig. 44-3).

The resistance associated with the length of wire in the winding suggests that one method of testing an inductor is to place an ohmmeter across the two terminals and compare the measured with the rated resistance. If they are the same, it may be assumed

Fig. 44-3. Series-equivalent form of an inductance.

Fig. 44-4. Circuits used to determine X_L.

in series with L, as in Fig. 44-4b. In a series circuit current is the same everywhere. Hence the current in R is the same as the current in L. The voltage V_R is then measured across R. Current is next determined by substituting the measured value of V_R and the known value of R in the equation

$$I = \frac{V_R}{R} \qquad (44\text{-}4)$$

Next, V_L is measured and X_L computed by substituting the values of V_L and I in Eq. (44-3).

that the winding is continuous, that is, it is neither open nor shorted. If the resistance is infinite, the inductor is open. If the resistance is very much smaller than the rated value, say 20 Ω instead of 500 Ω, it may be assumed that a large part of the coil winding is shorted. In practice, resistance checks of an inductor are made to determine if it is continuous or open. Special instruments are required to identify coils with shorted turns.

Measurement of X_L

The reactance of a coil (X_L) may be determined by measurement. In the circuit of Fig. 44-4, a sinusoidal voltage V causes a current I to flow. Ohm's law extended to ac circuits states that the voltage V_L across the coil is $V_L = I \times Z$. Where the resistance of the coil is small compared to X_L, $Z = X_L$ and

$$V_L = I \times X_L \qquad (44\text{-}2)$$

This equation may be written in the form

$$X_L = \frac{V_L}{I} \qquad (44\text{-}3)$$

Equation (44-3) can be used to determine the X_L of a coil at a given frequency F. This is accomplished by using an ac voltmeter to measure the voltage V_L across L, and by measuring the current I with an ac ammeter. The measured values of V_L and I are then substituted in Eq. (44-3), and X_L is computed.

Note that a slight error is introduced by this procedure, for L has resistance in addition to inductance. However, if the value of X_L is appreciably larger than the resistance R of the coil, R may be ignored.

An alternate method for measuring the current I does not require the use of an ammeter. Instead of the meter M, a resistor R of known value is placed

SUMMARY

1. Inductance (L) is that characteristic of a coil which opposes a *change* in current. The unit of inductance is the *henry* (H).
2. The amount of opposition that an inductance offers to a changing current is called inductive reactance (X_L), or simply reactance. Reactance is measured in ohms.
3. The inductive reactance of a coil is not a constant, but is directly proportional to the frequency (F) of the changing current. It is also directly proportional to the value of the inductance.
4. Inductive reactance may be computed from the formula

$$X_L = 2\pi FL$$

where X_L is in ohms, F in hertz, and L in henrys.
5. Every coil has some resistance R associated with it. The continuity of a coil winding may be determined by measuring the resistance of the coil. If the resistance measured is infinite, the coil winding is open.
6. Only the resistance of an inductance will limit dc current in a coil. The inductance of a coil will not affect dc current.
7. The ac voltage across a coil V_L is equal to the product of the ac current in the coil and the X_L of the coil, that is,

$$V_L = I \times X_L$$

8. If the ac voltage V_L across a coil and the ac current I in the coil are known, the inductive reactance of the coil may be computed from the equation

$$X_L = \frac{V_L}{I}$$

This equation assumes that the resistance R of the coil is small compared with the reactance of the coil.

SELF-TEST

Check your understanding by answering these questions:
1. Inductance of a coil is measured in ohms. _____ (true/false)
2. The _____ of a coil is directly _____ to the _____ of the current in the coil, if the L of the coil remains constant.
3. If $L = 2.5$ H and $F = 120$ Hz, $X_L =$ _____ Ω.
4. If the frequency F is doubled while L is held constant, X_L is _____.
5. The graph of X_L versus F is a _____ _____ if L is held constant.
6. The resistance R of a coil is associated with the resistance of the wire in the _____ of the coil.

7. If the ac voltage V_L across a coil is 4.5 V and the current I in the coil is 0.0225 A, then $X_L =$ _____ Ω.
8. If $F = 200$ Hz, and $X_L = 15,100$ Ω, then $L =$ _____ H.

MATERIALS REQUIRED

- Power supply: Line-isolation transformer and variable autotransformer operating from 120-V ac line
- Equipment: Variable regulated dc source; 0–25 mA ac meter or a multirange ac milliammeter covering this range; oscilloscope; EVM; 0–25 mA dc meter; variable-frequency AF sine-wave generator
- Resistor: ½-W 1,000-Ω; R as required for step 2
- Inductors: 8 H at 50 mA whose $R = 250$ Ω; 30 mH
- Miscellaneous: SPDT switch, SPST switch; fused line cord

PROCEDURE

Effect of an Inductance on DC and AC Current

1. Measure, and record in Table 44-1, the resistance R of the 8-H choke.
2. Connect the circuit of Fig. 44-5a. V is a variable regulated dc source. L is the 8-H choke; R a resistor equal in value to the resistance of the 8-H choke; S_2 a single-pole double-throw switch (which selects either L or R to complete the circuit); M a 0–25 mA dc meter; V_1 an EVM set on dc voltage; S_1 a switch to apply voltage to the circuit. S_1 is open. V is set for *zero* volts.
3. Throw S_2 to position 1. R is in the circuit. Close S_1. **Power on.** Adjust V until M measures 20 mA. Measure and record the voltage V across the resistor. *Do not vary the supply voltage.*
4. Throw S_2 to position 2. L is in the circuit. The voltage across L should be the same as across R. Measure the current I in L and record in

Table 44-1. **Power off.** Disconnect and remove the dc source, the milliammeter M, and the electronic voltmeter.
5. Connect the circuit of Fig. 44-5b. V is an ac source consisting of a variable autotransformer plugged into an isolation transformer. M is a 0–25 mA ac meter, V_1 is an EVM set on ac voltage. The other components are the same as in the preceding circuit. S_1 is *off*. V is set at 0 V.
6. Throw S_2 to position 1, placing R in the circuit. **Power on.** Adjust V until M measures 20 mA. Measure and record the voltage V across the resistor. *Do not vary the supply voltage.*
7. Throw S_2 to position 2, placing L in the circuit. Readjust V if necessary to place the same voltage across L as there was across R in step 6. Measure and record the current I in L.

TABLE 44-1.

Resistance, ohms	S_2 Position	Component in Circuit	dc		ac	
			V, volts	I, amperes	V, volts	I, amperes
L	1	R		0.02		0.02
R	2	L				

Fig. 44-5. Experimental circuit to demonstrate the effects of an inductance on direct and alternating current.

Reactance (X_L) of L at Line Frequency

8. The circuit of Fig. 44-5b is still connected. S_2 is in position 2 and L is in the circuit. M is the 0–25 mA range of the meter. Adjust V in turn for each of the currents shown in Table 44-2. Measure and record the voltage V at each current level. Compute the V/I ratio at each of the current levels and record as X_L in Table 44-2.

NOTE: The ratio of V/I is really the impedance Z. In this case we will ignore the resistance of L and consider the impedance at line frequency as a pure reactance, X_L. The error introduced by this procedure is small enough to be ignored.

Average the four values of X_L and record. Draw a graph of the volt-ampere characteristic of the inductance at line frequency. (Let V be the horizontal, I the vertical axis.)

Inductive Reactance (X_L) versus Frequency (F)

In this part of the experiment you will determine how X_L varies with frequency. X_L will not be measured directly, but will be computed as the ratio of V_L/I. The current I will not be measured directly, but will be computed from the voltage V_R measured across R, where $I = V_R/R$. The calculation for X_L can be made directly from V_R (the voltage across R), V_L (the voltage across L), and R, by substituting these measured values in Eq. (44-5).

For
$$X_L = \frac{V_L}{I} = \frac{V_L}{\dfrac{V_R}{R}}$$

Therefore
$$X_L = \frac{V_L}{V_R} \times R \qquad (44\text{-}5)$$

9. Connect the circuit of Fig. 44-6. V is a variable-frequency sine-wave generator capable of delivering a 20-V peak-to-peak ac voltage, variable from 10 Hz to 100 kHz. V_1 is an oscilloscope calibrated to measure peak-to-peak volts. Set the signal generator in turn to each of the frequencies shown in Table 44-3. Maintain the output V of the generator at 15 V peak-to-peak. Measure and record the voltage V_L and V_R for each frequency. Compute and record the value of I (V_R/R) in amperes, and the value of X_L

TABLE 44-2.

I, amperes	0.025	0.020	0.015	0.010	Average X_L at Line Frequency (ohms)
V, volts					
$X_L = \dfrac{V}{I}$ (ohms)					

Fig. 44-6. Circuit to determine X_L and variation of X_L versus F.

[use Eq. (44-5)] for each frequency. Show sample computations. Compute also, and record in Table 44-3, the formula value ($2\pi FL$) of X_L. Show a sample computation. Draw a graph of X_L (vertical axis) versus F (horizontal axis) from the data.

QUESTIONS

1. From the data in Table 44-1, is there any significant difference between alternating and direct current in L and R? Identify the difference and explain the reason for it.

2. Refer to the data in Table 44-2. Is X_L the same for each of the currents developed in the circuit? Should it be? Explain any unexpected result.
3. What does the volt-ampere characteristic (graph) tell you about X_L?
4. How does X_L vary with frequency? Substantiate your answer by referring to the data in Table 44-3. Refer also to the graph (Fig. 44-2).
5. Did your experiment verify the formula $X_L = 2\pi FL$? Substantiate your answer by reference to the data.
6. If X_L of a certain inductance is 5,000 Ω at frequency F, what is the X_L of the coil at $2F$? $5F$? nF?
7. Refer to Table 44-3. Are the computed values of I rms or peak-to-peak? Explain.

Answers to Self-Test

1. false
2. reactance; proportional; frequency
3. 1,884
4. doubled
5. straight line
6. winding
7. 200
8. 12 (approx.)

TABLE 44-3.

F, Hz	V, volts peak-to-peak	V_L, volts peak-to-peak	V_R, volts peak-to-peak	I, amperes $\left(\dfrac{V_R}{R}\right)$	X_L (ohms) $\dfrac{V_L}{V_R} \times R$	$2\pi FL$
2,000						
3,000						
5,000						
7,000						
8,000						
9,000						
10,000						

OBJECTIVES

1. To measure, by Lissajous patterns, the phase angle between voltage and current in a pure resistive circuit
2. To measure, by Lissajous patterns, the phase angle between voltage and current in an inductive circuit

INTRODUCTORY INFORMATION

Phase Relations between Voltage and Current in a Coil

In a dc circuit, as the voltage across a resistor increases, the current in R increases. This is also true in a resistive ac circuit, for if a sinusoidal voltage v is applied across R, the instantaneous variations of current i in R follow exactly the instantaneous changes in voltage v. Thus, at the instant v is going through zero, i is going through zero. When v is at maximum, i is at maximum. Voltage and current in R are said to be in phase. This relationship is shown graphically in Fig. 45-1.

In a pure inductive circuit voltage and current are *not* in phase, but current lags voltage by 90°. A graph of voltage and current (Fig. 45-2) shows this relationship. A phase meter or an oscilloscope acting as a phase indicator may be used to verify the phase relationship between voltage and current in an inductor. In this experiment an oscilloscope will be employed.

Phase Measurements by Means of Lissajous Patterns

In a previous experiment Lissajous patterns were used to measure frequency. Lissajous patterns may also be employed to determine the phase relationship between two sine waves of the *same* frequency. As in the case of frequency measurements, one of the signals is connected to the vertical, the other to the horizontal input of the oscilloscope. Internal oscilloscope sweep is disabled by setting the sweep selector switch to Ext.

If the signals have the same phase, the resulting pattern will be a line slanting upward from left to right. The slant angle will depend on the amplitude of the two signals. As the phase angle between the

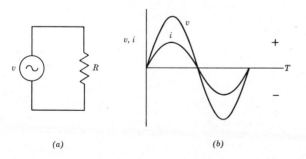

(a) *(b)*

Fig. 45-1. In a resistive circuit, voltage and current are in phase.

Fig. 45-2. In a pure inductive circuit, voltage and current are out of phase, and current lags voltage by 90°.

| 0° | 22.5° | 45° | 67.5° | 90° | 112.5° | 135° | 157.5° | 180° |
| 360° | 337.5° | 315° | 292.5° | 270° | 247.5° | 225° | 202.5° | |

Fig. 45-3. Lissajous patterns for two sinusoidal voltages of the same frequency but differing in phase (*U.S. Government Printing Office*).

two signals changes, the Lissajous figure will vary in the manner shown in Fig. 45-3. Thus as the phase angle increases from 0 to 90°, the line becomes a narrow ellipse, then a broader ellipse, and finally at 90° a circle. As the phase angle increases from 90° to 180°, the ellipse slants in the opposite direction, decreasing from a wide to a narrow ellipse. Finally at 180° it is a straight line slanting up from right to left. For phase shifts between 180° to 360° the Lissajous patterns are repeated in the reverse order. Thus the pattern for 225° is the same as for 135°. For 270° it is the same as for 90°, and so forth.

Figure 45-4 shows how the phase angle θ may be calculated. The pattern must be centered on the X and Y axes of the scope. The maximum Y distance, Y_M from the X axis, and the intercept Y_0 on the Y axis must be measured. The phase angle θ may then be computed by substituting the measured values Y_M and Y_0 in the formula

$$\sin \theta = \frac{Y_0}{Y_M} \qquad (45\text{-}1)$$

Thus
$$\theta = \text{arc } \sin \frac{Y_0}{Y_M} \qquad (45\text{-}2)$$

The method described will determine the phase difference if the following conditions are met:

1. The frequency of the two signals must be the same.
2. The amplitude of the two signals being compared must be the same. If they are not the same in amplitude, the vertical- and horizontal-gain controls of the oscilloscope must be adjusted to effect equal vertical and horizontal deflection.
3. The phase characteristics of the vertical and horizontal amplifiers of the oscilloscope must be the same for the frequency of the signal under study.
4. The waveform on the oscilloscope must be perfectly centered with respect to the X and Y axis.

It is possible to measure the phase relation between voltage and current in a coil by Lissajous patterns. However, an external resistor R must be connected in series with L, as in the circuit of Fig.

45-5. Care must be taken to select a resistor of such a value that the ratio of X_L/R is greater than 10. This circuit will act almost like a pure inductance. Current i in the series circuit of Fig. 45-5 is the same throughout the circuit. Moreover, the voltage v_R across R is in phase with the current i. Hence, by measuring the phase angle between voltage v_r across R, and the applied voltage v, we are in fact measuring the phase angle between v and i in the coil.

NOTE: If the ratio of X_L/R is equal to 10, the error in phase angle introduced by R is approximately 6°. As X_L/R becomes greater than 10, the error decreases.

SUMMARY

1. In a resistive circuit current and voltage (dc and ac) are in phase.
2. In an inductive circuit alternating current and voltage are *not* in phase. If the circuit is a pure inductance, that is, it has no resistance, then current lags voltage by 90°.
3. We can demonstrate this phase difference between ac voltage and current in an inductance by the method of Lissajous patterns, employing the circuit of Fig. 45-5.
4. In the circuit of Fig. 45-5, if $X_L/R > 10$, then the

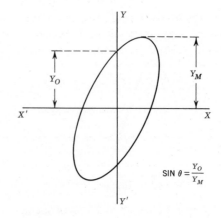

Fig. 45-4. Calculating the phase angle.

Fig. 45-5. Circuit for measuring phase angle between v and i in an inductive circuit.

Fig. 45-6. Circuit for measuring phase angle between v and i in a resistive circuit.

effect of R can be ignored and the circuit treated like a pure inductance.

5. In Fig. 45-5 the voltage v across the circuit is applied to the vertical input of the oscilloscope; the voltage across R (which is in phase with the current i in the circuit) is applied to the horizontal input of the scope. Controls are adjusted to equalize the height and width of the signal, and the waveform is centered about the origin. In a pure inductance the resulting pattern should be a circle, which demonstrates a 90° phase difference.

6. The phase angle between two voltages of the same frequency may also be measured by Lissajous patterns. Here if the resulting pattern is centered about the origin, as in Fig. 45-4, the sine of the phase angle θ is given by the formula

$$\sin \theta = \frac{Y_O}{Y_M}$$

where Y_O is the intercept of the curve on the Y axis, and Y_M is the maximum Y distance from the X axis (that is, Y_M is the projection on the Y axis of the point of *horizontal* tangency to the curve).

SELF-TEST

Check your understanding by answering these questions:

1. The phase angle between ac voltage and current in a pure resistive circuit is _____ degrees.
2. Current leads voltage by 90° in a pure inductance. _____ (true/false)
3. Phase differences between voltage waveforms of the _____ frequency may be measured by Lissajous patterns.
4. In measuring phase by Lissajous patterns, the internal sweep or time base of the oscilloscope must be _____ _____ .
5. In measuring phase the Lissajous patterns must be _____ about the _____ .
6. The phase angle θ of a Lissajous pattern whose y intercept, Y_O = two divisions, where Y_M = four divisions, is _____ degrees.

MATERIALS REQUIRED

- Equipment: Oscilloscope, AF sine-wave generator
- Resistors: ½-W two 1,000-Ω
- Inductor: 30 mH

PROCEDURE

Phase Relations in a Resistive Circuit

1. Connect the circuit of Fig. 45-6. The ac source is a sine-wave generator whose frequency $f =$ 50 kHz, and whose output voltage $v = 15$ V peak-to-peak. The vertical input leads of the oscilloscope are connected across the output

of the generator, points AC, with point C as the common ground. The horizontal oscilloscope leads are connected across R_2 the 1,000-Ω resistor, points BC, hot lead to B. The vertical is the voltage axis; the horizontal the current axis.

2. Disable the oscilloscope internal sweep or time base by setting the sweep selector on Ext.
3. Adjust the vertical- and horizontal-gain attenuators of the oscilloscope for equal vertical and horizontal deflection, respectively. Center the resulting pattern vertically with reference to the X axis and horizontally with respect to the Y axis (that is, the pattern is centered with reference to the origin).
4. Draw the resulting pattern in Table 45-1. Using Fig. 45-3 as a guide, record in Table 45-1 the phase difference between v and i.
5. Set the frequency of the signal generator to 10 kHz, still maintaining 15 V output. Oscilloscope controls remain as set in steps 1 through 3.
6. Draw the resulting pattern in Table 45-1 and record the phase difference between v and i.

Phase Relations in an Inductive Circuit

7. Replace R_1 with a 30-mH choke L as in Fig. 45-5. The oscilloscope is set for measuring the phase difference between v and i in the circuit of Fig. 45-5.
8. Set the frequency of the signal generator to 50 kHz. Maintain output v at 15 V peak-to-peak.
9. Adjust the vertical and horizontal-gain attenuators of the oscilloscope for equal vertical and horizontal deflection, respectively.

NOTE: If you have difficulty obtaining sufficient horizontal deflection, reverse the positions of the vertical and horizontal input leads, that is, connect the hot vertical lead to point B, hot horizontal lead to point A, and ground leads to C.

TABLE 45-1. Phase Relations in a Resistive Circuit

Generator Frequency	Lissajous Pattern	Phase Angle θ degrees
50 kHz		
10 kHz		

TABLE 45-2. Phase Relations in an Inductive Circuit

Generator Frequency	Lissajous Pattern	Y_O	Y_M	Phase angle θ degrees
50 kHz				
10 kHz				

Center the resulting pattern vertically and horizontally.

10. Draw the resulting pattern in Table 45-2. Indicate whether it is a circle, an ellipse, or a straight line.
11. Measure Y_O and Y_M and record in Table 45-2. (If the pattern is a circle, $Y_O = Y_M$.)
12. Calculate and record the phase angle θ. Show your computations.
13. Set the frequency control of the signal generator at 10 kHz, still maintaining the output level of the generator at 15 V peak-to-peak (same circuit connections).
14. Recenter the resulting pattern, if necessary.
15. Repeat steps 9 through 12.

QUESTIONS

1. In Fig. 45-5, with $f = 50$ kHz, what is the ratio of X_L/R? Show your computations.
2. (a) For the conditions in question 1 can we consider the circuit, including R a pure inductance? Why?
   (b) How is this demonstrated by the Lissajous pattern?
   (c) How is this demonstrated by the phase angle between v and i?
3. When $f = 10$ kHz, what is the ratio of X_L/R?
4. (a) For the conditions in question 3, can we consider the circuit a pure inductance? Why?
   (b) How is this demonstrated by the Lissajous pattern?
   (c) How is this demonstrated by the phase angle?
5. Refer to Table 45-1. What are the phase relations between v and i? How do you know?

Answers to Self-Test

1. zero
2. false
3. same
4. switched out
5. centered; origin
6. 30°

EXPERIMENT 46

Inductances in Series and in Parallel

OBJECTIVES

1. To verify experimentally that the total inductance L_T for two inductors L_1 and L_2 connected in series, where no mutual coupling exists, is

$$L_T = L_1 + L_2$$

2. To verify experimentally that L_T for two inductors connected in parallel, where no mutual coupling exists, is

$$\frac{1}{L_T} = \frac{1}{L_1} + \frac{1}{L_2}$$

INTRODUCTORY INFORMATION

Just as in the case of resistors, inductors may be connected in series or in parallel in a circuit. The effect of series-connected inductors is to offer greater opposition to alternating currents than either inductor can, acting alone. Measurement shows that series-connected inductors, where no mutual coupling exists, act like an equivalent single inductor whose inductance L_T is the sum of the individual inductances. That is,

$$L_T = L_1 + L_2 + L_3 + \cdots \qquad (46\text{-}1)$$

Two or more inductors connected in parallel, where no mutual coupling exists, act like an equivalent single inductor whose inductance L_T is given by the equation

$$\frac{1}{L_T} = \frac{1}{L_1} + \frac{1}{L_2} + \frac{1}{L_3} + \cdots \qquad (46\text{-}2)$$

It is apparent that the total inductance of parallel-connected inductors is less than the smallest inductor in the circuit.

Verification that the total inductance of series- and parallel-connected inductors, as given by Eqs. (46-1) and (46-2), respectively, may be obtained

by measuring L_T directly on an inductance bridge. Or L_T may be measured indirectly, as in this experiment.

Consider the circuit of Fig. 46-1. V is an ac source sending current through R and L. The reactance (X_L) of L may be determined, as in the preceding experiment, by measuring I and the voltage V_L across L, and substituting these values in the equation

$$X_L = \frac{V_L}{I} \qquad (46\text{-}3)$$

After X_L has been computed, L may be found from the equation

$$L = \frac{X_L}{2\pi F} \qquad (46\text{-}4)$$

In this equation, X_L is the value in ohms just obtained, F is the frequency in Hz of the alternating current, and L is the required inductance in henrys.

Example

What is the inductance of a coil whose reactance at 100 Hz is 3,000 Ω?

Solution. $L = \dfrac{X_L}{2\pi F} = \dfrac{3,000}{6.28(100)} = 4.78$ H

In the circuit of Fig. 46-1, the current I may be

Fig. 46-1. A series LR circuit may be used to determine the X_L of an inductor.

measured directly by an ac ammeter. Another method is to measure the resistance of R and the voltage V_R across R. The current may then be computed from the equation

$$I = \frac{V_R}{R}$$

Effect of Direct Current on the Inductance of an Iron-core Choke

In electronics iron-core coils, called chokes, are used as current-smoothing elements in power supplies. The inductance of these chokes is dependent on the direct current present in them. A filter choke rated 8 H at 50 mA direct current means that L is 8 H when a 50-mA direct current is in the choke. In the absence of the direct current, the choke will not measure 8 H. If there is less than 50 mA dc the choke will measure more than 8 H; when there is more than 50 mA dc, the choke will measure less than 8 H.

In this experiment you will verify the effect of direct current on the inductance of an iron-core choke by means of the experimental circuit (Fig. 46-2). V_{rms} is an isolated line-derived ac voltage. With switch S in position 1, current flows in the ac circuit consisting of the inductor L connected in series with R. There is no direct current in L. The inductance of L is determined by finding the ac current I in L, and the ac voltage V_{BC} across L. Now,

$$X_L = \frac{V_L}{I} \quad \text{and} \quad L = \frac{X_L}{2\pi F}$$

Next S is switched to position 2 and the output of the dc supply V_{dc} connected in series with V_{rms} is adjusted to the rated dc level of L. The inductance of L is again determined for comparison.

Fig. 46-2. Experimental circuit to determine the inductance of L.

SUMMARY

1. The total inductance L_T of series-connected inductors, where no mutual coupling exists, is given by the equation

$$L_T = L_1 + L_2 + L_3 + \cdots$$

where L_1, L_2, L_3, etc., are inductors connected in series.

2. The total inductance L_T of parallel-connected inductors, where no mutual coupling exists, is given by the formula

$$\frac{1}{L_T} = \frac{1}{L_1} + \frac{1}{L_2} + \frac{1}{L_3} + \cdots$$

3. It is apparent from summary statements 1 and 2 that the laws for L_T of inductors connected in combination (series or parallel), where no mutual coupling exists, are the same as the laws for R_T of resistors connected in combination.

4. The inductance of iron-core coils, or chokes, is rated for a particular level of dc current. Thus a choke rated 8 H at 50 mA dc will measure more than 8 H when there is less than 50 mA dc, and less than 8 H when there is more than 50 mA dc.

5. The inductance of coils and chokes may be measured directly by use of an inductance bridge.

6. The inductance of coils and chokes may also be determined indirectly by measuring the ac voltage V_L across the coil and the ac current I in the coil. From the measurements X_L can be computed, for $X_L = (V_L/I)$. Knowing X_L, it is possible to calculate L by the equation

$$L = \frac{X_L}{2\pi F}$$

SELF-TEST

Check your understanding by answering these questions:

1. Three chokes, 11-, 2.5-, and 8-H, are connected in series. No mutual coupling exists. Their total inductance is _____ H.

2. The same three chokes (question 1) are connected in parallel. Their total inductance is _____ H.

3. The ac voltage V_L across a coil L is 22 V. The frequency is 60 Hz. The ac current in the coil is 0.025 A. What is
   (a) $X_L = $ _____ Ω
   (b) $L = $ _____ H

4. A choke (L) and a 1-kΩ resistor are connected in series as in Fig. 46-3. The voltage across the resistor is 50 V and the voltage across the choke

Fig. 46-3. *LR* series circuit.

is 40 V. The frequency of the applied voltage is 60 Hz. What is the

(a) Current I in the circuit? _____ A

(b) X_L of the choke? _____ Ω

(c) L of the choke? _____ H

MATERIALS REQUIRED

- Power supply: Variable voltage-regulated dc source, line-isolated 6.3-V rms, 60-Hz line-derived source
- Equipment: Oscilloscope, EVM, 0–100 mA dc milliammeter
- Resistors: ½-W 10,000-Ω; 5-W 500-Ω
- Inductors: Two iron-core chokes, 8 H at 50 mA direct current, whose $R = 250\ \Omega$ (as in Experiment 45)
- Miscellaneous: SPST and SPDT switches; additional parts as required for step 8; fused line cord

---

PROCEDURE

Determining the Direct Current X_L of an Iron-core Choke

All ac voltages should be measured with an EVM and checked with an oscilloscope calibrated to measure rms values.

1. Connect the circuit of Fig. 46-2. *S* is in position 1. V_{rms} is the output of a variable autotransformer plugged into an isolation transformer deriving its power from the 120-V ac line. V_{rms} is set at *zero* volts. V_{dc} is set at *zero* volts. *M* is a 0–100 mA dc meter to measure direct current in the inductor.

2. With the EVM and oscilloscope connected across *R*, adjust the output of the ac supply until V_{AB}, the voltage across *R*, is 1 V. Determine the alternating current *I* in *R* and record in Table 46-1. Measure and record the voltage V_{BC} across *L*. Compute and record the X_L of *L*. Show your computations. Compute and record the inductance *L*. Show your computations. Record also

the rated value in henrys and the dc rating of *L*.

3. Set switch *S* to position 2. Adjust V_{dc} until there is 15 mA direct current in *L*. Measure the ac voltage across *R*. If it is not precisely 1 V rms, readjust V_{rms} until this voltage is obtained. Measure and record V_{BC}, the voltage across *L*. Determine and record *I* (alternating current), X_L, and *L*.

4. Repeat step 3 for 30 mA of direct current and finally for the rated value of direct current in *L*. *Be certain that the voltage V_{AB} across R is precisely 1 V. If it is not, readjust V_{rms} until there is precisely 1 V across R.*

Inductors in Series

All ac voltage measurements should be made with an EVM. These readings should be confirmed by measurement with an oscilloscope set on the ac voltage range and calibrated in rms values.

5. Determine the inductance of each inductor L_1

TABLE 46-1.

Direct current, mA	V_{AB}, volts	$I(ac)$, amperes	V_{BC}, volts	X_L, ohms	L, henrys	L, Rated Value, henrys
0	1					
15	1					
30	1					dc current
Rated value	1					

TABLE 46-2.

V_{L_1}, volts	V_R, volts	I_1, amperes	L_1, henrys
V_{L_2}, volts	V_R, volts	I_2, amperes	L_2, henrys

TABLE 46.3. Series Connection

V_{LT}	V_R	I	L_T

ages V_{LT} across L_1 and L_2 (A to B) and V_R across R.

c. Compute and record the current I and L_T.

Extra Credit

8. Design an experiment to verify the formulas for L_T for series- and parallel-connected RF chokes. Show circuit. Perform the experiment, tabulating your data. State your conclusions. Indicate equipment used.

QUESTIONS

1. Explain the method you used for determining the inductance of an inductor.
2. (*a*) What is the sum L_T of L_1 and L_2 recorded in Table 46-2?
   (*b*) How does this compare with the measured value L_T in Table 46-3?
   (*c*) Explain the difference, if any, between these two values.
3. (*a*) Determine L_T from the formula

$$\frac{1}{L_T} = \frac{1}{L_1} + \frac{1}{L_2}$$

   (*b*) How does this computed value of L_T compare with that obtained in Table 46-4?
   (*c*) Explain the difference, if any, between these two values.
4. Show the formula for L_T of three inductors in series, assuming no mutual coupling.
5. Show the formula for L_T of three inductors in parallel, assuming no mutual coupling.
6. Compare the formulas for R_T of resistors with those for L_T of inductors where no mutual coupling exists.

and L_2 using the following procedure:
a. Connect the circuit of Fig. 46-3.
b. Measure, and record in Table 46-2, the voltage V_{L_1} across the inductance L_1, and V_R across the resistor.
c. Replace L_1 by L_2 and repeat step **b**.
d. Compute the values of I_1 and I_2 and record them in Table 46-2. Show formulas used and computations.
e. Compute the values of L_1 and L_2 and record them in Table 46-2. Also show formulas used and computations.

6. Determine the total inductance L_T of L_1 and L_2 connected in series by the following method:
a. Connect the circuit of Fig. 46-4. Separate L_1 and L_2 as far as possible so that there is no coupling between them.
b. Measure, and record in Table 46-3, the voltages V_{LT} across L_1 and L_2 (i.e., from point A to B), and V_R across R.
c. Compute and record the current I and L_T.

Inductors in Parallel

7. Determine the total inductance L_T of L_1 and L_2 connected in parallel by the following method:
a. Connect the circuit of Fig. 46-5. Separate L_1 and L_2 as far as possible so that there is no coupling between them.
b. Measure, and record in Table 46-4, the volt-

Fig. 46-4. Effect of adding inductors in series.

Fig. 46-5. Effect of adding inductors in parallel.

226 *Basic Electricity*

TABLE 46-4. Parallel Connection

V_{LT}	V_R	I	L_T

7. Give a simple formula L_T for two inductors in parallel where no mutual coupling exists.

8. Why aren't the measured values of L_1 and L_2 in step 5 the same as their rated values?

Answers to Self-Test

Answers to slide-rule accuracy.

1. 21.5
2. 1.62
3. (a) 880; (b) 2.34
4. (a) 0.05; (b) 800; (c) 2.12

47 EXPERIMENT

Transformer Characteristics

OBJECTIVES

1. To determine experimentally the turns ratio of a given transformer
2. To measure the effect of an increase of current in the secondary winding (load current) on current in the primary (source current)
3. To resistance-test a transformer

INTRODUCTORY INFORMATION

Ideal Transformer

A transformer is a device for coupling ac power from a source to a load. The source of power is connected to the primary winding, the load to the secondary winding. In the process, certain transformations take place which are related to the construction, and to the primary to secondary turns ratio, of the transformer.

Transformers find many uses in electronics. There are power transformers, filament transformers (for heating the filaments of a vacuum tube), audio transformers, radio-frequency transformers, etc.

A conventional transformer consists of two or more windings on a core, magnetically coupled (Fig. 47-1). An ac voltage applied across the input or primary winding causes current in the primary. This sets up an expanding and collapsing magnetic field which cuts the turns of the secondary winding, inducing an ac voltage in the secondary. When a load is connected across the secondary, current flows in the load.

The core around which the primary and secondary are wound may be iron, as in the case of low-frequency power and audio transformers. An air core may be employed for coupling higher-frequency circuits. The core material and the geometry of the windings determine the characteristics of coupling.

An alternating current in a winding of an iron-core transformer magnetizes the core first in one direction and then in the other. It is this moving magnetic field in the core which cuts the windings of the other coil, inducing a voltage in it.

In an ideal iron-core transformer, that is, one in which there are no power losses, 100 percent source (primary) power would be delivered to the load. Equation (47-1) shows this power relationship.

$$V_p \times I_p = V_s \times I_s \qquad (47\text{-}1)$$

Here V_p and I_p are the primary voltage and current, respectively, while V_s and I_s are the secondary voltage and current. In a lossless transformer, the ratio between primary voltage and the voltage induced in the secondary is the same as the ratio (a) between the number of turns of the primary N_p and number of turns of the secondary N_s.

$$\frac{V_p}{V_s} = \frac{N_p}{N_s} = a \qquad (47\text{-}2)$$

Fig. 47-1. Stepdown transformer without load.

If $a = 1$, there are as many turns in the primary as there are in the secondary, and the voltages appearing across the primary and secondary are equal. This type of 1:1 transformer is called an isolation transformer.

If a is greater than 1, a lower voltage appears across the secondary than across the primary. This is called a voltage stepdown transformer.

If a is less than 1, a higher voltage appears across the secondary than across the primary. This is a voltage stepup transformer.

Equation (47-1) for an ideal transformer may be written also as

$$\frac{V_p}{V_s} = \frac{I_s}{I_p} \qquad (47\text{-}3)$$

From Eqs. (47-2) and (47-3) we have

$$\frac{V_p}{V_s} = \frac{I_s}{I_p} = a \qquad (47\text{-}4)$$

The last equation states that current and voltage in the windings of a transformer are inversely related. Therefore, a voltage stepup transformer is also a current stepdown transformer, whereas a voltage stepdown is a current stepup transformer.

Power Losses in a Transformer

The ideal transformer does not exist because there are power losses which do not permit 100 percent transfer of power from the source to the load. One such loss is related to the resistance of the windings and is the I^2R or heating loss. Thus, there are I^2R losses associated with the resistance of the primary winding (primary loss) and also with the resistance of the secondary (secondary loss). When there is no load on the secondary winding, there is no current, hence no power loss in the secondary. However, there is current and hence some I^2R loss in the primary.

Eddy currents are present in the core of an iron-core transformer. A circulating eddy current is induced in the iron core by the changing magnetic field. Eddy currents heat the transformer and act like an I^2R loss. Eddy currents thus rob the source of power and represent another power loss.

Another loss associated with a transformer core is *hysteresis loss*. The hysteresis effect results from the fact that magnetic lines of force lag behind the magnetizing force that causes them. Hysteresis can be understood by considering the fact that when a

Fig. 47-2. Primary current I_p increases when secondary current I_s increases, due to reflected impedance (R_1) from the secondary to the primary.

magnetizing force is removed from an iron-core magnetic circuit, a portion of the flux remains within the iron. This residual magnetism can be removed by applying to the iron a magnetizing force opposite in direction to that of the initial force. The energy required to demagnetize the iron acts as a core loss, which is associated with the reversal of magnetizing current in the winding.

One other loss must be mentioned. This is associated with the magnetic leakage which exists in a transformer. Not all the lines of magnetic force will link the turns of the secondary winding. The lines of force thus lost to the magnetizing circuit constitute magnetic leakage.

Effect of Load Current on Primary Current

Primary current in the transformer of Fig. 47-2 depends on the load current in the secondary, as is evident from the equation

$$\frac{I_s}{I_p} = a$$

or

$$I_p = \frac{I_s}{a} \qquad (47\text{-}5)$$

Therefore, as load current (I_s) increases due to a decrease in load resistance R_2, primary current must increase. The increase in primary current is explained by the assumption that the change in load impedance (as is evidenced by an increase in load current) appears also as a reflected impedance in parallel with the primary. Since power in the primary and secondary were assumed to be equal, and since power can be dissipated only in a resistance, the reflected impedance must be the resistance R_1 (Fig. 47-2).

In addition to the primary current I_p resulting from the reflected load impedance, the primary supplies the current for the iron-core losses and the magnetizing current. The vector sum of these two currents (the magnetizing current is out of phase with the voltage) is called the "exciting current." The exciting current is about 3 to 5 percent of the rated output of the transformer. This fact explains why primary current will normally measure more than that predicted by Eq. (47-5).

Resistance-testing Transformer Windings

Resistance (ohmmeter) tests of the individual windings of the small transformers used in electronics are used to determine the continuity of the windings. An ohmmeter test also establishes the resistance of each winding. The technician then compares the measured resistance with the rated value to determine if a suspected transformer is in fact defective.

The following considerations may be helpful in analyzing the results of resistance measurements.

Winding Measures Infinite Resistance. This winding is "open." The break may occur at the beginning or end of the winding where connection is made to the terminal leads. This type of break can be easily repaired by resoldering the leads to the winding. If the discontinuity is elsewhere, the transformer must be replaced.

Winding Resistance "Very" High. A winding whose resistance measures "very" high compared to its rated value may be open, or there may be a cold-solder joint at the terminal connections. If the condition cannot be corrected, the transformer must be replaced.

Winding Resistance "Very" Low Compared to Rated Value. Turns of the winding must be "shorted" somewhere on the transformer or the winding may be shorted to the frame. However, a small difference in resistance between the rated and measured values may be insignificant. For example, if a primary rated at 120 Ω measures 100 Ω, the difference may be attributable to the inaccuracy of the meter. In cases of doubt, other tests (which will not be discussed here) are indicated.

Resistance between Windings. The windings of transformers, other than autotransformers, are insulated from each other. There should be infinite resistance between insulated windings, as long as the transformer is not connected in a circuit. If the insulation between two windings breaks down there will be a measurable resistance between these windings, signifying a defective transformer.

Factors which Determine Resistance of a Winding. The resistance of a winding depends on the diameter of the wire and the number of turns. Thus, resistance varies inversely as the square of the diameter and directly as the number of turns. Large diameters are required for high-current windings; smaller diameters for windings which carry less current. The primary of a voltage stepdown transformer has more windings than the secondary. Moreover, a voltage stepdown is also a current stepup transformer. Hence the secondary must carry more current than the primary. Therefore, the resistance of the primary will be higher than that of the secondary. How much higher depends on the transformer ratio and on the diameter of the two wires.

SUMMARY

1. A transformer consists of two or more windings on a core. The core is iron for low-frequency applications or air for higher frequencies.
2. When an ac source is applied across the primary winding, a voltage is induced in the secondary. If there is a load on the secondary, current also flows in that winding.
3. A transformer is therefore a device which couples ac power from a source connected to the primary, to a load connected to the secondary winding.
4. In an ideal transformer all the source power would be delivered to the load. Thus in Fig. 47-2

$$V_p \times I_p = V_s \times I_s$$

where V_p and V_s are the primary and secondary voltages, respectively; I_p and I_s are the primary and secondary currents, respectively.
5. In an ideal transformer the ratio between the number of turns in the primary N_p and the number of turns in the secondary N_s is the same as the ratio of the voltage in the primary V_p to the voltage in the secondary V_s. Thus

$$\frac{N_p}{N_s} = a = \frac{V_p}{V_s}$$

6. In an ideal transformer the ratio of current in the secondary I_s to current in the primary I_p is also a. Thus

$$\frac{N_p}{N_s} = a = \frac{I_s}{I_p}$$

7. A voltage stepdown transformer is one in which the secondary voltage is lower than in the primary. A voltage stepdown transformer is also a current stepup transformer. In this type of transformer $a > 1$.

NOTE: The symbol $>$ means greater than; $<$ means smaller than.

8. If $a < 1$, the transformer is a voltage stepup, current stepdown device.
9. The power losses which occur in a transformer include:
   (a) I^2R or heating losses due to the resistance of the windings.
   (b) Eddy current losses present in an iron-core transformer.
   (c) Hysteresis losses, which result from the lag between the magnetic lines of force and the magnetizing force which induces them.
   (d) Magnetic leakage which occur because not all the magnetic lines of force link the turns of the secondary winding.
10. An increase in I_s will result in an increase in I_p, and a resulting decrease in primary impedance. The decrease in primary impedance is caused by the impedance reflected from the secondary into the primary. The reflected impedance is the resistance R_1 (Fig. 47-2) which acts in parallel with the primary winding.
11. The primary current supplies the power losses associated with a transformer. Therefore, because there are losses associated with every transformer, $I_p > I_s/a$.
12. Ohmmeter tests of the resistance of individual windings of a transformer are used to determine the continuity of the windings.
13. The separate windings of a transformer are insulated from each other and from the transformer frame. Therefore, the resistance between windings should be infinite, as long as the transformer is *not* connected in the circuit.
14. The resistance between an individual winding and the frame of the transformer should also be infinite.
15. The resistance of a transformer winding depends on the wire diameter and the number of turns. In a voltage step-up, current step-down transformer:
    (a) There are fewer turns in the primary than

secondary.
    (b) The secondary wire diameter is larger than that of the primary wire.
    (c) The resistance of the secondary is lower than the primary.

SELF-TEST

Check your understanding by answering these questions:

1. In a particular ideal transformer, the primary has 10 times as many turns as the secondary. The voltage across the primary is 120 V. The voltage across the secondary must be _____ V.
2. In the transformer of question 1, the current in the secondary is 3 A. The current in the primary must therefore be _____ A.
3. In the transformer of question 1, the turns ratio N_p/N_s is _____.
4. There is no *ideal* transformer for all practical purposes. _____ (true/false)
5. The power losses in a transformer include those due to:
   (a) _____ of the windings
   (b) _____ _____
   (c) _____
   (d) _____ _____
6. An increase in secondary current causes a(n) _____ (increase/decrease) in primary current.
7. The impedance reflected from the secondary into the primary is a _____ (resistance/inductance).
8. The power losses in a transformer are supplied by the _____ _____ _____.
9. In an iron-core power transformer which is not connected in a circuit the resistance measured between the primary and secondary windings is 10 Ω. This resistance _____ (is/is not) normal.
10. The primary and secondary windings must be _____ in the transformer in question 9.

MATERIALS REQUIRED

- Power supply: Source of 120-V 60-Hz
- Equipment: EVM; multirange ac milliammeter; decade-type rheostat; ohmmeter
- Transformer: AF output, 6 K6 to 3-Ω speaker
- Miscellaneous: SPST switch; fused line cord

PROCEDURE

Transformer Turns Ratio, a:

1. Connect the circuit of Fig. 47-1.

CAUTION: Be certain that the line voltage (120 V, alternating current) is applied to the *primary* of the voltage stepdown transformer. The secondary is not loaded.

2. Measure the voltages V_p (across the primary) and V_s (across secondary) and record in Table 47-1.
3. Compute and record the turns ratio a.

Effect of Load on Primary Current

4. Connect an ac milliammeter M in the primary circuit as in Fig. 47-2. The ac milliammeter should be set on the highest range, then adjusted to the proper range. R_G represents the internal resistance of the generator. V is the 120 V of the 60-Hz line. R_2 is a decade rheostat acting as the load resistor.
5. Measure and record the primary current I_p, the voltage across the primary V_p, and the voltage across the secondary V_s, in turn for each value of load resistance R_2 shown in Table 47-2. An EVM is used to measure V_p and V_s.
6. Compute and record the secondary current $(I_s = V_s/R_2)$. Substitute the value of I_s in the equation $I_{p_1} = I_s/a$. Compute and record I_{p_1}. (The computed value I_{p_1} is the current which would appear in the primary of an ideal transformer.)
7. Subtract I_{p_1} from I_p (the current measured in

TABLE 47-1.

V_p	V_s	$a = \dfrac{V_p}{V_s}$

the primary) and record in Table 47-2, under column labeled "ΔI."

8. Compute and record the power W_s dissipated by the load resistor R_2 for each value of R_2.

CAUTION: In all cases, be certain that the wattage rating of R_2 is high enough for the application.

Resistance Measurement of the Windings

9. Remove power and disconnect the transformer primary and secondary from the circuit.
10. Measure the resistance of the primary and secondary with an ohmmeter, and record in Table 47-3.
   Measure also and record the resistance between windings and between each winding and the frame.

QUESTIONS

1. Is the transformer you used a voltage stepup or a current stepup?
2. How is primary current affected by current in the secondary? Why?
3. Does ΔI appear to be relatively constant in Table 47-2? What does ΔI represent?
4. What losses, if any, are incurred in the transformer used in this experiment?

TABLE 47-2.

R_2, ohms	I_p, milli-amperes	V_p, volts	V_s, volts	$I_s = \dfrac{V_s}{R_2}$ milliamperes	$I_{p_1} = I_s/a$, milliamperes	$\Delta I = I_p - I_{p_1}$, milliamperes	W_s
15							
10							
5							
4							
3							
2							
1							

TABLE 47-3.

| | | Resistance, ohms | | |
		Primary to secondary	Primary to frame	Secondary to frame
Primary	Secondary			

5. Where does the power come from to supply the transformer losses?
6. What is the relationship between the resistance of a winding and the
   (a) Current rating of that winding?
   (b) Number of turns of the winding?
   Refer to your data to substantiate the answer.
7. Give the sources of error in the experimental data?
8. Compare the change in W_s as the load resistor (R_2) was varied from 15 to 4 Ω with the change that took place when R_2 was varied from 4 to 1 Ω. How do you account for the change?

Answers to Self-Test

1. 12
2. 0.3
3. 10
4. true
5. (a) heating; (b) eddy current; (c) hysteresis; (d) magnetic leakage
6. increase
7. resistance
8. input power source
9. is not
10. shorted

48

Capacitor Color Code and Testing Capacitors

OBJECTIVES

1. To find the capacitance value of a capacitor by means of the EIA color code
2. To use a capacitance checker to determine capacitance, leakage, and power factor
3. To use an EVM for testing shorted or leaky capacitors

INTRODUCTORY INFORMATION

Color Code

Presently, capacitance values of molded paper, mica, and disc ceramic capacitors are usually printed on the unit by the manufacturer. However, some equipment in use today may contain components marked according to several obsolescent systems. These obsolescent codes are shown here along with the codes which are presently being used for ceramic tubular-style capacitors.

The two types of molded capacitors, tubular and flat, are shown in Fig. 48-1. Color bands are used on the tubular-type and color dots on the flat type in accordance with the code shown in Fig. 48-2. Figures 48-1 and 48-2 explain how the code is applied. All values are in micromicrofarads (pF). It should be noted that the MIL code markings are somewhat different from the EIA markings. There are two types of molded mica, the flat postage-stamp type and the button silver mica, shown in Fig. 48-3. The accompanying obsolescent color code is shown in Figs. 48-3 and 48-4.

Figures 48-5 and 48-6 show an obsolescent system for color coding disc ceramic capacitors.

Figures 48-7 and 48-8 show the EIA color codes which are presently in use for ceramic tubular-style capacitors.

Capacitance Testers

Different types of capacitance testers are available to the technician. There is the tester which checks only the capacitance of a capacitor. This may be used to check the rated value of a capacitor or to determine the capacitance of a capacitor whose value is unknown. This type of tester will also detect open and shorted capacitors.

Other types, in addition to acting as capacitance bridges, also determine some or all of the following capacitor characteristics.

1. *Insulation Resistance for Paper, Mica, and Ceramic Capacitors.* Insulation resistance (R_l) is effectively a resistance in parallel with the capacitor. The value of insulation resistance considered normal varies with the type of capacitor. This characteristic is considered only for electrostatic and not electrolytic capacitors. Approximate values of insulation resistance are shown below.

    (*a*) *Molded Paper Capacitors.* Insulation resistance (R_l) is inversely proportional to capacitance. The product of $R_l \times \mu F$ should equal about 1,000 megohms-microfarads when new, depending on the ambient temperature.

    (*b*) *Mica Capacitors.* When new, these will have a value R_l greater than 3,000 MΩ.

    (*c*) *Ceramic Capacitors.* When new, these will have a value R_l greater than 7,500 MΩ.

2. *Leakage Current of Electrolytic Capacitors.* This test for electrolytic capacitors takes the place of the R_l test for electrostatic capacitors. Leakage current I is directly proportional to capacitance. Maximum leakage current permissible for electrolytics is usually included in the instruction manual of the capacitor checker.

Fig. 48-1. Molded capacitor types (*Centralab, a Div. of Globe-Union, Inc.*).

Color	Significant Figures	MOLDED PAPER TUBULAR CAPACITORS		
		Decimal Multiplier	Tolerance ±%	Voltage Volts
Black	0	1	20	—
Brown	1	10	—	100
Red	2	100	—	200
Orange	3	1000	30	300
Yellow	4	10000	40	400
Green	5	10^5	5	500
Blue	6	10^6	—	600
Violet	7	—	—	700
Gray	8	—	—	800
White	9	—	10_{EIA}	900
Gold	—	0.1	—	—
Silver	—	—	10_{EIA}	—
No Color	—	—	—	—

Fig. 48-2. Color code for molded capacitors (*Centralab, a Div. of Globe-Union, Inc.*).

Fig. 48-3. Molded mica capacitor types (*Centralab, a Div. of Globe-Union, Inc.*).

Color	Significant Figures	MOLDED MICA EIA Standard—MIL Specification Capacitors			
		Decimal Multiplier	Tolerance ±%	RMA Voltage Rating (All Capacitors)	Class
Black	0	1	20 (JAN RMA 1948)	—	A
Brown	1	10	—	100	B
Red	2	100	2	200	C
Orange	3	1000	3 (EIA)	300	D
Yellow	4	10000	—	400	E
Green	5	—	5 (EIA)	500	F (MIL)
Blue	6	—	—	600	G (MIL)
Violet	7	—	—	700	—
Gray	8	—	—	800	I (EIA)
White	9	—	—	900	J (EIA)
Gold	—	0.1	5 (MIL)	1000	—
Silver	—	0.01	10	2000	—
No Color	—	—	20 (old RMA)	500 (old RMA)	—

Fig. 48-4. Color code for molded mica capacitors (*Centralab, a Div. of Globe-Union, Inc.*).

Fig. 48-5. Disc ceramic capacitor types (*Centralab, a Div. of Globe-Union, Inc.*).

| Color | Significant Figures | Decimal Multiplier | CERAMICS EIA Standard—MIL Specification Capacitors | | Temperature Co-efficient Parts Per Million Per Degree Centigrade | T.C. for Extended Range TC HiCap | |
| | | | Tolerance | | | | |
			Capacity 10 mmf or Less	Capacity More Than 10 mmf		Significant Fig.	Multiplier
Black	0	1	±2.0 mmf$_{(MIL)}$	±20%	NPO	0	−1
Brown	1	10	±0.1 mmf	±1%	N33	—	−10
Red	2	100	—	±2%	N75	1	−100
Orange	3	1000	—	±2.5%$_{(EIA)}$	N150	1.5	−1000
Yellow	4	10000$_{(EIA)}$	—	—	N220	2.2	−10000
Green	5	—	±0.5 mmf	±5%	N330	3.3	+1
Blue	6	—	—	—	N470	4.7	+10
Violet	7	—	—	—	N750	7.5	+100
Gray	8	0.01	±0.25 mmf	—	+30	—	+1000
White	9	0.1	±1.0 mmf	±10%	N330±500	—	+10000
Gold	—	—	—	—	+100 (MIL)	—	—
Silver	—	—	—	—	Bypass & Coupling (EIA)	—	—
No Color	—	—	—	—	—	—	—

Fig. 48-6. Color code for disc ceramic capacitors (*Centralab, a Div. of Globe-Union, Inc.*).

Fig. 48-7. Color coding for general-purpose tubular capacitors (*Centralab, a Div. of Globe-Union, Inc.*).

Fig. 48-8. Color coding for temperature-compensating ceramic tubular capacitors (*Centralab, a Div. of Globe-Union, Inc.*).

Values of leakage current are read directly on the calibrated meter scale of the tester.

3. *Power Factor of Electrolytic Capacitors.* The power factor (P.F.) of an electrolytic capacitor is a measure of the power losses in that capacitor. The losses occurring in a capacitor may be indicated as a resistance R in series with the capacitor. The phase angle of this RC combination, therefore, indicates the losses of this capacitor. The cosine of this phase angle is defined as the power factor. Acceptable P.F. values for new electrolytic capacitors are usually given in the instruction manual accompanying the capacitance tester.

NOTE: A capacitor should be replaced if any of its measured characteristics do not come within the range of acceptable norms.

For proper testing, the entire capacitor, or at least one lead of the capacitor, should be disconnected from the circuit. An exception to the rule is for a tester on the market which checks shorts and opens of capacitors wired in the circuit.

The residual capacitance of a checker should be determined before it is used to measure the values of very small capacitances. Moreover, for accurate readings the pigtails of the capacitor should be placed directly across the terminals of the checker. Long clip leads should be avoided, since their capacitance will add to that of the capacitor under test.

The student or technician should be fully familiar with the operating instructions of the capacitance checker he is using in order to derive full benefit from it. Moreover, he should perform all the checks which the instrument is capable of making.

Testing Capacitors with an EVM

Completely shorted capacitors or very "leaky" capacitors (i.e., capacitors having a very low value of insulation resistance) may frequently be detected with a VOM, or EVM.

A simple resistance test in which the ohmmeter is placed across the terminals of the capacitor is used. If the capacitor is shorted, the meter will read zero resistance. Of course, this capacitor must be discarded. A low value of insulation resistance may also be detected by an ohmmeter test.

Occasionally a capacitor will become shorted or leaky only when a voltage is placed across it. This capacitor is said to break down under load. This defect cannot be detected with an ohmmeter but it

Fig. 48-9. Testing capacitor with an electronic voltmeter.

can be found with a dc voltmeter. In this "dynamic" test (see Fig. 48-9), a dc voltage is placed across the series combination of the voltmeter and capacitor. A good capacitor will show only a momentary charge on the voltmeter, after which zero voltage will be indicated. However, a capacitor breaking down under the voltage will give a voltage indication on the meter. The higher the leakage current, the higher the voltage will read. When checking for leakage in this fashion, the leakage resistance is in series with the meter and limits the current. If the leakage resistance is much greater than the input resistance of the voltmeter, the maximum current is mainly dependent upon the value of leakage resistance. The rated dc working voltage V of the capacitor should be used if possible for this test.

The input resistance of an EVM is much higher than that of a VOM on the low-voltage ranges. It is therefore more effective in measuring leakage of a nonelectrolytic capacitor. The best procedure to follow is to connect the EVM in series with the capacitor C, as in the circuit of Fig. 48-9. The EVM is first set to a high dc voltage range. If the voltage indication is low, after the capacitor has charged, the voltmeter is switched to the lowest range on which the meter is not overloaded. The voltage measured can be used to determine the insulation resistance R_I of the capacitor by means of the following equation:

$$R_I = R_{in} \frac{V - V_1}{V_1} \qquad (48\text{-}1)$$

where V = applied voltage

V_1 = measured voltage

R_{in} = input resistance of the meter. In the case of an EVM it includes both the resistance of the dc probe and the input resistance of the meter on the dc voltage ranges.

SUMMARY

1. Capacitance values of paper, mica, and disc ceramic capacitors are presently printed on the body of the capacitors.

2. Older nonelectrolytic capacitors used a color code for designating their values. The systems employed and the corresponding values are shown in Figs. 48-1 through 48-8.

3. Whichever system was used, these significant characteristics were normally included in the code or markings: (a) value in μF or pF; (b) tolerance in percent; (c) voltage rating.

4. Capacitance checkers are used to find the capacitance value of an unknown capacitor.

5. Capacitance testers may also check:
    (a) Insulation or leakage resistance of non-electrolytic paper, mica, and ceramic capacitors.
    (b) Leakage current of electrolytic capacitors.
    (c) Power factor of electrolytic capacitors.

6. A capacitor should be replaced if any of its measured characteristics are not in normal range.

7. Some capacitor checkers are especially designed to test for opens and shorts of capacitors connected in the circuit.

8. If the capacitance checker is not of the *go, no go* type used in summary step 7, it is *not* possible to check a capacitor directly *in* the circuit. It *is* necessary to *disconnect* one of the leads of the capacitor from the circuit before testing it.

9. In finding the capacitance value of very small valued capacitors, it is necessary to subtract the residual capacitance of the tester from the measured value. Moreover, long clip leads should be avoided, and the pigtails of the capacitor should be placed directly across the terminals of the checker.

10. In testing for shorted capacitors an ohmmeter test will reveal the short. For this type of test a VOM, or EVM may be used.

11. Some capacitors may break down or display excessive leakage only under load, that is, when their rated voltage appears across their terminals. To locate this type of trouble an EVM on the voltage ranges must be used.

12. The capacitor is connected in series with the electronic voltmeter, and the rated voltage V of the capacitor is placed across this series combination. The capacitor leakage resistance R_I can be determined from the voltage V_1 read on the voltmeter by means of the formula

$$R_I = R_{in} \times \frac{V - V_1}{V_1}$$

where R_I is the leakage resistance of the capacitor and R_{in} is the input resistance of the electronic voltmeter.

SELF-TEST

Check your understanding by answering these questions:

1. The characteristics of a nonelectrolytic capacitor which can be checked on a capacitance tester are:
    (a) _____ _____
    (b) _____ _____

2. The characteristics of an electrolytic capacitor which can be checked on a capacitance tester are:
    (a) _____ _____
    (b) _____ _____
    (c) _____ _____

3. When a color code is used on a nonelectrolytic capacitor, these are the capacitor characteristics which are usually designated:
    (a) _____ _____
    (b) _____
    (c) _____ _____

4. An _____ may be used to check a capacitor for shorts.

5. A dynamic test of a capacitor requires placing a voltage equal to the _____ _____ of the capacitor across a series combination including the capacitor and an _____.

6. Using the test circuit of Fig. 48-9 and Eq. (48-1), the insulation resistance of a capacitor, if $V = 600$ V, $R_{in} = 13 \times 10^6$ Ω, and $V_1 = 10$ V, is _____ Ω.

7. The insulation resistance of a capacitor is considered to be in _____ (series/parallel) with the capacitor.

8. A capacitor is considered defective if its insulation resistance is very _____ (high/low).

9. A capacitor is considered defective if the voltage measured in the test circuit of Fig. 48-9 is very low. _____ (true/false)

MATERIALS REQUIRED

- Power supply: dc voltage
- Equipment: EVM, capacitance checker
- Capacitors: Assortment of paper, mica, ceramic, and electrolytic, including 0.01 μF, 0.1 μF, 0.5 μF

PROCEDURE

1. Determine the value of each capacitor supplied, from its color code. Fill in the information required in Table 48-1.
2. Refer to the instruction manual of the capacitance checker for procedure to measure capacitance values.
3. Determine and record the residual capacitance of the checker.
4. Measure each capacitor and record the results in Table 48-1 in the row labeled "Measured value." If the capacitor is very small, compensate for the residual capacitance of the checker by subtracting residual capacitance from measured value.
5. If the checker has facilities, measure the insulation resistance of the molded mica and ceramic capacitors supplied, and record in Table 48-1.
6. If the checker has facilities, measure the leakage current and power factor of electrolytic capacitors supplied, and record in Table 48-1.
7. **a.** Place the leads of an ohmmeter across one of the nonelectrolytic capacitors. Note charging effect on meter.
   b. Reverse the ohmmeter leads across the capacitor. Describe the effect.
8. Measure each of the nonelectrolytic capacitors with an ohmmeter. Record the resistance of any whose insulation resistance is measurable.
9. Measure the resistance of each of the electrolytic capacitors supplied and record results in Table 48-2. Be sure to use the ohmmeter test-lead polarity which gives indications of greater resistance. In most cases this occurs when the positive ohmmeter test lead is connected to the positive capacitor terminal.
10. Connect a dc voltage across the series combination of a 0.01 μF capacitor and VOM as in Fig. 48-9. Set V at the rated voltage of the capacitor or to maximum voltage of the supply if the dc output of the power supply is not as high as the rated voltage of the capacitor. Record V in Table 48-3. Measure the voltage V_1 and record in Table 48-3. Record also the voltage range of the VOM and R_{in}. Compute and record R_l using Eq. (48-1).
11. Now, using the same capacitor, measure V_1 with an EVM and record in Table 48-3. Record also the applied voltage V, the meter range, and R_{in}. Compute and record R_l.
12. Repeat steps 10 and 11 for the other two capacitors in Table 48-3.

QUESTIONS

1. What is meant by residual capacitance of a capacitance checker?
2. When is it necessary to compensate for the residual capacitance of a checker?

TABLE 48-1.

Type	Capacitor									
	1	2	3	4	5	6	7	8	9	10
1st color										
2d color										
3d color										
4th color										
Coded value, μF										
Tolerance, %										
Measured value										
Insulation resistance										
Leakage current										
Power factor										

TABLE 48-2.

	Capacitor			
	1	2	3	4
Value				
dc working voltage				
Resistance				

3. How do the measured values of capacitance compare with the rated values (Table 48-1)? Explain any discrepancies.
4. What indication is there on your capacitance tester of (a) A shorted capacitor? (b) An open capacitor?
5. Is a resistance check of a capacitor usually a reliable test of capacitor leakage? Why?
6. Assuming the meter in Fig. 48-9 is a 1,000 Ω/V voltmeter, answer (a) through (d).
   (a) Replacing the 1,000 Ω/V voltmeter of Fig. 48-9 by its equivalent resistance on the 100-V range, draw the equivalent circuit of Fig. 48-9.
   (b) What would be the voltage shown on the meter on this range if the insulation resistance of the capacitor under test is 4 MΩ and $V = 600$ V? Compute the percentage deflection in terms of full-scale deflection.
   (c) Substitute the equivalent resistance of the 10-V range. What is the percentage of deflection? What would be the voltage shown on the meter?
   (d) How do the two percentages of deflection compare? Why?

7. Refer to the data in Table 48-3. Which is more effective in measuring the leakage resistance (R_l) of a capacitor, the VOM or the electronic voltmeter? Why?

Answers to Self-Test

1. (a) capacitance value; (b) insulation resistance
2. (a) capacitance value; (b) leakage current; (c) power factor
3. (a) capacitance value; (b) tolerance; (c) voltage rating
4. ohmmeter
5. voltage rating; EVM
6. 767×10^6
7. parallel
8. low
9. false

TABLE 48-3.

Meter Used	Capacitance (μF)	V (volts), Applied	V_1 (volts), Measured	Meter Range	R_{in} (ohms)	R_l (megohms), Computed
V O M	0.01					
	0.1					
	0.5					
E V M	0.01					
	0.1					
	0.5					

49 EXPERIMENT

RC Time Constants

OBJECTIVES

1. To determine experimentally the time it takes a capacitor to charge through a resistance
2. To determine experimentally the time it takes a capacitor to discharge through a resistance

INTRODUCTORY INFORMATION

Charging and Discharging a Capacitor

In electronic circuits, capacitors are used for many purposes. They are employed to store energy, to pass alternating current, and to block direct current. They act as filter elements, as components in "tuned" or resonant circuits, they are used in timing circuits, and they have many other functions.

Capacitors carry out their function by charging and discharging. A capacitor can store and hold a charge of electrons, a process known as charging. To charge a capacitor a voltage source is required, V in Fig. 49-1a. When switch S is closed, and the circuit is completed (Fig. 49-1b), electrons leave the negative terminal of the battery and enter the lower plate of capacitor C. Simultaneously, electrons leave the upper plate of C and return to the positive terminal of the battery. This process continues until the capacitor is fully charged, which occurs when the voltage across the capacitor equals the applied voltage V. When charged, the lower plate contains an excess of electrons and the upper plate a deficiency of electrons. The quantity of excess electrons on the lower plate is exactly equal to the quantity of electrons which left the upper plate. The voltage built up on the capacitor results from this difference in electron charge between the upper and lower plates. The polarity of charge on the capacitor plates is as shown in Fig. 49-1b. If, after charging, C is removed from the voltage source, as in Fig. 49-1c, it will continue to hold its charge as long as there is no complete circuit through which it can discharge.

If the charged capacitor is now placed across a resistor R (Fig. 49-2a), a discharge path is provided for C. Electrons now leave the negative plate of C, move through R, and enter the positive plate of C. This process continues until the excess electrons on the lower plate have replaced the deficiency of electrons in the upper. When this exchange of electrons has taken place, both plates return to their neutral or uncharged state, and the capacitor is said to be discharged (Fig. 49-2b). A voltmeter placed across C would now show *zero* voltage across the capacitor. The movement of electrons

Fig. 49-1. Charging a capacitor.

Fig. 49-2. Discharging a capacitor.

Fig. 49-3. *RC* circuit for charging a capacitor.

(charges) during the charge and discharge interval, in effect, constitutes electron current flow.

The relationship between the charge Q on a capacitor, the size C of capacitance, and the voltage V across C is expressed by

$$Q = C \times V \qquad (49\text{-}1)$$

In this formula Q is in coulombs, C in farads, and V in volts. It is apparent from Eq. (49-1) that the larger C is, the more coulombs it will store when fully charged from a given voltage source V.

Time Required to Charge a Capacitor

Consider the circuit of Fig. 49-3. We see a series circuit consisting of a capacitor C, a resistor R, a dc voltage source V, and an on-off switch S_1. A second switch S_2 is connected across the *RC* combination. Both S_1 and S_2 are open.

Assume that capacitor C has no charge on it. Now close S_1, leaving S_2 open, as in Fig. 49-4a. Closing S_1 completes the charging circuit for C. If a voltmeter were connected across C, the meter would read zero volts at the instant S_1 is closed. As C charges, the meter reads an increasing voltage. This experiment simply demonstrates that a capacitor does not charge instantaneously. It takes *time* to charge it. Our concern here is with the question, "How much time does it take to charge a capacitor?"

The manner in which C charges is shown in the accompanying graph (Fig. 49-4b). While S_1 was open, C did not charge because the battery voltage was not applied across the *RC* circuit. Thus, with S_1 open, the applied voltage $V = 0$, the charging current $i_c = 0$, the voltage across R, $v_R = 0$, and the voltage across C, $v_c = 0$. At the *instant* switch S_1 is closed, the battery "sees" only the resistance R in the circuit. That is, C acts like a short circuit. Hence, all the applied voltage V appears across R (at the instant S_1 is closed). There is maximum charging current limited only by the size of R, and there is zero voltage across C. But as S_1 remains closed, charging current causes a voltage to build

up across C, with the polarity shown in Fig. 49-4a. This is a bucking voltage which opposes the battery voltage and acts to reduce the charging current.

At any instant of time after the start of the charging interval, the voltage which causes current to flow in the circuit is the voltage source V minus the voltage v_c to which the capacitor has already charged. If we call this the active voltage V_A, then

$$V_A = V - v_c \qquad (49\text{-}2)$$

It is apparent that when $v_c = V$, the capacitor is fully charged, $V_A = 0$, and current in the circuit ceases.

The decay in voltage across the resistor follows the decay of charging current. Thus, at the end of the charging interval, though the battery is still applied across the *RC* circuit, the charging current i_c is zero, $v_R = 0$, and $v_c = V$.

Figure 49-4b shows that it took time for C to charge. From a mathematical analysis of this circuit, which is beyond the scope of this book, we find that it takes one time constant to charge C to 63.2 (approximately) percent of the applied voltage. The time required to charge a capacitor to 63.2 percent of the applied voltage is known as the *time constant* of the circuit. The value of the time constant t in seconds is equal to the product of the resistance R in ohms and the capacitance of C in farads. Thus

$$t = R \times C \qquad (49\text{-}3)$$

A "time constant" is not a fixed unit of time such as a second, but is a *relative* unit of time whose actual value in seconds is determined by the values of R and C in the charging circuit. Thus, in Fig. 49-4, if $R = 1$ MΩ and $C = 1$ μF, $t = 1$ s. But if $R = 100,000$ Ω and $C = 0.1$ μF, then $t = 0.01$ s.

The student may find the following relations a useful reference in calculating time constants:

(a)

(b)

Fig. 49-4. Capacitor C charging in an *RC* circuit.

R (in ohms) $\times$ C (in farads) = t (in seconds)
R (in megohms) $\times$ C (in microfarads)
$$= t \text{ (in seconds)}$$
R (in ohms) $\times$ C (in microfarads)
$$= t \text{ (in microseconds)}$$
R (in megohms) $\times$ C (in picofarads)
$$= t \text{ (in microseconds)}$$

Charge Rate of a Capacitor

An accurate graph of the rise of voltage across a capacitor which is charging is given in the "charge" graph of Fig. 49-5. This shows how the voltage rises across a capacitor C charging through a resistance R toward a voltage source V. The horizontal axis is the time axis set off in time constants (RC). The vertical axis is a percentage axis, with 100 percent representing the total voltage to which a capacitor can charge.

Examination of this graph shows that capacitor C charges approximately 63 percent in one time constant, 86 percent in two time constants, 95 percent in three time constants, 98 percent in four time constants, 99 percent (practically 100 percent) in five time constants. Let us consider these figures further. We note that at the end of the first time constant the "active voltage" is 37 percent of the original applied voltage. Also, during the second time constant the increase in charge of capacitor C is 23 percent (86 − 63 = 23). Now, this 23 percent rise during the second RC is 63 percent of 37. This suggests the rule which actually holds true for capacitor charge: In every time constant, a capacitor charges 63 (more accurately, 63.2) percent of the "active voltage." The active voltage can, of course, be computed using Eq. (49-2).

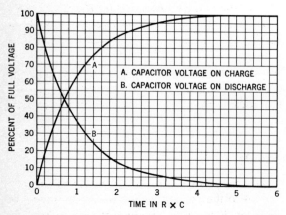

Fig. 49-5. Universal chart for charge and discharge of a capacitor.

Fig. 49-6. Capacitor discharging in an RC circuit.

Though a capacitor can theoretically "never" charge fully to the applied voltage, since there is always a remainder of 37 percent of the active voltage, to all intents and purposes we assume that C is fully charged at the end of five time constants.

Capacitor Discharge—Time Required

Now how does a capacitor discharge? Assume that C in Fig. 49-4 has charged to the full applied voltage. We now open S_1 and close S_2, as in Fig. 49-6a, permitting C to discharge through R. The variation of discharge current i_d, the change in voltage v_R across R, and the decay in voltage v_c across C are shown in the graphs of Fig. 49-6b. We note that the discharge current i_d is opposite in direction to the charge current i_c. Therefore, the voltage v_R across R has the opposite polarity to that in Fig. 49-4b.

Figure 49-5 shows the discharge curve for C. We note the following significant similarities between the discharge of C and the charge of C:

In one RC, C discharges 63 percent of its total voltage. In two RCs, C discharges 86 percent of its total voltage, etc. The rule for discharge is therefore similar to the rule for charge: A capacitor discharges 63 percent of its remaining voltage in one time constant.

If we refer to the graphs in Figs. 49-4b and 49-6b, we note that Kirchhoff's law applies both to charge and discharge circuits. In each case, at any instant of time, the applied voltage V is equal to the sum of the voltage across the capacitor and that across the resistor. That is,

$$V = v_c + v_R \qquad (49\text{-}4)$$

Before concluding our discussion of time constants, we should mention that we refer to the curves in Fig. 49-5 as exponential curves. That is, the equation which describes how v varies as a function of time involves t (time) as a variable exponent. Thus

$$v_c = V(1 - \varepsilon^{-t/RC}) \qquad (49\text{-}5)$$

where ε is a number (it is the base of the natural logs) equal to 2.71828.

Experimental Techniques in Observing Capacitor Charge and Discharge

We can set up an experiment using simple equipment to demonstrate how long it takes a capacitor to charge and discharge. However, our results will be approximate because we will not be using laboratory precision timing or measuring equipment.

Refer to Fig. 49-7a. By selecting the values of R and C, we can find the length of time it takes C to charge and discharge, using an ordinary watch with a seconds hand. In determining the charge time for C we use a large nonelectrolytic capacitor, say 1 μF, a 1-MΩ resistor, an EVM, a variable dc voltage source, and a watch. When we close switches S_1 and S_2, C will start charging toward the dc supply voltage. The EVM will read v_c, the voltage across C. We will find that in approximately 5 RCs (5 s in this case) the capacitor will be fully charged.

We should note here that C will not actually charge up to the total applied voltage, but to a value determined by the voltage divider, consisting of the internal resistance of the EVM and the value of R. Thus, if we use an EVM with an input resistance of 11 MΩ and if R is a 1-MΩ resistor, C will charge up to $^{11}/_{12}$ of the applied dc voltage. Thus if we wish to have C charge to 100 V, we would set the output of our dc supply to 109 V (approximately).

To observe the length of time it takes C to discharge, we open S_1. Now C will discharge through the input resistance of the meter which we assumed to be 11 MΩ. The discharge time constant is $C \times$ resistance of the meter and in this case would be 11 s. We can verify the discharge voltage at RC intervals, 11-s intervals in this case. Our results should roughly conform to the capacitor discharge curve (Fig. 49-5).

If, in the circuit of Fig. 49-7a, R is replaced by the EVM, as in Fig. 49-7b, another means is provided for determining the charge time of C. When switch S_1 is closed, C will charge through the input resistance R_{in} of the EVM. At the moment S_1 is closed, the full-battery voltage V appears across R_{in}, and the EVM measures the full voltage V. There is, of course, *zero* voltage across C at that instant. As C charges, the voltage v_c developed across C increases, while the voltage v_R developed across the input resistance of the EVM decreases. At the end of one RC, only 37 percent $(100 - 63)$ of the applied voltage V will be measured on the meter. At the end of two RCs, 14 percent $(100 - 86)$ will be measured by the EVM. At the end of the third, fourth, and fifth RCs, the EVM will show respectively, 5, 2, and 1 percent (approximately) of the applied voltage V. Of course the charging time constant in this circuit is $R_{in}C$.

Using this method to determine the length of time it takes C to charge, we would connect the circuit of Fig. 49-7b and set V to some value, say 100 V. The start of the timed interval would begin when S_1 is closed, and would end when the meter voltage dropped to 1 V (approximately). If the meter reading stabilizes at a measurable value of voltage, say 3 V, it does so because of capacitor leakage. In that event, readings should be discontinued just when the voltage reaches this stable level.

NOTE: As the voltage measured on the EVM falls, we can switch the meter to a lower range without disturbing the charging time constant.

By this method we can measure the charging time of C somewhat more accurately than by the preceding means. Hence it will be used in this experiment.

Fig. **49-7.** Experimental circuit for charge and discharge of a capacitor.

SUMMARY

1. The charge in coulombs Q on a capacitor is equal to the capacitance C in farads times the voltage V in volts to which the capacitor has charged. Thus, $Q = C \times V$.

2. The time required to charge a capacitor to 63.2 percent (approximately) of the applied voltage is called a *time constant*.
3. When a capacitor of C farads is charging through a resistance of R ohms, the time constant t in seconds of the charging circuit is $t = R \times C$.
4. If at any instant of time while a capacitor is charging, the active voltage V_A equals the difference between the charging source voltage V, and the voltage on the capacitor, that is,

$$V_A = V - V_C$$

then the added voltage to which the capacitor charges in one time constant is 63.2 percent of V_A.
5. It takes five time constants for a capacitor to charge to 99+ percent (approximately) of the applied voltage. We can say therefore that a capacitor charges to the applied voltage in five time constants.
6. When a capacitor of C farads is discharging through a resistor of R ohms, the discharge time constant t in seconds is $t = R \times C$.
7. When discharging, a capacitor will lose 63.2 percent of its charge in *one* time constant.
8. In each succeeding time constant the capacitor will lose an additional 63.2 percent of the voltage still across it.
9. It takes five time constants, approximately, to discharge a capacitor.
10. The graph of the charge and discharge of a capacitor is exponential and is shown in Fig. 49-5, the universal chart for the charge and discharge of a capacitor.
11. A time constant is a *relative*, not an absolute, measure of time.
12. The formula which gives the voltage V_C to which a capacitor charges at any instant of time t after the start of the charging cycle is

$$V_C = V(1 - \varepsilon^{-t/RC})$$

where V is the applied voltage and ε is the number 2.71828 (the base of the natural logs).

SELF-TEST

Check your understanding by answering these questions:
1. A 0.25-μF capacitor is charging through a 2.2-MΩ resistor toward an applied voltage of 50 V. In one time constant the capacitor will have charged to _____ V.
2. The capacitor in question 1 has a time constant of _____ s.
3. In _____ s, the capacitor in question 1 will have charged to 43 V.
4. A 0.05-μF capacitor charged to 100 V is permitted to discharge through a 220-kΩ resistor. At the end of _____ s the voltage across the capacitor will have dropped to 37 V (approximately).
5. It will take _____ time constants to discharge the capacitor almost completely.
6. The discharge time constant in question 4 is _____ s.
7. The charging or discharging curve shown in the universal time-constant chart (Fig. 49-5) is called an _____ curve.

MATERIALS REQUIRED

■ Power supply: Regulated variable direct current
■ Equipment: EVM
Optional for special credit: Square-wave generator, oscilloscope
■ Resistors: ½-W 1-MΩ; 10,000-Ω
■ Capacitors: 1 μF; 0.005 μF
■ Miscellaneous: Two SPST switches; watch with a seconds hand

PROCEDURE

Input Resistance of EVM

1. Connect the circuit of Fig. 49-7a. S_1 and S_2 are open. Set $V = 33$ volts (approximately).
2. Close S_1. If your EVM has an input resistance of 11 MΩ, the meter will read 30 V. If it does not, readjust V until the voltmeter reads 30 V.
3. With the EVM, measure the voltage V across the dc supply and record in Table 49-1.
4. Compute and record in megohms the input re-

TABLE 49-1. Discharge Time Constant of a Capacitor

V, volts	R_{in}, ohms		Discharge Time Constant, $R_{in}C$ _____ s	Volts across C
	Computed	Rated		
				30

Discharge Time, RC	v_c, volts				
	1st Trial	2d Trial	3d Trial	Average	Computed
1					
2					
3					
4					
5					
10					

sistance R_{in} of your EVM by substituting the measured value V in the formula

$$R_{in} = \frac{30}{V - 30} \times 1 \text{ M}\Omega \qquad (49\text{-}6)$$

Record also the rated value for R_{in}, including the resistance in the EVM probe. The computed and rated values should be the same (approximately).

5. Close S_2. Observe that the voltage indicated on the EVM falls to zero the moment S_2 is closed and then rises as C charges. After 10 s, approximately, C should be fully charged and the voltmeter should measure 30 V.

NOTE: Some meters are heavily damped and should be avoided in this experiment where instantaneous time is a factor.

Discharge Time Constant

6. Compute the discharge time constant $R_{in}C$ and record in Table 49-1. Use the rated value of R_{in}.
7. Open S_1. Observe and record v_c, the voltage across C, at every discharge time constant interval shown in Table 49-1. For example, if $R_{in}C = 11$ s, readings should be made at 11-s intervals and recorded in column labeled "v_c, 1st trial," until C is completely discharged.

NOTE: When the voltage read on the meter stabilizes at some level, you are reading a voltage produced by the dielectric strain within the capacitor. This may occur at about the tenth time constant.

8. Close S_1 and permit C to charge fully (30 V).
9. Repeat step 7. Record your readings in column labeled "v_c, 2d trial."
10. Repeat steps 8 and 9, recording your readings in column labeled "v_c, 3d trial."
11. Average the three values of v_c for one RC and record in column "v_c, average." Average also and record the value of v_c for two RCs, three RCs, etc. Draw a graph (in red pencil) of v_c (average) versus RC. Let RC be the horizontal axis.
12. Compute, and record in Table 49-1, the voltage v_c which should appear across C after one RC, two RCs, etc. For comparison with step 11, draw a graph (in green) of the computed values of v_c versus RC.

Charge Time Constant

13. Connect the circuit of Fig. 49-7b. Set the EVM on the 50-V or next higher range. Close S_1 and S_2 and set V at 30 V, as measured on the voltmeter.
14. Open S_2. Observe and record v_R, the voltage across the EVM, at every charge time-constant interval shown in Table 49-2.

NOTE: The charge time constant is the same as the discharge time constant, $R_{in}C$, recorded in Table 49-1.

Record readings in the column labeled "v_R, 1st trial." Compute the voltage v_c across the capacitor at every time-constant interval and re-

TABLE 49-2. Charge Time Constant of a Capacitor

Charge Time Constant, $R_{in}C$ ____ s	v_R, volts			v_c, volts				
	1st Trial	2d Trial	3d Trial	1st Trial	2d Trial	3d Trial	Average	Computed
1								
2								
3								
4								
5								
10								

cord in the column "v_c, 1st trial." Use the equation

$$v_c = V - v_R = 30 - v_R \qquad (49\text{-}7)$$

15. Close S_2, thus discharging capacitor C. V should be 30 V. Open S_2 and repeat step 14. Record the measured voltage in the column "v_R, 2d trial," and the computed values in the column labeled "v_c, 2d trial."

16. Repeat step 15. Record the measured voltage in the column "v_R, 3d trial," and the computed values in the column "v_c, 3d trial."

17. Average the three values of v_c at every time-constant interval and record in column "v_c, average."

18. Compute, and record in Table 49-2, the voltage v_c which should appear across C at the end of one RC, two RCs, etc. For comparison with the graph in step 17, draw a graph of the computed values of v_c versus RC (in ink).

QUESTIONS

1. What is meant by the term "time constant" for RC circuits?
2. How can we use the charging curve of a capacitor to measure time? Relate your answer to the universal time-constant chart.
3. What factors limit the accuracy of your measurements in this experiment?
4. What is the purpose of averaging the three trial values of v_c?
5. How do the average values of v_c compare with the computed values for:
   (a) The charge of a capacitor?
   (b) The discharge of a capacitor?
   Explain any discrepancies between computed and average values.
6. Do the graphs of v_c versus RC for charge, and v_c

versus RC for discharge, agree with the universal time-constant chart? Comment on any variations from the normal.
7. Compute the time constant for the following values of R and C:
   (a) $R = 220,000$ Ω, $C = 0.003$ μF
   (b) $R = 5,600$ Ω, $C = 0.25$ μF
   (c) $R = 2.2$ MΩ, $C = 250$ pF

Extra Credit

8. A periodic square wave (Fig. 49-8b) is applied across an RC circuit (Fig. 49-8a) and the voltages across R and C in turn are observed on an oscilloscope. Draw the waveforms which would appear across R and C. Explain how you determined what the waveforms would be.
9. Experimentally confirm your answer to question 8. Explain in detail the procedure you used and the results of your findings.

Answers to Self-Test

1. 31.6
2. 0.55
3. 1.1
4. 0.011
5. 5
6. 0.011
7. exponential

Fig. 49-8. Waveforms in an RC circuit.

50

Reactance of a Capacitor (X_C)

OBJECTIVE

To verify experimentally that the capacitive reactance X_C of a capacitor is given by the formula

$$X_C = \frac{1}{2\pi FC}$$

INTRODUCTORY INFORMATION

Reactance of a Capacitor

The capacitive reactance X_C of a capacitor is the amount of opposition it offers to current in an ac circuit. The unit of capacitive reactance is the ohm. However, like the X_L of a coil, the X_C of a capacitor cannot be measured with an ohmmeter. Rather, capacitive reactance must be measured indirectly from its effect on current in an ac circuit.

X_C is dependent on frequency and is given by the formula

$$X_C = \frac{1}{2\pi FC} \qquad (50\text{-}1)$$

Here X_C is in ohms, C in farads, and F is frequency in hertz (cycles per second). The value of C in microfarads (μF) may be substituted directly in the formula

$$X_C = \frac{10^6}{2\pi FC} \qquad (50\text{-}2)$$

For example, suppose it is required to find the reactance of a 0.1-μF capacitor at a frequency of 1,000 Hz. Substituting in Eq. (50-2),

$$X_C = \frac{10^6}{(6.28)(1,000)(0.1)} = 1,592 \ \Omega$$

It is apparent from Eqs. (50-1) and (50-2) that the higher the frequency, the smaller the reactance of a capacitor; and the lower the frequency, the higher the reactance. For direct current, where $F = 0$, X_C is infinite. Therefore, direct current will not flow through a capacitor.

The reactance of a capacitor may be determined by measurement. In the circuit of Fig. 50-1, a sinusoidal voltage V causes a current I to flow in the circuit. Ohm's law, extended to ac circuits, states that

$$I = \frac{V}{Z} \qquad (50\text{-}3)$$

where I is measured in amperes, V in volts, and Z in ohms. The term Z stands for impedance, which is the opposition to ac current. In Fig. 50-1, the reactance of C is the impedance[1] of the circuit. That is,

$$X_C = Z \qquad (50\text{-}4)$$

Hence, the capacitive current I is given by the equation

$$I = \frac{V}{X_C} = \frac{V_C}{X_C} \qquad (50\text{-}5)$$

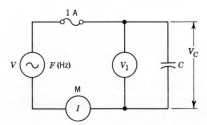

Fig. 50-1. A sinusoidal voltage V causes current I to flow in the circuit. We assume here that the resistance of the meter is zero. Therefore $Z = X_C$ and $V_C = V$.

[1] We assume here that the resistance of the current meter is so small, compared to X_C, that we may ignore its effect on ac current.

Fig. 50-2. Series *RC* circuit.

We see, therefore, that the amplitude of alternating current through a capacitor across which an ac voltage is connected is directly proportional to the amplitude of voltage (V_C) across the capacitor, and inversely proportional to the reactance (X_C) of the capacitor.

Equation (50-5) may be written in the form

$$X_C = \frac{V_C}{I} \qquad (50\text{-}6)$$

We may employ Eq. (50-6) in determining the reactance of a capacitor at a given frequency *F*. This is accomplished by measuring the voltage V_C across *C* with an ac voltmeter *V*, and by measuring the current *I* in the circuit with an ac ammeter. The values of V_C and *I* are then substituted in Eq. (50-6) and X_C is computed. Note in the experimental circuit of Fig. 50-1 that the voltmeter V_1 is connected directly across *C* and not across the voltage source *V*. Thus, we measure V_C and not *V*. The reason is that the ammeter *M* has resistance, across which there will be a voltage drop. If the resistance of *M* is high, its effect on impeding current would have to be considered if the applied voltage *V* were measured. By measuring V_C, Eq. (50-6) can be used to find X_C.

An alternate method for measuring alternating current does not require the use of an ac ammeter.

Consider the circuit of Fig. 50-2. This is a series circuit. Thus, the current flowing through *R* and *C* is the same. With this circuit arrangement it is possible to determine the value of X_C experimentally by measuring the voltage V_C across *C* and V_R across *R*. To determine the current *I* in amperes in the circuit, the known value of *R* in ohms and the measured value V_R in volts is substituted in Ohms' law, Eq. (50-7).

$$I = \frac{V_R}{R} \qquad (50\text{-}7)$$

Knowing *I*, it is possible to find X_C by substituting

for *I* and V_C in Eq. (50-6). It is also possible to find X_C without computing *I* by combining Eq. (50-6) and (50-7) and solving for X_C. We get

$$X_C = \frac{V_C}{V_R} \times R \qquad (50\text{-}8)$$

By substituting the measured values V_C, V_R, and *R* in Eq. (50-8), the value of X_C is determined. Having found X_C experimentally, it is possible to verify approximately the validity of the formula for X_C. This is done by comparing the experimental value of X_C with the formula value. The capacitance *C* must be known or measured.

SUMMARY

1. The amount of opposition that a capacitor offers to current in an ac circuit is called capacitive reactance (X_C) and is measured in ohms.
2. The capacitive reactance of a capacitor is not constant but is inversely proportional to the ac frequency (*F*). That is, the higher *F* is, the smaller is X_C; the lower *F* is, the greater is X_C.
3. X_C is also inversely proportional to the capacitance (*C*) of a capacitor.
4. Capacitive reactance of a capacitor *C*, in a circuit of frequency *F*, may be computed from the formula

$$X_C = \frac{1}{2\pi FC}$$

   where X_C is in ohms, *F* in hertz, and *C* in farads.
5. When *C* is given in microfarads (μF),

$$X_C = \frac{10^6}{2\pi FC}$$

6. X_C cannot be measured directly—only by its effect in an ac circuit. Thus in Fig. 50-1, we measure the voltage V_C across *C* and the current *I* in the circuit. Then, by extending Ohm's law to ac circuits, we can find X_C by substituting V_C and *I* in the formula

$$X_C = \frac{V_C}{I}$$

7. Another method of determining X_C is illustrated by the circuit of Fig. 50-2. Here the voltages V_R and V_C are measured across *R* and *C*, respectively. Then X_C is computed by substituting V_R and V_C in the formula

$$X_C = \frac{V_C}{V_R} \times R$$

Check your understanding by answering these questions:

1. The reactance of a capacitor is given in _____ .

2. The reactance of a capacitor depends on the _____ of the applied voltage source and the _____ of _____ .

3. The X_C of a capacitor whose value is 0.1 μF in a circuit where a 60-Hz signal is applied is _____ Ω.

4. If the frequency of the signal is changed to 6,000 Hz, the X_C of the 0.1 μF capacitor is _____ Ω.

5. The voltage across a capacitor is 9.5 V, and the current in the capacitor is 0.005 A. The capacitive reactance of the capacitor is _____ Ω.

6. A 1,000-Ω resistor and a 0.05 μF capacitor are connected in series. The voltage measured across the resistor is 3.5 V, and the voltage measured across the capacitor is 31.5 V. The X_C of the capacitor is _____ Ω.

7. The frequency of the applied voltage in question 6 is _____ Hz.

MATERIALS REQUIRED

- Power supply: Isolation transformer and variable autotransformer operating from the 120-V line
- Equipment: Electronic voltmeter, oscilloscope, 0-5 mA ac milliammeter, capacitance tester
- Resistors: ½-W, 5,600-Ω
- Capacitors: 0.5 μF, 0.1 μF
- Miscellaneous: Fused line cord

PROCEDURE

Determining X_C by Voltmeter-ammeter Method

1. With a capacitance tester, measure the values of C listed in Table 50-1 and record the figures in Table 50-1.
2. Connect the circuit (Fig. 50-1). Plug a variable autotransformer (ac supply) into the isolated outlet of an isolation transformer connected to the ac line. V_1 is an EVM set on the 25-volt ac voltage range. M is a 5-mA ac milliammeter. C is a 0.5-μF nonelectrolytic capacitor. Set the output of the ac supply at zero.
3. **Power on.** Increase the output of the ac supply until 2 mA of current is measured on the milliammeter. With the voltmeter, measure the voltage V_C across the capacitor and record in Table

50-1. Compute, and record in Table 50-1, the X_C of the capacitor using the measured values of V_C and I. Show your computations. Increase the ac voltage until $I = 3$ mA. Record V_C and the computed value of X_C in Table 50-1.
4. Repeat step 3 for $I = 4$ mA. Compute the value of X_C, substituting the measured value of C in Eq. (50-2). F is the ac line frequency. Show your computation. Repeat steps 3 and 4 for 0.1-μF capacitor, setting current to the levels shown in Table 50-1.

Determining X_C by Voltage Method

5. Measure the 5,600-Ω resistor listed in Table 50-2 and record its value. Record also the mea-

TABLE 50-1.

C (μF)				X_C (ohms)	
Rated	Measured	I (mA)	V_C (volts)	$X_C = \dfrac{V_C}{I}$	$X_C = \dfrac{10^6}{2\pi FC}$
0.5		2			
		3			
		4			
0.1		1			
		2			
		3			

TABLE 50-2.

C (μF)		R (ohms)		V_C, volts peak-to-peak	V_R, volts peak-to-peak	I_{mA}	X_C (ohms)	
Rated	Measured	Rated	Measured				$X_C = \dfrac{V_C \times R}{V_R}$	$X_C = \dfrac{10^6}{2\pi FC}$
0.5		5,600						
0.1		5,600						

sured values of the capacitors from Table 50-1. Record also the formula values of $X_C = 10^6/2\pi FC$, from Table 50-1.

6. Connect the circuit of Fig. 50-2. C is the 0.5-μF capacitor used previously. The line-isolated voltage supply V is set at 6.3 or 12.6 V rms 60 Hz or line frequency. Calibrate the oscilloscope for peak-to-peak ac voltage measurement. Measure, and record in Table 50-2, the peak-to-peak voltage V_C across C and the voltage V_R across R. Keep hot lead of scope at junction of R and C. Compute X_C by using (1) V_C, V_R, and R and substituting these values in Eq. (50-8), and (2) the measured value of C and the line frequency. Show your computations.

7. Repeat step 6 for $C = 0.1 \ \mu F$.

QUESTIONS

1. In your own words state Ohm's law for ac circuits.
2. (a) How would you apply Ohm's law to a pure capacitive circuit to determine the current in the capacitor?

(b) Which parameters (variables) must you know to determine current in a capacitor?
3. Upon what parameters does the reactance (X_C) of a capacitor depend? Explain.
4. Refer to the data in Tables 50-1 and 50-2. Does the experimental value of $X_C(V_C/I)$ remain constant in this experiment? Should it? Comment on any unexpected results.
5. How does the measured value of $X_C(V_C/I)$ compare with the formula value $10^6/2\pi FC$? Comment on any discrepancy.
6. List the factors which limit the accuracy of the measurements in this experiment.

Answers to Self-Test

Answers to slide-rule accuracy.

1. ohms
2. frequency; value; capacitance
3. 26,500
4. 265
5. 1,900
6. 9,000
7. 354

51

EXPERIMENT

Measuring Phase Angle between Voltage and Current in a Capacitive Circuit

OBJECTIVES

1. To measure the phase angle between voltage and current in a capacitive circuit by means of Lissajous patterns
2. To measure the phase angle between voltage and current in a capacitive circuit directly by using external synchronization or external triggering

INTRODUCTORY INFORMATION

Phase Relations between Voltage and Current in a Capacitor

In a circuit containing only capacitance, voltage and current are *not* in phase, but current *leads* voltage by 90°. A graph of voltage and current in a pure capacitive circuit, Fig. 51-1, shows this relationship. You will recall from Experiment 45 that this characteristic of a capacitor is just the opposite of that of an inductor, for in an inductor current *lags* voltage by 90°. It is apparent, then, that capacitors and inductors have opposite effects on voltage and

current in an ac circuit. These effects will be considered in later experiments.

Measuring Phase Relations between Voltage and Current in a Pure Capacitive Circuit by Lissajous Patterns

The phase relationship between voltage and current can be demonstrated with an oscilloscope, employing the Lissajous-pattern technique as in Experiment 45. Refer to Fig. 51-2, which shows a sinusoidal voltage source v applied to resistor R connected in series with a capacitor C. The value of resistance R selected is such that the capacitive reactance X_C is very much greater than R. If the ratio of X_C/R is greater than 10, the circuit will act almost like a pure capacitance. The voltage v_R across R in the circuit is in phase with the current i. Hence by measuring the phase angle between v_R and the applied voltage v, we are in fact measuring the phase angle between voltage and current in the capacitor, ignoring the small error introduced by the resistor R. The student is referred to Experi-

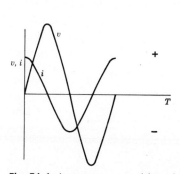

Fig. 51-1. In a pure capacitive circuit voltage and current are out of phase, and current leads voltage by 90°.

Fig. 51-2. Experimental circuit for demonstrating phase relationship between V and I in a capacitor.

Fig. 51-3. Oscilloscope connections for demonstrating directly phase relations between the applied voltage v and current i in a capacitive circuit. The sync/trigger selector of the oscilloscope is set to Ext. The applied voltage v is first displayed on the oscilloscope and acts as the reference waveform.

ment 45 for a discussion of the Lissajous-pattern method of measuring phase angles.

Direct Phase Measurements Using External Synchronization or Triggering

Phase angles between two waveforms having the same frequency, but appearing at different points in a circuit, may be measured directly with an oscilloscope. Either a triggered or nontriggered oscilloscope may be used. In the case of the triggered type, *triggering* must be external, that is, the trigger selector switch must be set to Ext. In the case of the nontriggered oscilloscope *synchronization* of the sweep must be external, that is, the sync selector switch must be set to Ext. Since it is anticipated that this text-laboratory manual will be used by students with both triggered and nontriggered oscilloscopes, in the discussion which follows the term "trigger/synchronize" will be employed to cover both types.

The technique requires that the sine-wave v applied to the series RC circuit (Fig. 51-3) be fed to the external sync/triggering jack of the oscilloscope and the sync/trigger selector switch set to Ext. The oscilloscope-sweep circuit is then synchronized/triggered at the same time, regardless of the point in the circuit where the waveform is being observed. The applied voltage v fed to the vertical input of the oscilloscope serves as the reference for phase measurements. The reference sine wave is conveniently centered on the graticule, so that the start of the sine wave coincides with a major vertical graph line A (Fig. 51-4).

If the oscilloscope sweep is linear, equal distances on the time base correspond to equal angular displacements. Thus in Fig. 51-4, one division is divided into five equal graduations. Since one alternation (180°) covers two graticule divisions, each graduation corresponds to 18°.

Once the reference waveform has been properly centered and its width in divisions noted, the number of degrees per graduation can be calculated. Then the vertical input (hot) lead of the oscilloscope is connected to point P (Fig. 51-3). The voltage waveform across R will now appear on the oscilloscope. The displacement (number of graduations) of the observed waveform, from the reference voltage, is measured on the time base. Multiplying the calculated *degrees per graduation* by the *number* of graduations of displacement, we get the phase difference between the applied voltage and current.

The position of the waveform across R relative to the reference waveform tells us whether the current i leads or lags the applied voltage v. Thus in Fig. 51-5, waveform 1 leads the reference waveform (ABC) by 90°. Waveform 2 lags by 90°.

NOTE: If a nontriggered oscilloscope is used, the width of the time base may be adjusted as desired for convenient measurements. For example the width of the sweep in Fig. 51-4 may be adjusted so that the waveform ABC is 2, 3, or 4 in. wide.

SUMMARY

1. In a capacitive circuit, alternating current and voltage are not in phase. If the circuit is a pure capacitance, that is, if it has no resistance, current leads voltage by 90°.
2. We can demonstrate this phase difference between applied voltage and current in a capacitive circuit by means of Lissajous patterns, using the circuit of Fig. 51-2.
3. In the circuit of Fig. 51-2, if $X_C/R > 10$, then the

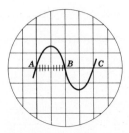

Fig. 51-4. Horizontal sweep used directly for phase measurements.

effect of R can be ignored and the circuit treated like a pure capacitance.

4. It is also possible to measure the phase difference between voltage and current in a capacitor directly, by using an oscilloscope set on Ext. sync/triggering, employing the circuit of Fig. 51-3.

5. The reference voltage waveform v is then centered on the graticule as in Fig. 51-4.

6. Once v is centered, the vertical hot lead of the oscilloscope is connected to the junction of C and R, point P in Fig. 51-3. The waveform across R now appears on the oscilloscope. This is the required current (i) waveform.

7. The horizontal displacement between v and i is measured on the oscilloscope graticule and changed into degrees.

8. Whether the current waveform leads or lags the reference waveform can be determined by noting if the current waveform is positive or negative, respectively, at the start of the reference waveform (that is, at the time when the positive alternation of the reference waveform is just starting).

SELF-TEST

Check your understanding by answering these questions:

1. In a pure capacitive circuit current _____ voltage by _____ degrees.

2. If the Lissajous-pattern method of determining phase difference is used, the pattern describing the phase relations between the applied voltage v and the current i in a pure capacitor will appear as a _____ on the oscilloscope, when the vertical and horizontal deflections are equalized.

3. In the Lissajous-pattern method the Int. sweep of the oscilloscope must be _____

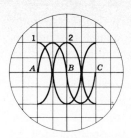

Fig. 51-5. Waveform 1 leads waveform ABC by 90°. Waveform 2 lags waveform ABC by 90°.

_____. The reference voltage is then applied to the _____ input of the oscilloscope, and the current waveform is applied to the _____ input of the scope.

4. In the direct method of measuring phase relations, the sync/triggering switch of the oscilloscope must be set to _____.

5. In the direct method of measuring phase difference, the width of the complete reference sine waveform on the oscilloscope is two divisions. Each division is divided into 10 graduations. The displacement between the start of the waveform being measured and the start of the reference waveform is three graduations. The phase difference is _____ degrees.

6. In question 5 the displaced waveform is in its negative alternation, just at the time that the reference waveform is starting to go positive. The displaced waveform therefore _____ the reference waveform.

MATERIALS REQUIRED

■ Equipment: Oscilloscope, AF signal generator
■ Resistors: ½-W 3,300-Ω
■ Capacitors: 0.05 μF

PROCEDURE

Measuring Phase Relations in a Capacitor by Lissajous Patterns

1. Disable the internal sweep of the oscilloscope by setting the sweep selector switch on Ext.

2. Connect the circuit of Fig. 51-2. A sine-wave AF generator, set at 60 Hz maximum output, is the voltage source. The output of the generator

is the reference voltage v_{AG} applied to the vertical input terminals of the oscilloscope.

NOTE: v_{AG} may be considered the voltage across C because $X_C/R > 10$ in this circuit. The voltage v_{BG} across R has the same phase as the current in R. Hence v_{BG} has the same phase as the current in C.

The voltage v_{BG} is connected to the horizontal input of the oscilloscope. G, common to both inputs, is connected to the ground terminal of the oscilloscope.

3. Adjust the vertical- and horizontal-gain controls for equal vertical and horizontal deflection. The Lissajous pattern you see should be a circle, showing that there is a 90° phase difference between voltage and current in a capacitor.

NOTE: If you cannot get sufficient horizontal deflection, reverse the vertical and horizontal input leads and repeat procedural step 3.

Direct-Phase Measurement Using External Sync/Triggering

4. Connect the circuit of Fig. 51-3. Set the sync/trigger selector switch to Ext. Feed the AF generator output signal to the Ext. sync/trigger jack on the oscilloscope. Adjust the generator output 60-Hz signal to 20 V peak-to-peak. Set oscilloscope vertical-gain controls to two or three divisions of vertical deflection. Set oscilloscope controls for a two- or three-cycle display.

5. Center the waveform vertically with respect to the horizontal graticule axis. Center the waveform horizontally so that the start of the positive alternation coincides with one of the major vertical graph lines on the graticule, as in Fig. 51-4. Determine the degree per (horizontal) graduation factor of the time base _____.

6. Now connect the vertical input lead to point P, the junction of R and C in Fig. 51-3. Readjust the vertical-gain controls for two or three divisions of vertical deflection on the screen. Ob-

serve the waveform. Does it lead or lag the reference waveform? _____

7. Measure the number of graduations of horizontal displacement of the waveform across R_1 from the reference and convert into degrees of phase difference. _____ degrees

QUESTIONS

1. Explain in detail, illustrating with diagrams, the Lissajous-pattern method you used for measuring the phase relationship between voltage and current in a capacitor.
2. (a) Did the current in a capacitor lead or lag the voltage applied?
   (b) By how many degrees?
3. (a) What difficulties, if any, did you encounter using the direct method of phase measurement?
   (b) How did you overcome these difficulties?
4. How did the results of the two measurements compare as to degrees of phase difference? Explain any difference in results.
5. In the experimental circuits used (Figs. 51-2 and 51-3) identify the
   (a) Voltage waveform across the capacitor
   (b) The current waveform in the capacitor
6. In Figs. 51-2 and 51-3, what is the value of the ratio X_C/R at 60 Hz? Show your computations.
7. Why are the applied voltage and the voltage across C assumed to be the same in Figs. 51-2 and 51-3?

Answers to Self-Test

1. leads; 90
2. circle
3. turned off; vertical; horizontal
4. Ext.
5. 54
6. lags

Total Capacitance of Capacitors in Series and in Parallel

OBJECTIVES

1. To verify experimentally that the total capacitance C_T of capacitors connected in parallel is $C_T = C_1 + C_2 + C_3 + \cdots$
2. To verify experimentally that the total capacitance C_T of capacitors connected in series is $1/C_T = 1/C_1 + 1/C_2 + 1/C_3 + \cdots$

INTRODUCTORY INFORMATION

Total Capacitance of Parallel-Connected Capacitors

Series- and parallel-connected capacitor combinations are frequently used in electronics. Consider the circuit of Fig. 52-1. In this circuit C_1 and C_2 are connected in parallel. The ac voltage V is applied equally across C_1 and C_2. There are two paths for current, namely, through C_1 and C_2. Current I_1 flows through C_1 and I_2 flows through C_2. I_1 and I_2 combine to form the total line current I_T in the circuit. As far as the ac generator V is concerned, it sees a capacitance C_T through which I_T is flowing. C_T is therefore the equivalent capacitance resulting from the parallel combination. The formula for C_T is

$$C_T = C_1 + C_2 \qquad (52\text{-}1)$$

The formula may be extended for more than two capacitors connected in parallel, and is

$$C_T = C_1 + C_2 + C_3 + \cdots \qquad (52\text{-}2)$$

Total Capacitance of Series-connected Capacitors

In the series-connected circuit of Fig. 52-2, the same line current I flows through C_1 and C_2. The effect of connecting these capacitors in series is to increase the total capacitive reactance and thus reduce the line current which would flow if either capacitor were in the circuit alone.

Since the reactance of the series combination is greater than the reactance of either capacitor considered alone, the total capacitance C_T must be smaller than the capacitance of either C_1 or C_2. The formula for total capacitance C_T for two or more capacitors connected in series is

$$\frac{1}{C_T} = \frac{1}{C_1} + \frac{1}{C_2} + \frac{1}{C_3} + \cdots \qquad (52\text{-}3)$$

The total capacitance C_T of capacitor combinations may be determined experimentally in a number of ways. The simplest is to connect the required combination and measure C_T by means of a capaci-

Fig. 52-I. Capacitors connected in parallel in ac circuit.

Fig. 52-2. Capacitors connected in series in ac circuit.

tance tester. If a laboratory-standard capacitance bridge is used, C_T can be measured to a high degree of accuracy.

A second method, used in this experiment, is to determine the capacitive reactance X_{CT} of the combination. Once X_{CT} is determined, C_T may be found from the equation

$$C_T = \frac{1}{2\pi F X_{CT}} \qquad (52\text{-}4)$$

The reactance of the capacitor combination may be found by measuring the current I_C and the voltage V_C across the combination. X_{CT} is then computed from the equation

$$X_{CT} = \frac{V_C}{I_C} \qquad (52\text{-}5)$$

The capacitive current may be measured with an ac ammeter. Or the ac voltage V_R across a series-connected resistor may be measured and the series current, which is also the capacitive current I_C, may be computed from the equation

$$I_C = \frac{V_R}{R} \qquad (52\text{-}6)$$

SUMMARY

1. In a circuit consisting of two or more capacitors connected in parallel, the total current I_T is equal to the sum of the branch currents. Therefore the total current is greater than the branch current in any one of the capacitors in the circuit.
2. In a circuit consisting of two or more parallel-connected capacitors, the total capacitance C_T is equal to the sum of each of the individual capacitors. That is,

$$C_T = C_1 + C_2 + C_3 + \cdots$$

3. A simple way to remember the above formula is to note that the method of computing C_T of parallel-connected capacitors is the same as the method of computing R_T of series-connected resistors.
4. In series-connected capacitors, the current is the same in each of the capacitors. However, the total reactance X_{CT} of series-connected capacitors is greater than the reactance of any one capacitor in the circuit. Therefore, the total current I_T in a circuit consisting of series-connected capacitors is less than the current would be if any one of the capacitors were in the circuit by itself.

5. In a circuit consisting of two or more series-connected capacitors, the total capacitance C_T of the combination is

$$\frac{1}{C_T} = \frac{1}{C_1} + \frac{1}{C_2} + \frac{1}{C_3} + \cdots$$

6. The total capacitance of *series-connected capacitors* is computed in the same way as the total resistance R_T of *parallel-connected resistors*.
7. One method of determining experimentally the total capacitance C_T of capacitor combinations is to measure the C_T of the combination with a capacitance bridge.
8. Another method is to measure the total alternating current I_T of the combination and the voltage V_C across the combination, and to determine X_{CT} using the formula

$$X_{CT} = \frac{V_C}{I_T}$$

9. Now, knowing X_{CT} and frequency F of the alternating current, find C_T by substituting X_{CT} and F in the formula

$$C_T = \frac{1}{2\pi F X_{CT}}$$

SELF-TEST

Check your understanding by answering these questions:

1. In Fig. 52-1, the current I_1 in C_1 is 0.05 A and the current I_2 in C_2 is 0.1 A. The total current I_T in the circuit is _____ A.
2. In the circuit of Fig. 52-1, $C_1 = 0.2$ μF and $C_2 = 0.4$ μF. The total capacitance C_T of the combination is $C_T =$ _____ μF.
3. If the applied voltage in the circuit of question 1 is 10 V, the reactance
   (a) $X_{C1} =$ _____ Ω
   (b) $X_{C2} =$ _____ Ω
   (c) $X_{CT} =$ _____ Ω
4. In the circuit of Fig. 52-2, $C_1 = 0.25$ μF and $C_2 = 0.05$ μF. The total capacitance C_T of this combination is $C_T =$ _____ μF.
5. If the frequency F of the applied voltage V is 100 Hz, then in the circuit of question 4,
   (a) $X_{C1} =$ _____ Ω
   (b) $X_{C2} =$ _____ Ω
   (c) $X_{CT} =$ _____ Ω
6. In the circuit of Fig. 52-3, $I_T = 0.015$ A, $V_C = 6.3$ V, and $F = 60$ Hz. Then

(a) $X_{CT} =$ _____ Ω
(b) $C_T =$ _____ μF

MATERIALS REQUIRED

- Power supply: Isolated 6.3-V rms (18-V peak-to-peak) ac source
- Equipment: Oscilloscope, capacitance tester, ac milliammeter (0–5 mA), EVM
- Capacitors: 0.5 μF, two 0.1 μF, 0.05 μF
- Resistor: ½-W 56,000-Ω
- Miscellaneous: One SPST switch; fused line cord

Fig. 52-3. Voltage-current method for determining X_{CT} and then C_T of parallel-connected capacitors.

PROCEDURE

Total Capacitance of Capacitors Connected in Parallel

Method 1

1. With a capacitance tester, measure the capacitance of each of the capacitors listed in Table 52-1 and record their values.
2. Measure also, and record in Table 52-1, the total capacitance C_T of each of these parallel combinations of capacitors:
   a. Capacitors C_1, C_2, C_3, C_4
   b. Capacitors C_2, C_3, C_4
   c. Capacitors C_1, C_2, C_3
   d. Capacitors C_1, C_2
   Compute, and record in Table 52-1, the C_T (formula value) of each of these combinations using the measured values of the individual capacitors for your computations. Show a sample computation.
   Example. Assume capacitor C_1 measures 0.47 μF and C_2, 0.11 μF. Then

$$C_T = C_1 + C_2 = 0.47 + 0.11 = 0.58 \ \mu F$$

Method 2

3. Connect the circuit of Fig. 52-3. The capacitors are combination **c.** from step 2. M is an ac 0–5 mA meter. The EVM is set on the 10-V range to measure the voltage across the capacitors. V is an isolated 6.3-V rms, line-derived source. Measure, and record in Table 52-2, the total current I_T and the voltage V_C across the capacitors. Compute and record the value of X_{CT} applying Eq. (52-5) to the measured values of V_C and I_T.
   Substituting the computed value of X_{CT} in Eq. (52-4), compute and record the total capacitance C_T of combination **c.** Show your computations.
   Sample computation. Assume V_{CT} is 7.8 V, $I = 0.0021$ A, and $F = 60$ Hz. Then

$$X_{CT} = \frac{V_{CT}}{I_T} = \frac{7.8}{0.0021} = 3,720 \ \Omega$$

$$C_T = \frac{1}{2\pi F X_{CT}} = \frac{1}{6.28(60)(3,720)} \ F$$

TABLE 52-1.

Capacitor No.	C_1	C_2	C_3	C_4	C_T: Parallel Combinations	
					Measured (Method 1)	Formula Value
Rated value, μF	0.5	0.1	0.1	0.05	a.	
					b.	
Measured value, μF					c.	
					d.	

TABLE 52-2.

Parallel Combination	I_T, mA	V_C, volts	X_{CT}, ohms	C_T (μF)	
				Computed (Method 2)	Formula Value
c.					
d.					

$$C_T = \frac{10^6}{6.28(60)(3{,}720)} \ \mu F$$

$$C_T = 0.715 \ \mu F$$

From Table 52-1 transcribe the formula value of combination **c** to Table 52-2.

4. Repeat step 3 for combination **d**.

Total Capacitance of Capacitors Connected in Series

Method 1

5. With a capacitance tester, measure and record in Table 52-3 the total capacitance of each of the combinations of the following series-connected capacitors:

NOTE: For rated value of capacitors, refer to Table 52-1.

a. Capacitors C_1, C_2
b. Capacitors C_1, C_2, C_3
c. Capacitors C_1, C_2, C_3, C_4
d. Capacitors C_1, C_4

Compute, and record in Table 52-3, the C_T of each of these combinations, using the measured values of the individual capacitors listed in Table 52-1.

Example. Assume capacitor 1 measures 0.47 μF and capacitor 2 measures 0.11 μF. Then for combination **a**:

and

$$\frac{1}{C_T} = \frac{1}{C_1} + \frac{1}{C_2}$$

$$\frac{1}{C_T} = \frac{1}{0.47} + \frac{1}{0.11} = 11.2$$

$$C_T = \frac{1}{11.2} = 0.089 \ \mu F$$

Method 2

6. Connect the circuit of Fig. 52-4. With an oscilloscope calibrated to measure rms volts, measure and record in Table 52-3 the voltage V_R across R and the voltage V_{CT} across the series capacitors (points A to B). Solve, and record in Table 52-3, the current I_T in this circuit by substituting the measured values of V_R and R in the equation

$$I_T = \frac{V_R}{R}$$

Compute and record the value of X_{CT} by applying Eq. (52-5) to the measured values of V_{CT} and I_T. Substituting the computed value of X_{CT} in Eq. (52-4), compute and record the total capacitance C_T of the series combination **a** shown in Fig. 52-4. (See sample computations, step 3.)

7. Add C_3, a 0.1-μF capacitor in series with C_1 and C_2, and repeat step 6. Record for combination **b**.

QUESTIONS

1. (*a*) Refer to Fig. 52-1. What would be the effect on line current (total current I_T) of adding another capaci-

TABLE 52-3.

Series Combination	C_T (μF)		V_R, volts	V_{CT}, volts	I_T, mA	X_{CT}, ohms	C_T μF (method 2)
	Method 1 Measured	Computed					
a.							
b.							
c.			X	X	X	X	X
d.			X	X	X	X	X

Fig. 52-4. Voltage-current method for determining X_{C_T} and then C_T of series-connected capacitors.

tor C_3 in parallel with C_1 and C_2?

(b) Explain why.

2. (a) Refer to Fig. 52-2. What would be the effect on line current of adding another capacitor C_3 in series with C_1 and C_2?

(b) Explain why.

3. Refer to Table 52-1. How do the measured values of C_T (using method 1) compare with the formula values of C_T [computed by using Eq. (52-1)]? Explain any discrepancies.

4. Refer to Table 52-2. How do the computed values of C_T (using method 2 for determining the total capaci-

tance of parallel-connected capacitors) compare with the formula values? Explain any discrepancies.

5. In measuring C_T of parallel-connected capacitors, which method, 1 or 2, appears to give the more accurate results? Why?

6. State in your own words the formula for finding total capacitance of capacitors connected in parallel.

7. Refer to Table 52-3. What verification, if any, do you have for the formula for total capacitance of series-connected capacitors? Explain.

8. Which of the two methods seems more accurate for measuring C_T of series-connected capacitors? Explain why?

Answers to Self-Test

Answers to slide-rule accuracy.

1. 0.15
2. 0.6
3. (a) 200; (b) 100; (c) 66.6
4. 0.0416
5. (a) 6,360; (b) 31,800; (c) 38,160
6. (a) 420; (b) 6.33

53

The Capacitive Voltage Divider

OBJECTIVES

1. To derive analytically that the voltage V_1 across a capacitor C_1 in a series-connected capacitive voltage divider is given by the formula

$$V_1 = V \times \frac{C_T}{C_1}$$

where V = voltage applied across all the capacitors and C_T = total capacitance of the series-connected capacitors

2. To confirm experimentally the formula in objective 1

INTRODUCTORY INFORMATION

AC Voltage across a Capacitor

Ohm's law extended to ac circuits states that the current I in a circuit is the ratio of the applied voltage V to the total opposition to ac current. We call this total opposition *impedance* and label it Z. Thus

$$I = \frac{V}{Z} \qquad (53\text{-}1)$$

Determination of ac impedances can be quite complex. However, in a circuit containing only capacitance C, such as that in Fig. 53-1, the total opposition to alternating current is the reactance X_C of capacitor C. Therefore, in the circuit of Fig. 53-1, Z and X_C are the same and

$$I = \frac{V}{X_C} \qquad (53\text{-}2)$$

From Eq. (53-2) it is quite evident that the voltage drop V_C across capacitor C is equal to the product

of the current I in the capacitor and the reactance X_C of the capacitor.

$$V_C = I \times X_C \qquad (53\text{-}3)$$

Capacitive Voltage Divider

If an ac voltage is applied across capacitors connected in series, as in Fig. 53-2, voltage division takes place across the capacitors. There is a voltage drop across each capacitor which is proportional to the reactance of the capacitor. Now since in a series circuit, the current I is everywhere the same in the circuit, the voltage V_1 across C_1 is

$$V_1 = I \times X_{C1}$$

Similarly $\qquad V_2 = I \times X_{C2} \qquad (53\text{-}4)$

Since capacitive reactance is inversely proportional to capacitance, the smaller the capacitor, the greater the voltage drop across it in a capacitive voltage divider. Moreover, since the reactance of a 0.05 μF capacitor is twice the reactance of a 0.1 μF capacitor, $V_1 = 2V_2$ in Fig. 53-2. If the applied voltage V were 18 V in that circuit, then $V_1 = 12$ V and $V_2 = 6$ V, with the sum of V_1 and V_2 being equal to the applied voltage V.

The relationships just discussed may be generalized. Thus in a three-capacitor divider (Fig. 53-3),

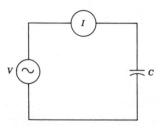

Fig. 53-1. The total opposition to alternating current in this circuit is the reactance (X_C) of C.

Fig. 53-2. Voltage distribution in a capacitive voltage divider.

$$V_1 = I \times X_{C1} = I \times \frac{1}{2\pi F C_1}$$

$$V_2 = I = \frac{1}{2\pi F C_2} \qquad (53\text{-}5)$$

$$V_3 = I \times \frac{1}{2\pi F C_3}$$

Dividing V_1 by V_2 leads to

$$\frac{V_1}{V_2} = \frac{C_2}{C_1}$$

Similarly

$$\frac{V_2}{V_3} = \frac{C_3}{C_2} \qquad (53\text{-}6)$$

and

$$\frac{V_3}{V_1} = \frac{C_1}{C_3}$$

Now consider the total reactance X_{CT} of the series-connected capacitors in Fig. 53-3.

$$V = I \times X_{CT} \qquad (53\text{-}7)$$

That is, the applied voltage V equals the product of I and the total reactance of the series-connected capacitors.

Since $V_1 = I \times X_{C1}$, dividing V_1 by V leads to

$$\frac{V_1}{V} = \frac{I \times X_{C1}}{I \times X_{CT}} = \frac{X_{C1}}{X_{CT}} = \frac{\dfrac{1}{2\pi F C_1}}{\dfrac{1}{2\pi F C_T}}$$

and

$$\frac{V_1}{V} = \frac{C_T}{C_1}$$

Therefore

$$V_1 = V \times \frac{C_T}{C_1} \qquad (53\text{-}8)$$

Equation (53-8) states that the voltage drop across any capacitor C_1, in a series-connected capacitive voltage divider, equals the product of the applied voltage and the ratio of the total capacitance C_T to

the capacitance of C_1. Note that this is similar to the formula for voltage distribution in a series-connected resistive voltage divider, with the exception that in a resistive divider V_1 (across R_1) = $V \times (R_1/R_T)$.

That is, in a series capacitive divider the positions of C_1 and C_T (see Eq. 53-8) are reversed, as compared to the positions of R_1 and R_T in a series-resistive divider.

It is now possible to apply Eq. (53-8) to the circuit of Fig. 53-2. In that circuit

$$\frac{1}{C_T} = \frac{1}{C_1} + \frac{1}{C_2} = \frac{1}{0.05} + \frac{1}{0.1}$$

and

$$C_T = \frac{1}{30} \ \mu F$$

Substituting $C_T = 1/30$, $C_1 = 1/20$, and $V = 18$ in Eq. (53-8) gives

$$V_1 = 18 \times \frac{1/30}{1/20} = 12 \ V$$

Similarly $V_2 = 6 \ V$

These are the same results as were obtained in our earlier solution of this circuit.

We can now also verify Eqs. (53-6), for according to that equation,

$$\frac{V_1}{V_2} = \frac{C_2}{C_1}$$

and

$$\frac{12}{6} = \frac{0.1}{0.05} = \frac{2}{1}$$

as expected.

SUMMARY

1. The voltage drop V_1 across a capacitor C_1 in a capacitive voltage divider is given by the formula $V_1 = I \times X_{C1}$, where I is the current in the capacitor and X_{C1} is the reactance of the capacitor.

Fig. 53-3. Capacitive voltage divider.

2. In a series-connected capacitive voltage divider the ratio $V_1/V_2 = C_2/C_1$, where V_1 and V_2 are the voltage drops across C_1 and C_2, respectively, and C_1 and C_2 are measured in units of capacitance (that is, farads, microfarads, etc.). Similarly $V_2/V_3 = C_3/C_2$, $V_3/V_4 = C_4/C_3$, etc.
3. The total current I in a capacitive voltage divider where the applied voltage is V and the total capacitance is C_T is

$$I = \frac{V}{X_{CT}}$$

4. The voltage V_1 across a capacitor C_1 in a series-connected voltage divider with applied voltage V is

$$V_1 = V \times \frac{C_T}{C_1}$$

SELF-TEST

Check your understanding by answering these questions:

1. In the circuit of Fig. 53-1, $V = 10$ V and $X_C = 1{,}000\ \Omega$. The current I in the circuit is _____ A.
2. In the circuit of Fig. 53-2, $X_{C1} = 1{,}000\ \Omega$. Therefore (a) $X_{C2} =$ _____ Ω; (b) $X_{CT} =$ _____ Ω; (c) $I =$ _____ A.
3. In the capacitive voltage divider (Fig. 53-3), $C_1 = 0.1\ \mu F$, $C_2 = 0.05\ \mu F$, $C_3 = 0.5\ \mu F$, and $V = 100$ V.
   (a) $V_1 =$ _____ V
   (b) $V_2 =$ _____ V
   (c) $V_3 =$ _____ V

MATERIALS REQUIRED

- Power supply: Isolated 6.3-V rms (18-V peak-to-peak) ac source
- Equipment: Capacitance tester, electronic voltmeter
- Capacitors: 0.5 μF, two 0.1 μF, 0.05 μF
- Miscellaneous: Fused line cord

PROCEDURE

1. With the capacitance tester check the value of each capacitor and record in Table 53-1.
2. Connect the circuit of Fig. 53-2. V is a 6.3-V rms (18 V peak-to-peak) source. Measure and record in Table 53-2 the source voltage V.
3. Measure also and record in Table 53-2 the voltages V_1 and V_2 across C_1 and C_2, respectively.
4. Using the measured values of C_1 and C_2 from Table 53-1, compute and record the voltages V_1 and V_2 which should appear across C_1 and C_2, respectively. Show your computations.
5. Add V_1 and V_2 (measured) and record in Table 53-2.

6. Connect the circuit of Fig. 53-3 using the values of C noted in Table 53-1. Measure and record V, V_1, V_2, and V_3. Add V_1, V_2, and V_3 (measured) and record in Table 53-2. Compute and record the voltages V_1, V_2, and V_3 which should appear in the circuit.
7. In Fig. 53-3, replace C_1 with C_4 and repeat step 6.

QUESTIONS

1. State in your own words the rule for computing the voltage V_A, across a capacitor C_A in a series-connected capacitive voltage divider. Also write the formula for C_A.
2. Refer to Table 53-2. How do the measured voltage values compare with the computed values? Explain any differences.
3. Do the results of your measurements confirm the formula in question 1? Explain.
4. What conclusion can you reach about the sum of the voltages in a series-connected capacitive divider?

TABLE 53-1.

	C_1	C_2	C_3	C_4
Rated Value (μF)	0.05	0.1	0.1	0.5
Measured Value				

TABLE 53-2.

Step	Measured Voltages					Computed Voltages				Sum of Measured Voltages
	V	V_1	V_2	V_3	V_4	V_1	V_2	V_3	V_4	
2, 3, 4, 5				X	X			X	X	
6					X				X	
7		X				X				

Prove your answer by referring specifically to the data in Table 53-2.

5. Three capacitors are connected in series across a 100-V line. The capacitances of the capacitors are in the ratio 1:2:3. Find the voltage across each of the capacitors. Show your computations.

Answers to Self-Test

1. 0.01
2. (a) 500; (b) 1,500; (c) 0.0042
3. (a) 31.25; (b) 62.5; (c) 6.25

54

Impedance of a Series RL Circuit

OBJECTIVES

1. To verify experimentally that the impedance Z of a series RL circuit is given by the equation

$$Z = \sqrt{R^2 + X_L^2}$$

2. (Optional) To verify experimentally that the relationship between Z, R, and X_L is given by the equation

$$Z = \frac{R}{\cos \theta} = \frac{X_L}{\sin \theta}$$

where θ is the phase angle between R and Z

INTRODUCTORY INFORMATION

Impedance of a Series RL Circuit

The total opposition to alternating current in an ac circuit is called the impedance of the circuit. Consider the circuit of Fig. 54-1. If it is assumed that the choke coil (L) through which alternating current flows has zero resistance, the current is impeded only by the X_L of the choke. That is, $Z = X_L$. In this case, if $L = 8$ H and $F = 60$ Hz,

$$X_L = 2\pi FL = 6.28(60)(8) = 3,015 \ \Omega$$

How much current I will there be in the circuit if $V = 6.3$ V? Apply Ohm's law,

$$I = \frac{V}{X_L} \qquad (54\text{-}1)$$

where I is in amperes, V in volts, and X_L is in ohms. Therefore, $I = 6.3/3015 = 2.09$ mA.

If there is resistance R associated with the inductance L, or if L is in series with a resistor of, say, 3,000 Ω (Fig. 54-2), there will be less than 2.09 mA

of current. How much current will flow, assuming the same X_L as previously computed? Measurement shows that there is 1.485 mA in the circuit. It is evident that the total impedance of the resistor R connected in series with L is *not* simply the arithmetic sum of R and X_L. It can be demonstrated mathematically that Z is the result of the vector sum of X_L and R.

The vector diagram of Fig. 54-3 shows that R and X_L are at right angles to each other and that X_L leads R by 90°. We say that there is a 90° phase difference between R and X_L. Vector addition is achieved by drawing a line through A parallel to OB, and another line through B parallel to OA. C is the intersection of these two lines and is the terminus of the resultant vector OC. In this case, the vector AC is the same as X_L. Hence, OC or Z is the resultant of the vector sum of X_L and R.

In Fig. 54-3, Z is seen to be the hypotenuse of the right triangle of which X_L and R are the legs. Applying the Pythagorean theorem to this right triangle, we note that

$$Z = \sqrt{R^2 + X_L^2} \qquad (54\text{-}2)$$

Equation (54-2) makes it possible to determine the impedance of an RL circuit if R and X_L are known. For example, if $R = 40 \ \Omega$ and $X_L = 25 \ \Omega$, then

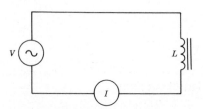

Fig. 54-1. The current I in a pure inductance is limited by the inductive reactance of the circuit.

Fig. 54-2. The impedance of L in series with R is greater than that of L alone.

$$Z = \sqrt{40^2 + 25^2} = \sqrt{1600 + 625}$$
$$= \sqrt{2225} = 47.17 \ \Omega$$

Equation (54-2) and the illustrative problem point up a very important difference between the mathematics of ac circuits and that of dc circuits. In a dc circuit consisting of series-connected resistors, the total opposition to current R_T is simply the arithmetic sum of R_1, R_2, etc. In an ac circuit consisting of a series-connected resistor and inductance the total opposition to current, the impedance Z, is *not* the arithmetic sum of R and X_L, but *is the vector sum of R and X_L.*

Impedance calculations based on Eq. (54-2) can become cumbersome, since this equation does not lend itself easily to slide-rule computations. Trigonometry provides a more effective means of analysis of ac circuits, and trigonometry is therefore extensively employed in ac computations.

If R and X_L are known in Fig. 54-2, the angle θ between R and Z can be determined. For

$$\tan \theta = \frac{X_L}{R}$$

and

$$\theta = \text{arc tan} \frac{X_L}{R} \qquad (54\text{-}3)$$

We can now calculate Z. For

$$\frac{X_L}{Z} = \sin \theta$$

and

$$Z = \frac{X_L}{\sin \theta} \qquad (54\text{-}4)$$

Similarly

$$Z = \frac{R}{\cos \theta} \qquad (54\text{-}5)$$

The process of finding Z is resolved, therefore, to the following simple operations:

1. Find θ, the phase angle between Z and R, by applying Eq. (54-3).

2. Find Z by substituting the values of R, X_L, and θ in Eq. (54-4) or (54-5).

Example

If $R = 40 \ \Omega$ and $X_L = 25 \ \Omega$, find Z trigonometrically.

Solution. (to slide-rule accuracy) Draw the vector triangle (see Fig. 54-3).

1.
$$\tan \theta = \frac{X_L}{R} = \frac{25}{40} = 0.625$$
$$\theta = 32°$$

2.
$$Z = \frac{X_L}{\sin \theta} = \frac{25}{\sin 32°} = 47.2 \ \Omega$$

or
$$Z = \frac{R}{\cos \theta} = \frac{40}{\cos 32°} = 47.2 \ \Omega$$

The results agree substantially with those obtained when we used the right-triangle method.

Ohm's Law Applied to a Series RL Circuit

Ohm's law extended to ac circuits states that the current I in a circuit equals the ratio of the applied voltage V and impedance Z.

$$I = \frac{V}{Z} \qquad (54\text{-}6)$$

and
$$Z = \frac{V}{I} \qquad (54\text{-}7)$$

Equation (54-7) provides the means for verifying the relationships among Z, X_L, and R. Thus, in the circuit of Fig. 54-2 we can measure V and I and compute Z by substituting the measured values of V and I in Eq. (54-7). Z is then calculated by the right-triangle formula [Eq. (54-2)] or by the trigo-

Fig. 54-3. Vector relations in a series *RL* circuit. Impedance is a vector quantity.

nometric formulas [Eqs. (54-4) and (54-5)]. If the results agree, i.e., if the value of Z is substantially the same for each of these methods, then we have verification of the theoretical formulas for Z.

Example

Assume that in the series circuit (Fig. 54-2), $X_L = 25\ \Omega$, $R = 40\ \Omega$, and the applied voltage $V = 47$ V. An ac ammeter is inserted in the circuit (the circuit must be broken and the ammeter inserted in series with R and L) and it measures 1 A. The impedance of the circuit can now be determined. For

$$Z = \frac{V}{I} = \frac{47}{1} = 47\ \Omega$$

The current measurement and subsequent computation for Z is one verification of Eqs. (54-2), (54-4), and (54-5).

Another technique for determining the current in the ac circuit of Fig. 54-2, without measuring current directly, is to measure the ac voltage V_R developed across the resistor R. The current in R may then be found from the formula

$$I = \frac{V_R}{R} \qquad (54\text{-}8)$$

Since current is everywhere the same in a series circuit, the computed value of I is the current in the circuit. We may then solve for the impedance of the circuit by substituting I in the equation $Z = V/I$, where V is the known measured value of the applied voltage and I is the computed current.

SUMMARY

1. In an ac circuit the total opposition to current is called the impedance (Z) of the circuit.
2. In a series RL circuit, the impedance Z is the vector sum of R and X_L, where X_L leads R by an angle of 90° (Fig. 54-3).
3. From the impedance right triangle in Fig. 54-3 the numerical value of Z can be found from the equation

$$Z = \sqrt{X_L{}^2 + R^2}$$

4. In Fig. 54-3, if θ is the phase angle between Z and R, then

$$\theta = \text{arc tan } \frac{X_L}{R}$$

5. Z may be computed trigonometrically if X_L and R are known. Thus

$$Z = \frac{X_L}{\sin\ \theta}$$

or

$$Z = \frac{R}{\cos\ \theta}$$

where the angle θ is computed from the equation in summary step 4.
6. Ohm's law extended to ac circuit states that the current I equals the ratio of the applied voltage V and impedance Z. Thus

$$I = \frac{V}{Z}$$

SELF-TEST

Check your understanding by answering these questions. Students who have not studied trigonometry may skip questions 2, 3, and 4.
1. In the series RL circuit (Fig. 54-2), $R = 1,000\ \Omega$ and $X_L = 2,000\ \Omega$. $Z = $ _____ Ω.
2. In the same circuit as in question 1, $\theta = $ _____ degrees.
3. In the same circuit as in question 1,

   (a) $\dfrac{R}{\cos\ \theta} = $ _____ Ω

   (b) $\dfrac{X_L}{\sin\ \theta} = $ _____ Ω

4. The ratio $R/\cos\ \theta$ or $X_L/\sin\ \theta$, where θ is the phase angle between Z and R, also gives the _____ of the circuit of Fig. 54-2.
5. In a series RL circuit, $R = 500\ \Omega$, $X_L = 500\ \Omega$, and the applied voltage $V = 100$ V. The current I in the circuit is $I = $ _____ A.
6. In the circuit of Fig. 54-2, $V_R = 15$ V. The current in the inductor L is _____ A.

MATERIALS REQUIRED

■ Power supply: Isolation transformer and variable autotransformer operating from the 120-V 60-Hz line
■ Equipment: EVM, 0–25 mA ac meter
■ Resistors: 2-W 5,000-Ω; ½-W 2,700-Ω
■ Inductor: 8 H at 50-mA direct current, whose resistance is 250 Ω
■ Miscellaneous: SPST switch; fused line cord

Fig. 54-4. Determining impedance Z of an RL circuit.

PROCEDURE

1. Connect the circuit of Fig. 54-4. S is open. R is a 2-W 5,000-Ω resistor. L is an iron-core choke rated 8 H at 50-mA direct current. V_A is a variable autotransformer plugged into T, a 1:1 isolation transformer, which derives its power from the 117-volt 60-Hz line. M is a 0–25 mA ac meter, and V is an EVM set on the ac voltage range.

2. Close S. **Power on.** Adjust the output of the ac supply until the meter measures 15 mA (0.015 A). Measure the rms voltage V applied to L and R and record in Table 54-1. Measure also and record V_R, the voltage across R, and V_L, the voltage across L. Compute and record X_L and Z. ($X_L = V_L/I$; $Z = V_{AB}/I$.)

3. Substitute the computed value of X_L and R (5 k Ω) in the equation $Z = \sqrt{R^2 + X_L^2}$. Find Z and record in Table 54-1. Show computations.

   Confirm the measured value of I by comput-ing I from the equation $I = V_R/R$. Record in Table 54-1. **Power off.**

4. Replace R with a 2,700-Ω resistor. **Power on.** Repeat the measurements in steps 2 and 3 and record in Table 54-1.

Extra Credit

5. From Table 54-1 transcribe to Table 54-2 the values of $Z = V_{AB}/I$ and $X_L = V_L/I$ corresponding to $R = 5,000\ \Omega$ and $R = 2,700\ \Omega$. For each value of X_L and R, determine and record $\tan\theta$ and θ using Eq. (54-3). Next, compute and record the corresponding values of Z, using Eqs. (54-4) and (54-5). Show computations.

QUESTIONS

1. Refer to Table 54-1. Are the two values of Z corresponding to $R = 5,000\ \Omega$ equal? Should they be? Are

TABLE 54-1.

R, ohms	I, amperes	V_{AB}, volts	V_R, volts	V_L, volts	$X_L = \dfrac{V_L}{I}$ ohms	$Z = \dfrac{V_{AB}}{I}$ ohms	$Z = \sqrt{R^2 + X_L^2}$ ohms	$I = \dfrac{V_R}{R}$ amperes
5,000	0.015							
2,700	0.015							

TABLE 54-2.

R, ohms	$Z = \dfrac{V_{AB}}{I}$ ohms	$X_L = \dfrac{V_L}{I}$ ohms	$\tan\theta = \dfrac{X_L}{R}$	θ, degrees	Z (ohms) = $\dfrac{R}{\cos\theta}$	Z (ohms) = $\dfrac{X_L}{\sin\theta}$
5,000						
2,700						

the two values of Z corresponding to $R = 2,700 \ \Omega$ equal?

2. From the data in Table 54-1, what conclusion can you draw concerning the formula $Z = \sqrt{R^2 + X_L^2}$? Support your answer.
3. Refer to Table 54-1. Is the metered value of I (using milliammeter) the same as the computed value $I = V_R/R$? Explain why it should be the same.
4. Define Z.
5. What are the possible sources of error in this experiment?
6. In Fig. 54-4, would it make any difference if the voltmeter V were connected across the points CD instead of AB? Explain why.

Extra Credit

7. Do the results of the data in Table 54-2 confirm Eqs. (54-4) and (54-5)? Explain.

Answers to Self-Test

Answers to slide-rule accuracy.

1. 2,240
2. 63.4
3. (a) 2,240; (b) 2,240
4. impedance
5. 0.1414
6. 0.005

55

EXPERIMENT

Voltage Relationships in a Series RL Circuit

OBJECTIVES

1. To measure the phase angle θ between the applied voltage V and the current I in a series RL circuit
2. To verify experimentally that the relationships among the applied voltage V, the voltage V_R across R, and the voltage V_L across L are described by the equations

$$V = \sqrt{V_R^2 + V_L^2}$$
$$V_R = V \cos \theta = V \times \frac{R}{Z}$$
$$V_L = V \sin \theta = V \times \frac{X_L}{Z}$$

INTRODUCTORY INFORMATION

Phase Angle between Applied Voltage and Current in a Series RL Circuit

Since current is the same in every part of a series circuit, current I is shown as the reference vector in considering the phase relations among V_R, V_L, and V in the series RL circuit (Fig. 55-1a). Current and voltage in a resistor are in phase. Hence the voltage V_R across R is drawn as an extension of the vector I (Fig. 55-1b). Current in an inductance lags the voltage across the inductance by 90°. Therefore the vector V_L is shown leading I by 90°. It is evident, therefore, that V_L leads V_R by 90°.

Vector Sum of V_R and V_L Equals Applied Voltage V

V_R and V_L are, respectively, the voltages across R and L in the series RL voltage divider (Fig. 55-1a). Does the sum of V_R and V_L equal the applied voltage V? Obviously it *cannot* be the arithmetic sum, since V_R and V_L are 90° out of phase. But V *is* the *vector* sum of V_R and V_L. This fact is shown in Fig. 55-1b, where V also appears as the hypotenuse of the right triangle of which V_R and V_L are the legs. Applying the Pythagorean theorem,

$$V = \sqrt{V_R^2 + V_L^2} \qquad (55\text{-}1)$$

what is the phase relationship between the applied voltage V and the current I in the inductive circuit? In Fig. 55-1b, I is seen to lag the applied voltage V by an angle θ. Is this angle θ the same as the angle θ between Z and R in the impedance vector diagram in the previous experiment? They are the same, as will be evident from Fig. 55-2.

Figure 55-2a is a redrawing of Fig. 55-1b, with the composition of each of the voltages as shown. Thus, V_R is the product of I and R, V_L is the product of I and X_L, and V is the product of I and Z. Since

(a)   (b)

Fig. 55-1. Phase relations in a series RL circuit.

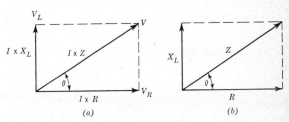

(a)   (b)

Fig. 55-2. Phase angle θ between V and I is the same as phase angle θ between Z and R.

I is a common factor in each of these products, it may be eliminated, leaving the impedance vector diagram (Fig. 55-2b). It is evident, therefore, that the angle θ is the same in Figs. 55-1b and 55-2b.

Further analysis of Fig. 55-1b reveals the relationship between V, V_R, V_L, and the phase angle θ. Since

$$\frac{V_R}{V} = \cos\theta \qquad (55\text{-}2)$$

$$V_R = V\cos\theta \qquad (55\text{-}3)$$

Similarly it follows that

$$V_L = V\sin\theta \qquad (55\text{-}4)$$

From the impedance triangle we know that

$$\cos\theta = \frac{R}{Z}$$

and

$$\sin\theta = \frac{X_L}{Z}$$

Substituting these values in Eqs. (55-3) and (55-4), we find that

$$V_R = V \times \frac{R}{Z} \qquad (55\text{-}5)$$

and

$$V_L = V \times \frac{X_L}{Z} \qquad (55\text{-}6)$$

Equations (55-3), (55-4), (55-5), and (55-6) may be used to compute the voltages V_R and V_L in a series RL circuit where the applied voltage V, the resistance R, and the inductive reactance X_L are known.

Example

If 47 V is applied across a circuit consisting of a 40-Ω resistor in series with an inductance whose reactance is 25 Ω, what is the value of:

1. θ, the phase angle between V and I?
2. V_R, the voltage across R?
3. V_L, the voltage across L?

Solution 1. (to slide rule accuracy) Using Eqs. (55-3) and (55-4),

1. $$\tan\theta = \frac{X_L}{R} = \frac{25}{40} = 0.625$$

$$\theta = 32°$$

2. $$V_R = V\cos\theta$$
$$= 47\cos 32°$$
$$= 40 \text{ V}$$

3. $$V_L = V\sin\theta$$
$$= 47\sin 32°$$
$$= 25 \text{ V}$$

Solution 2. Using Eqs. (55-5) and (55-6),

1. $$\tan\theta = \frac{X_L}{R} = \frac{25}{40} = 0.625$$

$$\theta = 32°$$

2. $$Z = \frac{R}{\cos\theta} = \frac{40}{\cos 32°} = 47\ \Omega$$

$$V_R = V \times \frac{R}{Z}$$

$$= 47 \times \left(\frac{40}{47}\right)$$

$$= 40 \text{ V}$$

3. $$V_L = V \times \frac{X_L}{Z}$$

$$= 47 \times \frac{25}{47}$$

$$= 25 \text{ V}$$

We can check solutions 1 and 2 by use of Eq. (55-1). By substituting the computed values of V_R and V_L in Eq. (55-1) we find that

$$V = \sqrt{40^2 + 25^2}$$

$$= 47 \text{ V}$$

The computed value of V is substantially the same as the given applied voltage V and our solutions are therefore correct.

SUMMARY

1. In a series RL (inductive) circuit the current I lags the applied voltage V by some angle θ, called the phase angle.
2. The phase angle θ between V and I is the same as the angle θ between Z and R in the impedance vector diagram of the RL circuit (Experiment 54). The angle θ is also the same as the angle between V and V_R in Fig. 55-1b.
3. The value of θ depends on the relative values of X_L, R, and Z, and may be computed from *any one* of the following formulas:

$$\theta = \text{arc tan}\ \frac{X_L}{R}$$

$$\theta = \text{arc sin}\ \frac{X_L}{Z}$$

$$\theta = \text{arc cos } \frac{R}{Z}$$

4. In a series RL circuit the voltage drop V_L across the inductance leads the voltage drop V_R across the resistor by 90°.
5. V_R and V_L are added vectorially to give the applied voltage V in the circuit. That is, V is the vector sum of V_R and V_L.
6. The relationship between the applied voltage V, the voltage drop V_R across R, and the voltage drop V_L across L is given by the formula:

$$V = \sqrt{V_R{}^2 + V_L{}^2}$$

7. If the applied voltage V and the phase angle θ are known, then V_R and V_L may be calculated from these formulas:

$$V_R = V \cos \theta$$
$$V_L = V \sin \theta$$

8. If the applied voltage V and R, X_L, and Z are known in a series RL circuit, then V_R and V_L may be calculated from these formulas:

$$V_R = V \times \frac{R}{Z}$$

$$V_L = V \times \frac{X_L}{Z}$$

9. To verify that V is the vector sum of V_R and V_L, we measure V, V_R, and V_L in a series RL circuit and substitute the measured values V_R and V_L in $\sqrt{V_R{}^2 + V_L{}^2}$. If the numerical result is equal to the measured voltage V, then we have the confirmation we require.

SELF-TEST

Check your understanding by answering these questions:

Fig. 55-3. Measuring phase angle between V and I.

1. In the circuit of Fig. 55-1, the measured values of V_R and V_L are, respectively, 30 and 20 V. The applied voltage V must then equal _____ V.
2. The phase angle between V and V_R is _____ _____ as the phase angle between the applied voltage V and the current I in a series RL circuit.
3. The phase angle θ between V and I in the problem of question 1 is _____ degrees.
4. In the circuit of Fig. 55-1, $X_L = 600 \ \Omega$, $R = 800 \ \Omega$, and $V = 120$ V. In this circuit
    (a) $Z = $ _____ Ω
    (b) $\theta = $ _____ degrees
    (c) $V_R = $ _____ V
    (d) $V_L = $ _____ V
5. In a series RL circuit arc sin $X_L/Z = $ arc sin V_L/V. _____ (true/false)

MATERIALS REQUIRED

- Power supply: Isolated 6.3-V rms source
- Equipment: EVM, oscilloscope
- Resistors: ½-W 4,700-Ω, 1,200-Ω
- Inductor: 8 H at 50-mA direct current, whose resistance is 250 Ω
- Miscellaneous: SPST switch; fused line cord

PROCEDURE

Measuring the Phase Angle θ

In Fig. 55-1a the voltage V_R across R is in phase with the current I. The angle θ by which the applied voltage V leads the current I can therefore be found by measuring the angle between V and V_R by the method of Lissajous patterns described in Experiment 45.

1. Connect the circuit of Fig. 55-3a. V is a 6.3-V rms line-isolated source. L is an iron-core choke rated 8 H at 50 mA direct current, whose resistance is 250 Ω. Since the circuit value of $R = $ 4,700 Ω, we can ignore the resistance of L. An insignificant error is introduced as a result. Connect the oscilloscope as shown. Switch the

TABLE 55-1.

R, ohms	Y_M, in	Y_O, in	θ (degrees) = arc sin $\frac{Y_O}{Y_M}$	V_R, volts	$I = \frac{V_R}{R}$ amperes	V_L, volts	$X_L = \frac{V_L}{I}$ ohms	θ (degrees) = arc tan $\frac{X_L}{R}$
4,700								
1,200								

sweep selector switch to Ext. sweep. **Power on.**

2. Adjust the vertical gain of the oscilloscope for 1 division of vertical deflection and the horizontal gain for 1 division of horizontal deflection. Center the resulting pattern vertically with respect to the X axis and horizontally with respect to the Y axis. The pattern will now also be centered with respect to the origin, as in Fig. 55-3b.

NOTE: The coordinate axes refer to the etched faceplate, the graticule, on the screen.

Measure, and record in Table 55-1, the maximum Y distance from the X axis, Y_M, and the intercept Y_O on the Y axis. Compute the phase angle θ by substituting Y_M and Y_O in the equation

$$\theta = \text{arc sin } \frac{Y_O}{Y_M} \qquad (55\text{-}7)$$

Show your computations.

3. To confirm that the measured angle θ above is the same as the angle between Z and R proceed as follows: With an EVM, measure and record the rms voltage V_R across R. Compute I in the circuit by substituting the measured value V_R and the rated value of R in the equation

$$I = \frac{V_R}{R}$$

Show your computation. Now measure V_L rms across L. Compute and record the reactance X_L of L ($X_L = V_L/I$). Determine the angle θ by substituting the computed value X_L and the rated value of R in the equation

$$\theta = \text{arc tan } \frac{X_L}{R}$$

Show your computations. The two values of θ should be the same or reasonably close.

4. Replace R with a 1,200-Ω resistor. Repeat the measurements and computations, steps 2 and 3.

Verifying the Formula $V = \sqrt{V_R{}^2 + V_L{}^2}$

5. Transcribe from Table 55-1 to Table 55-2 the measured rms voltages V_R and V_L for each value of resistance. For the corresponding values of V_R and V_L evaluate $\sqrt{V_R{}^2 + V_L{}^2}$ and record in Table 55-2.

Optional: Extra Credit

6. From the data in Table 55-1 and 55-2 verify the relationships in Eqs. (55-3), (55-4), (55-5), and (55-6). List the transcribed data and computed data in tabular form. Explain the method used to verify and show your computations.

QUESTIONS

1. Explain why the two values of θ corresponding to $R = 4,700 \ \Omega$ (Table 55-1) should be equal. If they are not, explain why. For $R = 1,200 \ \Omega$?
2. What are the factors which may contribute to errors in the Lissajous-pattern method of determining phase angle, as used in this experiment?
3. Do the results of your experiment confirm the relationship $V = \sqrt{V_R{}^2 + V_L{}^2}$? Substantiate your answer by referring to the proper data.
4. Ascertain the missing values in Table 55-3, which

TABLE 55-2.

R, ohms	V applied, volts rms	V_R, volts	V_L, volts	$\sqrt{V_R{}^2 + V_L{}^2}$ volts
4,700	6.3			
1,200	6.3			

TABLE 55-3.

R, ohms	L, henrys	V Applied, volts	F, Hz	V_L, volts	V_R, volts	θ, degrees	X_L, ohms	Z, ohms
10,000	0.03	100	100,000					
	0.05	80	50,000	60				

is based on measurements in a series *RL* circuit. Show your computations.

Extra Credit

5. What conclusions do you draw from procedural step 6? Substantiate your answer by referring to your computations.

Answers to Self-Test

Answers to slide-rule accuracy.

1. 36.1
2. the same
3. 33.7
4. (*a*) 1,000; (*b*) 36.8; (*c*) 96; (*d*) 72
5. true

56

(Optional) The j Operator in the Analysis of LC Circuits

OBJECTIVES

1. To define the meaning of the j operator as used in electrical circuit analysis
2. To transform the phasors $a + jb$ into $\rho\angle\theta$ and $\rho\angle\theta$ into $a + jb$
3. To verify experimentally that the impedance Z of a series-connected RL circuit is

$$Z = R + jX_L = |Z|\angle\theta$$

4. To verify experimentally that the applied voltage V of a series-connected RL circuit is

$$V = V_R + jV_L = |V|\angle\theta$$

INTRODUCTORY INFORMATION

The j Operator

In Experiment 54 we learned that in a series RL circuit, inductive reactance X_L can be represented as leading the resistance R by 90° and that the impedance Z of the circuit is equal to the vector sum of X_L and R. Moreover, in Experiment 55 it was demonstrated that the voltage V_L across L leads by 90° the voltage V_R across R in an LR circuit, and that the applied voltage V is the vector sum of V_R and V_L. To describe these relationships the j operator is used in the mathematics of ac circuits. The symbol j represents the imaginary number $\sqrt{-1}$.

The symbol j as defined above is called a mathematical operator, because it represents a rotation of 90°. Consider the horizontal axis in Fig. 56-1 as being the axis of all *real* numbers, that is, all positive and negative numbers between $+\infty$ and $-\infty$. Then consider the j axis as the axis of all imaginary numbers between $+j\infty$ and $-j\infty$. The number $j5$ is the result of rotating a real number $+5$, 90°

counterclockwise. The number $j5$, then, falls along the j axis, and is on the axis of imaginaries.

What would be the result of rotating the number $+j5$ another 90° counterclockwise? Graphically it is evident that the resultant number should be the real number -5. It can be shown that this graphical representation is also mathematically correct. For $j5$, rotated 90° CCW, is written as $j(j5)$ or simply j^25. Since $j = \sqrt{-1}$, $j^2 = -1$ and j^25 then becomes -5. Similarly, if the number -5 is rotated 90° CCW, it becomes $-j5$, and if $-j5$ is rotated 90° CCW, it becomes $+5$. That is, a number which has undergone four 90° rotations returns to its original starting position.

It should be quite evident, then, that the j operator can be treated as an algebraic symbol which obeys the laws of algebra. For

$$j = \sqrt{-1}$$
$$j^2 = -1$$
$$j^3 = -1j$$
$$j^4 = +1$$
$$j^5 = j, \text{ etc.}$$

The Complex Number a + jb (Rectangular Form)

Expressions involving *real* numbers and imaginaries are called complex numbers and are written in the form $a + jb$ where a and b are real numbers. The complex numbers $a + jb$ can be represented graphically as a point P in the complex plane (Fig. 56-1). Take, for example, the number $15 + j10$ ($a = 15$, $b = 10$). To find the point P that this represents, proceed to $+15$ on the axis of reals (Fig. 56-1) and erect a perpendicular to the horizontal axis at this point. Also at $+j10$, on the axis of imaginaries, erect a perpendicular to the vertical axis. The intersection of these two perpendiculars is the point P, and it represents the expression $15 + j10$. To reach the point $15 + j10$ starting at the origin requires moving 15 units to the right along the horizontal axis (the axis of reals) and 10 units up along a perpendicular parallel to the j axis. Similarly, for *any* complex number $a + jb$, a represents the component along the horizontal axis, and b represents the component along the vertical axis. That is, a is the projection of point P on the horizontal axis, b is the projection of point P on the vertical axis.

The complex number $a + jb$ is said to be in the *rectangular form* because a and b are the rectangular coordinates of point P. The complex number $a + jb$ can represent the point P or the radius vector drawn from the origin to P.

The Complex Number $\rho\angle\theta$ (Polar Form)

Polar coordinates can also be used to locate the point $15 + j10$. Refer to Fig. 56-2. The point P can be found if the length P of the radius vector $0P$ and if the angle θ are known. Since the triangle $0QP$ is a right triangle, ρ can be computed by the Pythagorean theorem. Thus $\rho = \sqrt{15^2 + 10^2} = 18.03$. Moreover, $\tan \theta = {}^{10}/_{15} = 0.667$, and therefore $\theta = 33.7°$. Thus the polar form of $15 + j10$ is $18.03 \angle 33.7°$.

In the form $\rho\angle\theta$ (Fig. 56-3a), ρ is the distance of the point P from the origin. Therefore ρ is the length of a radius, with center at the origin 0 describing a circular arc about the origin. The positive angle θ is the number of degrees of the arc, as the radius moves from the horizontal in a counter-clockwise direction. The distance ρ is always positive. However, the angle θ may be negative. Thus the number $\rho\angle-\theta$ in Fig. 56-3b is obtained by rotating the radius $\theta°$ in a clockwise direction.

Transforming a Number in the Rectangular Form into Its Equivalent in Polar Form

To transform the complex number $a + jb$ into the polar form $\rho\angle\theta$, use the following general formula:

$$\rho = \sqrt{a^2 + b^2}$$
$$\theta = \arctan \frac{b}{a} \tag{56-1}$$

The length of the radius vector ρ can also be found by trigonometry. Thus, from Fig. 56-2:

$$\rho = \frac{b}{\sin \theta} = \frac{a}{\cos \theta} \tag{56-2}$$

where
$$\theta = \arctan \frac{b}{a}$$

AXIS OF IMAGINARY NUMBERS

Fig. 56-1. Representation of numbers in the complex plane. P is the number $15 + j10$.

Fig. 56-2. Representation of a number in polar coordinates.

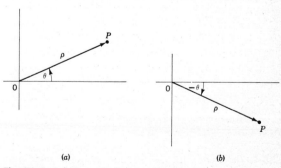

Fig. 56-3. (a) The number $\rho \angle\theta$; (b) the number $\rho \angle-\theta$.

Transforming from Polar Form into Rectangular Form

If a complex number is given in the polar form $\rho \angle \theta$, it can be transformed into the rectangular form $a + jb$. Consider the complex number $10 \angle 36.85°$ in Fig. 56-4. If a perpendicular is drawn from P to the horizontal axis, a right triangle is formed with the arms a and b.

Since
$$a = 10 \cos 36.85°$$
$$a = 8$$

Moreover
$$b = 10 \sin 36.85°$$

Therefore
$$b = 6$$

The number $10 \angle 36.85°$ can therefore be written as $8 + j6$. Similarly, any number in polar form $\rho \angle \theta$ can be written in the form $a + jb$, where

$$a = \rho \cos \theta$$
$$b = \rho \sin \theta \tag{56-3}$$

and

Therefore
$$\rho \angle \theta = \rho \cos \theta + j\rho \sin \theta \tag{56-4}$$

Impedance Z of an RL Circuit Expressed in Rectangular Form

Consider the series RL circuit in Fig. 56-5a. The impedance diagram for this circuit is shown in Fig. 56-5b. Since the impedance Z of this circuit is the vector sum of X_L and R, and since X_L and R are at right angles, impedance Z may be written in the rectangular form, using the j operator. Thus if X_L is the inductive reactance of L, and R is the resistance of the resistor, then

$$Z = R + jX_L \tag{56-5}$$

Equation (56-5) defines impedance as a vector in the complex plane, with horizontal component equal to R ohms and a vertical component equal to X_L ohms. This equation does not give the numerical value $|Z|$ of the impedance. If $|Z|$ is required, it

Fig. 56-4. The vector $10 \angle 36.85°$ can be transformed into the complex number $8 + j6$.

Fig. 56-5. (a) series RL circuit, (b) impedance diagram of series RL circuit: $Z = R + jX_L$.

may be computed from any one of the equations in (56-6):

$$|Z| = \sqrt{R^2 + X_L^2}$$
$$|Z| = \frac{R}{\cos \theta} \tag{56-6}$$
$$|Z| = \frac{X_L}{\sin \theta}$$

where
$$\theta = \arctan \frac{X_L}{R}$$

It is evident that Eqs. (56-6) are the same as those found in Experiment 54.

A numerical example will illustrate the manner in which Eqs. (56-5) and (56-6) are applied.

Example

In Fig. 56-5, $R = 80 \ \Omega$ and $X_L = 60 \ \Omega$.
(a) Write an equation for Z using the j operator.
(b) Find the numerical value $|Z|$.
(c) Find the phase angle θ.

Solution.
(a) $Z = R + jX_L = 80 + j60$
(b) $|Z| = \sqrt{R^2 + X_L^2} = \sqrt{80^2 + 60^2} = 100$
(c) $\theta = \arctan \dfrac{X_L}{R} = \arctan \dfrac{60}{80} = 36.85°$

Impedance Z of an RL Circuit Expressed in Polar Form

Since it is possible to express Z in rectangular form, it is also possible to express it in polar form. Thus in Fig. 56-5,

$$Z = |Z| \angle \theta$$

where
$$Z = \sqrt{X_L^2 + R^2} = \frac{R}{\cos \theta} = \frac{X_L}{\sin \theta} \tag{56-7}$$

and
$$\theta = \arctan \frac{X_L}{R}$$

If $R = 80$ Ω and $X_L = 60$ Ω, applying Eq. (56-7) we get

$$Z = 100 \angle 36.85°$$

V, V_R and V_L in a Series RL Circuit Expressed in Rectangular Form; in Polar Form

In the series LR circuit of Fig. 56-5a, the relationship between the applied voltage V, the voltage drop V_R across R, and the voltage drop V_L across L can be represented by the phasor diagram in Fig. 56-6. (When rotating radius vectors are used in electric circuit analysis, they are called *phasors;* this is the terminology we will use.) Applying our definition of j we can write the phasor V in rectangular form:

$$V = V_R + jV_L$$

The numerical value $|V|$ can be found from the equation

$$|V| = \sqrt{V_R{}^2 + V_L{}^2} = \frac{V_R}{\cos \theta} = \frac{V_L}{\sin \theta} \quad (56\text{-}9)$$

where $\qquad \theta = \text{arc tan} \dfrac{V_L}{V_R}$

The phasor V can also be written in polar form:

$$V = |V| \angle \theta$$

where $\quad |V| = \sqrt{V_R{}^2 + V_L{}^2} = \dfrac{V_R}{\cos \theta} = \dfrac{V_L}{\sin \theta} \quad (56\text{-}10)$

and $\qquad \theta = \text{arc tan} \dfrac{V_L}{V_R}$

If we know the values of $|V|$ and θ, we can simply write V in the polar form:

$$V = |V| \angle \theta$$

We can then transform the polar into the rectangular form by writing

$$V = |V| \cos \theta + j|V| \sin \theta \quad (56\text{-}11)$$

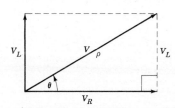

Fig. 56-6. $V = V_R + jV_L$ or $V = \rho \angle \theta$.

It is evident then that $|V| \cos \theta = V_R$, and $|V| \sin \theta = V_L$.

SUMMARY

1. The symbol j represents the imaginary number $\sqrt{-1}$. The symbol j is used as an operator in electrical circuit analysis to indicate a counterclockwise rotation of 90°. The symbol $-j$ represents a clockwise rotation of 90°.
2. The sum of a real number a and an imaginary number jb is called a *complex* number and is written $a + jb$.
3. The complex number $a + jb$ represents a point P in the complex plane (Fig. 56-1), or the radius vector OP, where a is the horizontal component and b the vertical component of OP. The format $a + jb$ is called the rectangular form.
4. The complex number $a + jb$ can also be written in the polar form $\rho \angle \theta$, that is, $\rho \angle \theta = a + jb$, where

$$\rho = \sqrt{a^2 + b^2} = \frac{a}{\cos \theta} = \frac{b}{\sin \theta}$$

and $\qquad \theta = \text{arc tan} \dfrac{b}{a}$

5. If a vector is given in the polar form, that is, if ρ and θ are known, then the vector can be transformed into the rectangular form by the equation

$$\rho \angle \theta = \rho \cos \theta + j\rho \sin \theta$$

6. Rotating radius vectors, used in electrical circuit analysis, are called phasors.
7. The impedance Z of a series RL circuit is a vector quantity. X_L and R have a 90° phase difference. If the values of X_L and R are known, their impedance Z can be written in the rectangular form by the equation

$$Z = R + jX_L$$

8. Z can also be written in the polar form by the equation $Z = \rho \angle \theta$ where

$$\rho = \sqrt{R^2 + X_L{}^2} = \frac{R}{\cos \theta} = \frac{X_L}{\sin \theta}$$

and $\qquad \theta = \text{arc tan} \dfrac{X_L}{R}$

9. If the impedance of a series RL circuit is given in the polar form, that is, if the values of ρ and θ are known, then Z can be expressed in rectangular form by the equation

$$Z = \rho \cos \theta + j\rho \sin \theta$$

In this equation, $\rho \cos \theta = R$, and

$$\rho \sin \theta = X_L$$

10. In a series RL circuit the phasor relationship between the applied voltage V, the voltage drop V_R across R, and V_L across L is given by the formula:

$$V = V_R + jV_L$$

or $$V = |V| \angle \theta$$

where $$|V| = \sqrt{V_R{}^2 + V_L{}^2} = \frac{V_R}{\cos \theta} = \frac{V_L}{\sin \theta}$$

and $$\theta = \arc \tan \frac{V_L}{V_R}$$

SELF-TEST

Check your understanding by answering these questions:

1. In the circuit of Fig. 56-5a $V = 20$ V, $R = 150$ Ω, and $X_L = 100$ Ω. The impedance Z, in rectangular form, is $Z = $ _____.
2. In the circuit of question 1, the phase angle θ between Z and R is _____ degrees.
3. In the circuit of question 1, impedance Z written in polar form is $Z = $ _____.
4. In the circuit of question 1, the applied voltage V, that is, the phasor V, written in polar form is $V = $ _____.
5. In the circuit of question 1, the phasor V written in rectangular form is $V = $ _____.
6. From the equation in question 5, we know that the voltage V_R across R is _____ V, while the voltage V_L across L is _____ V.
7. In the circuit of Fig. 56-5a, $Z = 100$ Ω, $V_R = 60$ V, and $V_L = 80$ V. The phasor V written in rectangular form is $V = $ _____.
8. In the circuit of question 7, the phase angle θ between V and V_R is _____ degrees.
9. In the circuit of question 7, the phasor V written in polar form is $V = $ _____.
10. In the circuit of question 7, the impedance Z written in polar form is $Z = $ _____.
11. In the circuit of question 7, the impedance Z written in rectangular form is $Z = $ _____.
12. From the equation in question 11, we know that $R = $ _____ Ω and $X_L = $ _____ Ω.

MATERIALS REQUIRED

- Power supply: Isolation transformer and variable autotransformer operating from the 120-V 60-Hz line
- Equipment: Electronic voltmeter
- Resistors: 2-W 5,000 Ω
- Inductor: 8 H at 50-mA direct current whose resistance is 250 Ω
- Miscellaneous: SPST switch; fused line cord

PROCEDURE

Impedance Z of a Series RL Circuit

1. Connect the circuit of Fig. 56-7. S is open. R is a 2-W 5,000-Ω resistor. L is an iron-core choke rated 8 H at 50 mA dc. V_A is a variable transformer plugged into T, a 1:1 isolation transformer which receives power from the 120-V 60-Hz line. Connect the electronic voltmeter, set on ac volts, across R.
2. Close S. **Power on.** Adjust the output of the ac supply until the voltage V_R across R measures 75 V.
3. Measure also and record in Table 56-1 the voltage, V_L across L, and the voltage V_{AB} which is the numerical value $|V|$ of the voltage applied across the series-connected RL circuit.
4. Open switch S. **Power off.** Measure the resist-

ance of R and record in Table 56-1.
5. Using the measured values of V_R and R compute and record in Table 56-1 the current I in the circuit.
6. Compute and record the numerical value $|Z|$ of the impedance in the circuit using the equation $|Z| = |V|/I$, where $|V|$ is the measured value.

Fig. 56-7. Experimental circuit for determining the impedance Z of an RL circuit.

TABLE 56-1.

Measured Values				Computed Values			
V_R volts	V_L volts	$V_{AB} = \lvert V \rvert$ volts	R ohms	I amp	$\lvert Z \rvert$ ohms	θ degrees	X_L ohms
75							
Z in polar form				Z in rectangular form			
$Z =$				$Z =$			

7. Compute and record the phase angle θ from the equation

$$\theta = \arctan \frac{V_L}{V_R}$$

8. Compute and record in Table 56-1 the value of X_L using the relationship

$$X_L = \frac{V_L}{I}$$

9. Using the computed values $\lvert Z \rvert$ and θ write the vector Z in polar form in Table 56-1.
10. Transform the polar form of Z into its rectangular form, using the relationship

$$Z = \lvert Z \rvert \cos \theta + j \lvert Z \rvert \sin \theta$$

and record in Table 56-1.

Voltage Relationships in a Series RL Circuit

11. From the measured values V_R and V_L in Table 56-1, write the phasor V (applied voltage) in rectangular form using the equation $V = V_R + jV_L$ and record in Table 56-2.
12. Compute the numerical value $\lvert V \rvert$ of the phasor V using any one of the equations

$$\lvert V \rvert = \sqrt{V_R{}^2 + V_L{}^2} = \frac{V_R}{\cos \theta} = \frac{V_L}{\sin \theta}$$

and record in Table 56-2. Employ the values V_R, V_L, and θ from Table 56-1.
13. From Table 56-1 record in Table 56-2 the measured value $V_{AB} = \lvert V \rvert$.
14. Using the value of $\lvert V \rvert$ from step 12, and the

value of θ from Table 56-1, write the phasor V in polar form in Table 56-2.

QUESTIONS

1. Refer to the equation for Z in rectangular form, Table 56-1 and to step 10 in procedure.
   (a) Does the measured value of $R = \lvert Z \rvert \cos \theta$?
   (b) Should it?
   (c) If the two values are not the same, explain why they are not.
2. Refer to the equation for Z in rectangular form, Table 56-1 and to step 10 in procedure.
   (a) Does the computed value of $X_L = \lvert Z \rvert \sin \theta$?
   (b) Should it?
   (c) If the two values are not the same, explain why not.
3. Have you confirmed experimentally (approximately) the analytical statement that $Z = R + jX_L$? Refer specifically to your data to show such confirmation or lack of it.
4. If your data does not confirm (approximately) the statement $Z = R + jX_L$ explain why not.
5. Refer to Table 56-2 and to step 12. Does the computed value of $\lvert V \rvert$ equal the measured value of $\lvert V \rvert$? If not, why not?
6. Have you confirmed experimentally (approximately) the analytical statement that $V = V_R + jV_L$? Refer specifically to your data to show such confirmation or lack of it.
7. If your data does not confirm (approximately) that $V = V_R + jV_L$ explain why not.

Answers to Self-Test

Answers to slide-rule accuracy.

1. $150 + j100$
2. 33.7

TABLE 56-2.

V in rectangular form	$\lvert V \rvert$ computed	$\lvert V \rvert$ measured	V in polar form
$V =$			$V =$

3. $180\,\underline{/33.7°}$
4. 20
5. $16.64 + j11.1$
6. 16.64; 11.1
7. $60 + j80$

8. 53.1
9. $100\,\underline{/53.1°}$
10. $100\,\underline{/53.1°}$
11. $60 + j80$
12. 60; 80

57

Impedance of a Series RC Circuit

OBJECTIVES

1. To verify experimentally that the impedance Z of a series RC circuit is given by the equation

$$Z = \sqrt{R^2 + X_C^2}$$

2. (Optional) To verify experimentally that the relationship between Z, R, and X_C is given by the equation

$$Z = \frac{R}{\cos \theta} = \frac{X_C}{\sin \theta}$$

where θ is the phase angle between Z and R

INTRODUCTORY INFORMATION

Analysis of the impedance of RC circuits, Fig. 57-1, is very similar to analysis of the impedance of RL circuits (see Experiment 54). It can be demonstrated mathematically that the impedance Z of an RC circuit is a *vector* quantity resulting from the *vector* sum of X_C and R.

Figure 57-2 is a vector diagram which shows that R and X_C are at right angles to each other. Note, however, that X_C *lags* R by 90°, in contradistinction to X_L, which leads R by 90°.

In Fig. 57-2, Z is shown as the vector sum of R and X_C. Z is therefore the hypotenuse of a right triangle of which R and X_C are the legs. Applying the

Pythagorean theorem to R, X_C, and Z, we see that

$$Z = \sqrt{R^2 + X_C^2} \qquad (57\text{-}1)$$

Equation (57-1) makes it possible to determine the impedance of an RC circuit if R and X_C are known. For example, if $R = 30 \ \Omega$ and $X_C = 40 \ \Omega$, then $Z = \sqrt{30^2 + 40^2} = 50 \ \Omega$.

Refer again to Fig. 57-2. If R and X_C are known, the angle θ between R and Z can be determined.

For

$$\tan \theta = \frac{X_C}{R} \qquad (57\text{-}2)$$

and

$$\theta = \text{arc} \tan \frac{X_C}{R} \qquad (57\text{-}3)$$

We can now calculate Z. Thus

$$\frac{X_C}{Z} = \sin \theta \qquad (57\text{-}4)$$

and

$$Z = \frac{X_C}{\sin \theta} \qquad (57\text{-}5)$$

Similarly

$$Z = \frac{R}{\cos \theta} \qquad (57\text{-}6)$$

The process of finding Z, therefore, is resolved to the following simple operations:

1. Find θ by applying Eqs. (57-2) and (57-3).
2. Find Z by applying either Eq. (57-5) or (57-6).

Fig. 57-1. Series *RC* circuit.

Fig. 57-2. Impedance of a series *RC* circuit.

Example

If $R = 30\ \Omega$, $X_C = 40\ \Omega$, find Z trigonometrically.

Solution. (to slide rule accuracy) Draw the vector triangle (see Fig. 57-2).

1.
$$\tan \theta = \frac{X_C}{R} = \frac{40}{30} = 1.33$$

$$\theta = 53.1°$$

2.
$$Z = \frac{R}{\cos \theta} = \frac{30}{\cos 53.1°} = 50\ \Omega$$

or
$$Z = \frac{X_C}{\sin \theta} = \frac{40}{\sin 53.1°} = 50\ \Omega$$

This result agrees with that obtained when we used the right-triangle formula.

Ohm's Law Applied to a Series RC Circuit

Ohm's law states that the current I in a circuit equals the ratio of the applied voltage V and impedance Z.

$$I = \frac{V}{Z} \qquad (57\text{-}7)$$

and
$$Z = \frac{V}{I} \qquad (57\text{-}8)$$

Equation (57-8) provides the means of verifying the relationship between Z, X_C, and R. Thus, in the circuit of Fig. 57-1, we can measure V and I and compute Z by substituting the measured values of V and I in Eq. (57-8). We can then calculate Z, using Eq. (57-1), (57-5), or (57-6). If the results agree, that is, if the value of Z is substantially the same for each of these methods, then we have verification of the right-triangle relationship between Z, X_C, and R and also verification of Eqs. (57-1), (57-5), and (57-6).

Example

Assume that in the series circuit (Fig. 57-1), $X_C = 40\ \Omega$, $R = 30\ \Omega$, and the applied voltage $V = 10$ V. An ac milliammeter is inserted in the circuit and it measures 0.2 A. We can now determine the impedance for

$$Z = \frac{V}{I} = \frac{10}{0.2} = 50\ \Omega$$

The current measurement and subsequent computation for Z is one verification of Eqs. (57-1), (57-5), and (57-6).

Another technique for determining the current in the ac circuit of Fig. 57-1 without measuring current directly is to measure the ac voltage V_R developed across the resistor R. The current in R may then be found from the equation

$$I = \frac{V_R}{R} \qquad (57\text{-}9)$$

Since current is everywhere the same in a series circuit, the computed value of I *is* the current in the circuit. We may then solve for the impedance of the circuit in the manner described in the preceding example, that is, by substituting I in the equation

$$Z = \frac{V}{I}$$

where V is the known or measured value of applied voltage and I is the computed current.

SUMMARY

1. In a series-connected RC circuit the impedance Z is the vector sum of R and X_C, where X_C lags R by 90° (Fig. 57-2).
2. From the impedance right triangle in Fig. 57-2, the numerical value of Z can be found from the equation

$$Z = \sqrt{R^2 + X_C{}^2}$$

3. In Fig. 57-2, if θ is the phase angle between Z and R, then,

$$\theta = \text{arc tan } \frac{X_C}{R}$$

4. Z may be computed trigonometrically if X_C and R are known. Thus

$$Z = \frac{X_C}{\sin \theta}$$

or
$$Z = \frac{R}{\cos \theta}$$

where θ is the angle computed from the equation in summary step 3.
5. The impedance Z of an RC circuit may also be computed, if the applied voltage V and the current I in the circuit are known from the basic equation

$$Z = \frac{V}{I}$$

6. If the value of Z is substantially the same for each of the methods (equations), we have veri-

284 **Basic Electricity**

fication of the right-triangle relationship between Z, X_C, and R, and also verification of the equations which define Z.

SELF-TEST

Check your understanding by answering these questions. Students who have not studied trigonometry may skip questions 2, 3, and 4.

1. In the series RC circuit (Fig. 57-1), $R = 3,000\ \Omega$, $X_C = 1,200\ \Omega$, $Z =$ _____ Ω.
2. In the same circuit as in question 1 $\theta =$ _____ degrees.
3. In the same circuit as in question 1
   (a) $\dfrac{R}{\cos \theta} =$ _____ Ω
   (b) $\dfrac{X_C}{\sin \theta} =$ _____ Ω
4. The ratio $R/\cos \theta$ or $X_C/\sin \theta$, where θ is the phase angle between Z and R, also gives the _____ of the circuit of Fig. 57-1.
5. In a series RC circuit, $R = 1,200\ \Omega$, $X_C = 1,500\ \Omega$, and the applied voltage $V = 50$ V. The current I is $I =$ _____ A.

Fig. 57-3. Experimental circuit for determining the impedance Z of a series RC circuit.

6. In the circuit of question 1, $V_R = 45$ V. The current I in the capacitor C is _____ A.

MATERIALS REQUIRED

- Power supply: Line-isolation transformer and variable autotransformer operating from 120-V line
- Equipment: 0–5 mA and 0–25 mA ac milliammeters, or a multirange milliammeter covering these ranges; EVM
- Resistors: 1-W 5,000-Ω; ½-W 22,000-Ω
- Capacitors: 0.5 μF, 0.1 μF
- Miscellaneous: SPST switch; fused line cord

PROCEDURE

1. Connect the circuit of Fig. 57-3. S is open. R is a 5,000-Ω resistor; C is an 0.5-μF capacitor. V_A is a variable autotransformer plugged into T, a 1:1 isolation transformer. M is an ac milliammeter set on the 0–25 mA range. V is an EVM set to the 100-V ac range.
2. Close S. Adjust the output of the autotransformer until M measures 10 mA (0.01 A). With the voltmeter, measure the voltage V across the RC circuit and record in Table 57-1. Compute Z by substituting the measured values of V and I in Eq. (57-8). Record Z in the column V/I.
3. Compute X_C using the formula

$$X_C = \frac{1}{2\pi FC}$$

where F is the power-line frequency.
Record X_C in Table 57-1. Calculate and record Z using Eq. (57-1).
4. Measure and record V_R, the voltage across R. Compute I, using Eq. (57-9). Record I in the column V_R/R.
5. **Power off.** Replace C by a 0.1-μF capacitor and R by a 22,000-Ω resistor. Switch the milliammeter to the 0–5 mA range. Repeat steps 2, 3, and 4, but set the output of the variable autotransformer for 2 mA (0.002 A) of current.

TABLE 57-1.

R, ohms	C, μF	I, amperes	V, volts	$Z = \dfrac{V}{I}$ ohms	X_C, ohms	$Z = \sqrt{R^2 + X_C^2}$ ohms	V_R, volts	$I = \dfrac{V_R}{R}$ amperes
5,000	0.5	0.01						
22,000	0.1	0.002						

TABLE 57-2.

R, ohms	C, μF	I, amperes	$Z = \dfrac{V}{I}$ ohms	X_C, ohms	$\tan \theta = \dfrac{X_C}{R}$	θ, degrees	Z (ohms) = $\dfrac{R}{\cos \theta}$	Z (ohms) = $\dfrac{X_C}{\sin \theta}$
5,000	0.5	0.01						
22,000	0.1	0.002						

Extra Credit

6. From Table 57-1 transcribe to Table 57-2 the values of $Z = V/I$, and X_C corresponding to $C = 0.5$ and 0.1 μF. For each C and R determine $\tan \theta$ and θ using Eqs. (57-2) and (57-3) and record in Table 57-2. Next compute the corresponding values of Z using Eqs. (57-5) and (57-6).

QUESTIONS

1. If the two values of Z in Table 57-1, corresponding to any one RC circuit, are equal, what conclusion can you draw concerning the formula $Z = \sqrt{R^2 + X_C^2}$? Are the values equal?
2. What are the possible sources of error in the data in Table 57-1?
3. For a specific Z in Table 57-1, is the value of I mea-sured with a milliammeter the same as the value of I computed from the equation $I = V_R/R$? Should they be the same? Explain any discrepancies.
4. Define Z.

Extra Credit

5. Do the results of your data in Table 57-2 confirm Eqs. (57-5) and (57-6)? Explain.

Answers to Self-Test

Answers to slide-rule accuracy.

1. 3,230
2. 21.8
3. (a) 3,230; (b) 3,230
4. impedance
5. 0.026
6. 0.015

58

EXPERIMENT

Voltage Relationships in a Series RC Circuit

OBJECTIVES

1. To measure the phase angle θ between the applied voltage V and the current I in a series RC circuit
2. To verify experimentally that the relationships among the applied voltage V, the voltage V_R across R, and the voltage V_C across C are described by the equations

$$V = \sqrt{V_R^2 + V_C^2}$$

$$V_R = V \cos \theta = V \times \frac{R}{Z}$$

$$V_C = V \sin \theta = V \times \frac{X_C}{Z}$$

INTRODUCTORY INFORMATION

There is a striking similarity between analysis of RC and RL circuits (see Experiment 55). The difference is that in an inductive circuit, current I lags

the applied voltage V, whereas in a *capacitive* circuit, current I *leads* the applied voltage V.

Phase Relations between the Applied Voltage and Current in a Series RC Circuit

The current I is common in every part of a series circuit. Therefore, current is used as the reference phasor in considering the phase relations between V_R and V_C in the series RC circuit of Fig. 58-1a. Since current and voltage in a resistor are in phase, the voltage V_R is shown in Fig. 58-1b as an extension of the current phasor. But *current* in a *capacitor leads the voltage across* the capacitor by 90°. Therefore, the phasor V_C is shown lagging the current I by 90°. It is evident, therefore, that V_C lags V_R by 90°.

V_R and V_C are, respectively, the voltages across R and C in the series RC voltage divider (Fig. 58-1a). As in the case of an LR circuit, the applied voltage V is the vector sum of V_R and V_C. This fact is shown in Fig. 58-1b. We also see that V is the hypotenuse of a right triangle of which V_R and V_C are the legs. Therefore, applying the Pythagorean theorem,

$$V = \sqrt{V_R^2 + V_C^2} \qquad (58-1)$$

Figure 58-1b also reveals the phase relationship between the applied voltage V and the current I in the series RC circuit. The current I is seen to lead the applied voltage V by the angle θ.

The question that suggests itself is: Is the angle θ, by which current leads the applied voltage in a series RC circuit, the same as the angle θ between the impedance vector Z and the resistance R in the previous experiment? They are the same, as can

(a)   (b)

Fig.58-1. Phase relations in a series RC circuit.

be seen from Fig. 58-2. Part (*a*) is a redrawing of Fig. 58-1*b* and shows the phase relations between V, V_R, and V_C. But in addition, the composition of each of these voltages is indicated. Thus, V_R is the product of I and R, V_C the product of I and X_C, and V the product of I and Z. Since I is a common factor in each of these products, it may be eliminated, leaving the impedance vector diagram (Fig. 58-2*b*).

It is evident, therefore, that the angle θ is the same in Figs. 58-1*b* and 58-2*b*.

Analysis of Fig. 58-1*b* reveals the relationships among V, V_R, V_C, and the phase angle θ. Thus, since

$$\frac{V_R}{V} = \cos \theta \qquad (58\text{-}2)$$

$$V_R = V \cos \theta \qquad (58\text{-}3)$$

Similarly it follows that

$$V_C = V \sin \theta \qquad (58\text{-}4)$$

Referring to the impedance triangle, it is also evident that

$$\cos \theta = \frac{R}{Z}$$

and

$$\sin \theta = \frac{X_C}{Z}$$

By substituting these values of the trigonometric functions in Eqs. (58-3) and (58-4), we find that

$$V_R = V \times \frac{R}{Z} \qquad (58\text{-}5)$$

and

$$V_C = V \times \frac{X_C}{Z} \qquad (58\text{-}6)$$

Equations (58-5) and (58-6) state that in a series RC circuit the voltage V_R across R may be found by multiplying the applied voltage V by the ratio of the resistance and impedance (R/Z), and the voltage V_C across C is found by multiplying the applied voltage V by the ratio of capacitive reactance and impedance (X_C/Z).

(a)   (b)

Fig. 58-2. The phase angle θ between V and I is the same as the angle between Z and R in a series RC circuit.

Example

If 100 V is applied across a circuit consisting of a 30-Ω resistor in series with a capacitor whose reactance is 40 Ω, what is the value of:

1. θ, the phase angle between V and I?
2. V_R, the voltage across R?
3. V_C, the voltage across C?

Solution 1. (to slide rule accuracy)

1.
$$\tan \theta = \frac{X_C}{R} = \frac{40}{30} = 1.33$$

$$\theta = 53.1°$$

2.
$$V_R = V \cos \theta$$
$$= 100 \cos 53.1°$$
$$= 60 \text{ V}$$

3.
$$V_C = V \sin \theta$$
$$= 100 \sin 53.1°$$
$$= 80 \text{ V}$$

Solution 2. (to slide rule accuracy)
1. As in the previous case $\theta = 53.1°$
2. Find Z by the right triangle method.

$$Z = \sqrt{R^2 + X_C^2} = \sqrt{30^2 + 40^2} = 50 \ \Omega$$

$$V_R = V \times \frac{R}{Z} = 100 \times \frac{30}{50} = 60 \text{ V}$$

3.
$$V_C = V \times \frac{X_C}{Z} = 100 \times \frac{40}{50} = 80 \text{ V}$$

Check: $\sqrt{V_R^2 + V_C^2} = \sqrt{60^2 + 80^2} = 100 \text{ V} = V$.

Since the applied voltage V is 100 V, it is evident that the solution to the problem is correct.

SUMMARY

1. In a series RC (capacitive) circuit the current I *leads* the applied voltage V by some angle θ, called the phase angle.
2. The phase angle θ between V and I is the same as the angle θ between Z and R in the impedance vector diagram (Fig. 58-1*b* and 58-2*b*). The angle θ is also the same as the angle between V and V_R (Fig. 58-1*b*).
3. The value of θ depends on the relative values of X_C, R, and Z, and may be computed from any one of the following formulas:

$$\theta = \text{arc tan } \frac{X_C}{R}$$

$$\theta = \text{arc sin } \frac{X_C}{Z}$$

$$\theta = \arccos \frac{R}{Z}$$

4. In a series RC circuit the voltage drop V_C across the capacitor *lags* the voltage drop V_R across the resistor by 90°.

5. V_C and V_R are phasors and are added vectorially to give the applied voltage V.

6. The relationship between the applied voltage V and the voltage drops V_R and V_C across R and C, respectively, is given by the formula

$$V = \sqrt{V_R{}^2 + V_C{}^2}$$

7. If the applied voltage value V and the phase angle θ are known, then V_R and V_C may be calculated from these formulas:

$$V_R = V \cos \theta$$
$$V_C = V \sin \theta$$

8. If the applied voltage V and R, X_C, and Z are known in a series RC circuit, then V_R and V_C may be calculated from these formulas:

$$V_R = V \times \frac{R}{Z}$$

$$V_C = V \times \frac{X_C}{Z}$$

9. To verify that V is the phasor sum of V_R and V_C in a series RC circuit, we measure V, V_R, and V_C and substitute the measured values of V_R and V_C in $\sqrt{V_R{}^2 + V_C{}^2}$. If the numerical result is equal to the measured value V, then we have the required confirmation.

SELF-TEST

Check your understanding by answering these questions:

1. In the circuit of Fig. 58-1a, the measured values of V_R and V_C are, respectively, 50 and 120 V. The applied voltage V must then equal _____ V.

2. The phase angle between voltages V and _____ is the same as the phase angle between V and I in the circuit of Fig. 58-1a.

3. The phase angle θ between V and I in question 1 is _____ degrees.

4. In the circuit of Fig. 58-1a, $X_C = 200$ Ω, $R = 300$ Ω, and $V = 60$ V. In this circuit
   (a) $Z =$ _____ Ω
   (b) $\theta =$ _____ degrees
   (c) $V_R =$ _____ V
   (d) $V_C =$ _____ V

5. In a series RC circuit arc tan $X_C/R =$ arc cos R/Z. _____ (true/false)

MATERIALS REQUIRED

- Power supply: Line-isolation transformer and variable autotransformer operating from 120-V ac line
- Equipment: Oscilloscope, EVM
- Resistors: ½-W 12,000-Ω; 5,600-Ω
- Capacitor: 0.5 µF
- Miscellaneous: SPST switch; fused line cord

PROCEDURE

Measuring Phase Angle

In Fig. 58-1a the voltage V_R across R is in phase with current I in the circuit. The phase angle θ, by which the applied voltage V lags the current I, can therefore be found by measuring the angle between V and V_R, by the method of Lissajous patterns employed in Experiment 55.

1. Connect the circuit of Fig. 58-3a. V is the output of the variable autotransformer which is plugged into a 1:1 isolation transformer in the manner of Fig. 40-3. Switch S is *off*. Set the output of V at 20 V, approximately. Switch the sweep selector of the oscilloscope to Ext. sweep. **Power on.**

2. Adjust the vertical-gain control of the oscilloscope for two divisions of vertical deflection, and the horizontal-gain control for two divisions of horizontal deflection. *Center* the resulting pattern vertically with reference to the X axis, and horizontally with reference to the Y axis. The pattern will now also be centered with reference to the origin, as in Fig. 58-3b.

   Measure, and record in Table 58-1, the maxi-

Fig. 58-3. Measuring phase angle between V and I.

TABLE 58-1.

R, ohms	C, μF	Y_M, div.	Y_O, div.	θ (degrees) = arc sin $\dfrac{Y_O}{Y_M}$	X_C, ohms	θ (degrees) = arc tan $\dfrac{X_C}{R}$	Z, ohms
5,600	0.5						
12,000	0.5						

mum Y distance Y_M from the X axis, and the intercept Y_O on the Y axis. Compute the phase angle θ by substituting Y_M and Y_O in the equation

$$\theta = \text{arc sin } \frac{Y_O}{Y_M} \qquad (58\text{-}7)$$

Show your computation.

3. Compute and record X_C of the capacitor C in Fig. 58-3a. For comparison, determine the phase angle θ by substituting the values of X_C and R in the equation

$$\theta = \text{arc tan } \frac{X_C}{R}$$

Show your computation. The computed and measured values of θ should be very close to each other. Also compute and record the impedance Z of the circuit. Show your computation.

4. **Power off.** In the circuit of Fig. 58-3a, replace R with a 12,000-ohm resistor. As in the preceding steps, measure and record the new phase angle between V and I. Also compute and record the phase angle, as in step 3, for comparison. Also

compute and record the impedance Z of the circuit.

Verifying the Formula $V = \sqrt{V_R{}^2 + V_C{}^2}$

5. Remove the oscilloscope connections from the circuit. Set the output of the isolated supply at 30 V, as measured on an EVM. With the voltmeter, measure the voltages V_R and V_C and record in Table 58-2. Calculate the value of $\sqrt{V_R{}^2 + V_C{}^2}$ and record in Table 58-2. Show your computations.

6. Repeat step 5 for each value of supply voltage V shown in Table 58-2.

7. **Power off.** Replace the 12,000-Ω resistor with a 5,600-Ω resistor. Repeat and record the measurements for V_R and V_C for each value of applied voltage shown in Table 58-2. In each case calculate and record the value of $\sqrt{V_R{}^2 + V_C{}^2}$.

Extra Credit

8. From the data in Tables 58-1 and 58-2, verify the relationships in Eqs. (58-3), (58-4), (58-5),

TABLE 58-2.

R, ohms	C, μF	V, volts	V_R, volts	V_C, volts	$\sqrt{V_R^2 + V_C^2}$ volts
12,000	0.5	30			
		40			
		50			
5,600	0.5	30			
		40			
		50			

and (58-6). Show your computations. Record the required data in tabular form.

QUESTIONS

1. What conclusion can you draw from the measured and computed results of θ in Table 58-1? Explain.
2. What are the sources of error in the measurements for θ (Table 58-1)?
3. What conclusion can be drawn from the data in Table 58-2 concerning V, V_R, and V_C? Explain.
4. Find the missing values in Table 58-3, which is based

on measurements in a series RC circuit. Show your computations.

Answers to Self-Test

Answers to slide-rule accuracy.

1. 130
2. V_R
3. 67.4°
4. (a) 360; (b) 33.7; (c) 50; (d) 33.2
5. true

TABLE 58-3.

	R, ohms	C, μF	V, volts	F, Hz	V_C, volts	V_R, volts	θ, degrees	X_C, ohms	Z, ohms
a	1,200	0.10	80	1,000					
b		0.25	100	500	50				
c				100	20	30		10,000	

EXPERIMENT 59

(Optional) The j Operator in the Analysis of RC Circuits

OBJECTIVES

1. To verify experimentally that the impedance Z of a series-connected RC circuit is

$$Z = R - jX_C = |Z| \underline{/-\theta}$$

2. To verify experimentally that the applied voltage V of a series-connected RC circuit is

$$V = V_R - jV_C = |V| \underline{/-\theta}$$

INTRODUCTORY INFORMATION

The use of the j operator (Experiment 56) facilitates analysis and computations of impedance, voltage, and current in RC circuits. In this experiment we will therefore learn how to write Z of an RC circuit in rectangular form using the j operator. Moreover, the relationship between the applied voltage V and the voltage drops V_R across R and V_C across C will be written in rectangular form, using the j operator. Z and V will also be written in the polar form.

Impedance Z of an RC Circuit Expressed in Rectangular Form

Consider the series RC circuit in Fig. 59-1a. The impedance diagram for this circuit is shown in Fig. 59-1b. Since the impedance Z is a vector whose components are X_C and R, and since X_C and R are at right angles, the impedance Z may be written in the rectangular form, using the j operator. It should be noted, however, that X_C lags R (considered as reference) by 90°. Therefore, since R is a positive real number lying on the axis of real numbers in the complex plane, X_C must lie on the axis of imagi-

nary numbers (j) in the complex plane, and specifically on the *negative* side of the j axis. Thus if X_C is the capacitive reactance of C, and R is the resistance of the resistor in Fig. 59-1a, then

$$Z = R - jX_C \qquad (59\text{-}1)$$

Equation 59-1 states that the impedance of a series-connected RC circuit is a vector, and when written in the rectangular form, the positive horizontal component is the value of resistance R, and the negative vertical component is the value of X_C. This equation does not give the *numerical* value $|Z|$ of the impedance. If $|Z|$ is required it may be computed from the equation

$$|Z| = \sqrt{R^2 + X_C^2} \qquad (59\text{-}2)$$

Refer again to Fig. 59-1b. Note that the phase angle is negative, $-\theta$, since Z lags R. The numerical value Z of the impedance can also be found from the equations

$$|Z| = \frac{R}{\cos(-\theta)} = \frac{R}{\cos \theta}$$

or $\qquad |Z| = \frac{-X_C}{\sin(-\theta)} = \frac{X_C}{\sin \theta}$ $\qquad (59\text{-}3)$

A numerical example will illustrate the manner in which Eqs. (59-2) and (59-3) are applied.

Fig. 59-1. (a) Series-connected RC circuit; (b) Vector diagram showing relationship between Z, R, and X_C. Note the phase angle is negative, $-\theta$, since Z lags R.

Fig. 59-2. $V = V_R - jV_C$ or $V = |V| \underline{/-\theta}$

Example

In Fig. 59-1a, $R = 30\ \Omega$, $X_C = 40\ \Omega$.
1. Write an equation for Z using the j operator.
2. Find the numerical value $|Z|$.
3. Find the phase angle θ.

Solution
1. $Z = R - jX_C = 30 - j40$
2. $|Z| = \sqrt{R^2 + X_C{}^2} = \sqrt{30^2 + 40^2} = 50$
3. $\theta = \text{arc tan } \dfrac{X_C}{R} = \text{arc tan } \dfrac{40}{30} = 53.1°$

Impedance Z of an RC Circuit in Polar Form

The polar form of the impedance Z of a series-connected RC circuit follows directly from the rectangular form. Thus in Fig. 59-1b

$$Z = |Z| \underline{/-\theta}$$

where $\quad |Z| = \sqrt{R^2 + X_C{}^2} = \dfrac{R}{\cos \theta} = \dfrac{X_C}{\sin \theta} \quad$ (59-4)

and $\qquad \theta = \text{arc tan } \dfrac{X_C}{R}$

Now if $R = 30\ \Omega$ and $X_C = 40\ \Omega$, applying the Eqs. (59-4), we get

$$Z = 50 \underline{/-53.1°}$$

V, V_R and V_C in a Series RC Circuit Expressed in Rectangular Form; in Polar Form

In the series RC circuit of Fig. 59-1a, the relationship between the applied voltage V and the voltage drop V_R across R and the voltage drop V_C across C can be represented by the phasor diagram in Fig. 59-2. The phasor V in rectangular form is written

$$V = V_R - jV_C \qquad (59\text{-}5)$$

The numerical value V can be found from the equations

$$|V| = \sqrt{V_R{}^2 + V_C{}^2} = \dfrac{V_R}{\cos(-\theta)} = \dfrac{-V_C}{\sin(-\theta)}$$

or $\qquad |V| = \dfrac{V_R}{\cos \theta} = \dfrac{V_C}{\sin \theta}$

(59-6)

where $\qquad \theta = \text{arc tan } \dfrac{V_C}{V_R}$

In polar form, the phasor V is written

$$V = |V| \underline{/-\theta} \qquad (59\text{-}7)$$

where $|V|$ is defined in exactly the same way as it is in Eq. (59-6).

If we know the values of $|V|$ and θ, for a series RC circuit, we can simply write V in polar form

$$V = |V| \underline{/-\theta}$$

We can then transform the polar into the rectangular form

$$V = |V| \cos \theta - j|V| \sin \theta \qquad (59\text{-}8)$$

Comparing Eqs. (59-5) and (59-8), we see that

$$V \cos \theta = V_R$$

and $\qquad V \sin \theta = V_C$

SUMMARY

1. The impedance Z of a series RC circuit is a vector. X_C lags R by 90°. If the values of X_C and R are known, Z can be written in the rectangular form

$$Z = R - jX_C$$

2. Z can also be written in the polar form by the equation

$$Z = |Z| \underline{/-\theta}$$

where $|Z| = \sqrt{R^2 + X_C{}^2} = \dfrac{R}{\cos \theta} = \dfrac{X_C}{\sin \theta}$

and $\qquad \theta = \text{arc tan } \dfrac{X_C}{R}$

3. If the impedance of a series RC circuit is given in the polar form, that is, if the numerical values $|Z|$ and θ are known, then Z can be expressed also in the rectangular form using the equation

$$Z = |Z| \cos \theta - j|Z| \sin \theta$$

In this equation $|Z| \cos \theta = R$, and $|Z| \sin \theta = X_C$.
4. In a series RC circuit the phasor relationship between the applied voltage V and the voltage drops V_R across R and V_C across C is given by the formulas:

$$V = V_R - jV_C$$

The j Operator in the Analysis of RC Circuits **293**

or
$$V = |V| \, \underline{/-\theta}$$

where
$$|V| = \sqrt{V_R{}^2 + V_C{}^2} = \frac{V_R}{\cos \theta} = \frac{V_C}{\sin \theta}$$

and
$$\theta = \text{arc tan} \, \frac{V_C}{V_R}$$

Fig. 59-3. Experimental circuit for determining the impedance Z of an RC circuit.

SELF-TEST

Check your understanding by answering these questions:

1. In the circuit of Fig. 59-1a, $V = 30$ V, $R = 220$ Ω, and $X_C = 150$ Ω. The impedance Z written in rectangular form is $Z = $ _____.
2. In the circuit of question 1, the phase angle θ between Z and R is _____ degrees.
3. In the circuit of question 1, the impedance Z written in polar form is $Z = $ _____.
4. In the circuit of question 1, the phasor V written in polar form is $V = $ _____.
5. In the circuit of question 1, the phasor V written in rectangular form is $V = $ _____.
6. From the equation in question 5 we know that the voltage V_R across R is _____ V, while the voltage V_C across C is _____ V.
7. In the circuit of Fig. 59-1a, $Z = 1{,}000$ Ω, $V_R = 60$ V, $V_C = 80$ V. The phasor V written in rectangular form is $V = $ _____.
8. In the circuit of question 7, the phase angle θ between V and V_R is _____ degrees, and V _____ (leads/lags) V_R by this angle.
9. In the circuit of question 7, the phasor V written in polar form is $V = $ _____.
10. In the circuit of question 7, the vector Z written in polar form is $Z = $ _____.
11. In the circuit of question 7, the vector Z written in rectangular form is $Z = $ _____.
12. From the equation in question 11 we know that $R = $ _____ Ω and $X_C = $ _____ Ω.

MATERIALS REQUIRED

- Power supply: Isolation transformer and variable autotransformer operating from the 120-V ac line
- Equipment: Electronic voltmeter
- Resistors: 1/2-W 5,600-Ω
- Capacitor: 0.5 μF
- Miscellaneous: SPST switch; fused line cord

PROCEDURE

Impedance Z of a Series RC Circuit

1. Connect the circuit of Fig. 59-3. S is open. V_A is a variable autotransformer plugged into T, a 1:1 isolation transformer which receives power from the 120-V ac line. The voltmeter V_1 set on ac volts is connected across R.
2. Close S. **Power on.** Adjust the output of the ac supply until the voltage V_R, measured across R, $= 36$ V.
3. Measure also and record in Table 59-1 the voltage V_C across C and the voltage V_{AB} which is the numerical value $|V|$ of the voltage V applied across the series-connected RC circuit.

4. Open switch S. **Power off.** Measure the resistance of R and record in Table 59-1.
5. Using the measured values V_R and R, compute and record in Table 59-1 the current I in the circuit.
6. Compute and record the numerical value $|Z|$ of the impedance in the circuit using the equation $|Z| = |V|/I$, where $|V|$ is the measured value.
7. Compute and record the phase angle θ from the equation

$$\theta = \text{arc tan} \, \frac{V_C}{V_R}$$

TABLE 59-1.

Measured Values				Computed Values			
V_R volts	V_C volts	$V_{AB} = \|V\|$ volts	R ohms	I amp	$\|Z\|$ ohms	θ degrees	X_C ohms
36							

Z in polar form	Z in rectangular form
$Z =$	$Z =$

8. Compute and record in Table 59-1 the value of X_C using the relationship

$$X_C = \frac{V_C}{I}$$

9. Using the computed value of Z and θ, write the vector Z in polar form in Table 59-1.
10. Transform the polar form of Z into the rectangular form using the relationship

$$Z = |Z| \cos \theta - j|Z| \sin \theta$$

and record in Table 59-1.

Voltage Relationships in a Series RC Circuit

11. From the measured values V_R and V_C in Table 59-1, write the phasor V (applied voltage) in rectangular form and record in Table 59-2.
12. Compute the numerical value $|V|$ of the phasor V using any one of the equations

$$|V| = \sqrt{V_R{}^2 + V_C{}^2} = \frac{V_R}{\cos \theta} = \frac{V_C}{\sin \theta}$$

and record in Table 59-2. Employ the values V_R, V_C, and θ from Table 59-1.
13. From Table 59-1 record in Table 59-2, the measured value $V_{AB} = |V|$.
14. Using the value of $|V|$ from step 12 and the value of θ from Table 59-1, write the phasor V in polar form in Table 59-2.

QUESTIONS

1. Refer to the equation for Z in rectangular form, Table 59-1 and to step 10 in procedure.
   (a) Does the measured value of $R = |Z| \cos \theta$?

(b) Should it?
   (c) If the two values are not the same, explain why they are not.
2. Refer to the equation for Z in rectangular form, Table 59-1 and to step 10 in procedure.
   (a) Does the computed value of $X_C = |Z| \sin \theta$?
   (b) Should it?
   (c) If the two values are not the same, explain why they are not.
3. Have you confirmed experimentally (approximately) the analytical statement that $Z = R - jX_C$? Refer specifically to your data to show that you have or have not confirmed it.
4. If your data does not confirm (approximately) that $Z = R - jX_C$, explain why not.
5. Refer to Table 59-2 and to step 12. Does the computed value of $|V|$ equal the measured value of $|V|$? If not, why not?
6. Have you confirmed experimentally (approximately) the analytical statement that $V = V_R - jV_C$? Refer specifically to your data to show that you have or have not confirmed it.
7. If your data does not confirm that $V = V_R - jV_C$ explain why not.

Answers to Self-Test

Answers to slide-rule accuracy.

1. $220 - j150$
2. 34.2
3. $267 \angle{-34.2°}$
4. $30 \angle{-34.2°}$
5. $24.8 - j16.9$
6. $24.8; 16.9$
7. $60 - j80$
8. 53.1; lags
9. $100 \angle{-53.1°}$
10. $1000 \angle{-53.1°}$
11. $600 - j800$
12. $600; 800$

TABLE 59-2.

V in Rectangular Form	$\|V\|$ Computed Volts	$\|V\|$ Measured Volts	V in Polar Form
$V =$			$V =$

60 EXPERIMENT

Frequency Response of a Reactive Circuit

OBJECTIVES

1. In a series-connected RL circuit to measure the effect on impedance and on current of a change in frequency of the voltage source
2. In a series-connected RC circuit to measure the effect on impedance and on current of a change in frequency of the voltage source

INTRODUCTORY INFORMATION

Impedance of a Series RL Circuit

The impedance of a series-connected RL circuit is given by

$$Z = \sqrt{R^2 + X_L{}^2} \qquad (60\text{-}1)$$

It is apparent from Eq. (60-1) that if R remains constant in value, a change in X_L will affect Z. Thus, as X_L increases, Z also increases. As X_L decreases, Z decreases. Since

$$X_L = 2\pi FL \qquad (60\text{-}2)$$

we can change X_L either by increasing or decreasing the size of L, with F remaining constant. Or we can change X_L by increasing or decreasing F, with L remaining constant.

Let us consider the effect on X_L of varying F, with L constant. It is evident from Eq. (60-2) that as F increases, X_L increases. As F decreases, X_L decreases. Z will therefore also increase or decrease respectively as F increases or decreases.

An example will illustrate this variation. Suppose the X_L of L in Fig. 60-1 changes with frequency as shown in Table 60-1. If $R = 30$, Z computed will take on the values shown in Table 60-1 (to slide

rule accuracy). Note that the effect of R on Z decreases as X_L increases. Figure 60-2 is a graph of Z versus F in this RL circuit.

Current versus Frequency in an RL Circuit

The current in an ac circuit is given by the equation

$$I = \frac{V}{Z} \qquad (60\text{-}3)$$

Current varies inversely with Z. Since Z varies directly with F in a series RL circuit, current will also vary inversely with F. Thus, current in a series RL circuit will increase as F decreases, and will decrease as F increases.

If we assume that the voltage V applied to the circuit of Fig. 60-1 is 100 V, then the current takes on the values shown in Table 60-1. In Fig. 60-2 we also see a graph of I versus F. This is the frequency response of the circuit.

Analyzing the two graphs, it becomes evident that the impedance of an RL circuit varies directly with frequency, while the current varies inversely.

Impedance of a Series RC Circuit

The impedance of a series RC circuit is given by the equation

Fig. 60-1. Series RL circuit to determine effect of change in frequency F on impedance Z and on current I.

TABLE 60-1.

F, Hz	X_L, ohms	$Z = \sqrt{R^2 + X_L^2}$ ohms	$I = \dfrac{V}{Z}$ (amperes)
0	0	30	3.33
100	40	50	2.00
150	60	67.2	1.488
200	80	85.5	1.17
250	100	104.5	0.956
300	120	124	0.806
400	160	163	0.614
500	200	202	0.495

$$Z = \sqrt{R^2 + X_C^2} \qquad (60\text{-}4)$$

Though Eq. (60-4) and (60-1) are the same, except for the substitution of X_C for X_L, the variation of Z with frequency is very much different for the RC circuit. The reason is that X_C varies inversely with frequency. Thus

$$X_C = \frac{1}{2\pi FC} \qquad (60\text{-}5)$$

Hence, as F increases, X_C decreases. As F decreases, X_C increases. Therefore, the impedance of a series RC circuit increases with a decrease in frequency but decreases with an increase in frequency. That is, Z varies inversely with frequency.

Current versus Frequency in an RC Circuit

In a series RC circuit, as F decreases, X_C increases, Z increases, and I decreases. As F increases, X_C decreases, Z decreases, and I increases. It is evident, therefore, that in a series RC circuit, current varies directly with frequency F. Again this relationship is opposite to that of a series RL circuit.

The foregoing analysis shows that the effects of a capacitor and an inductor on current in series RC and RL circuits are opposite. This fact is related to the phase relationship between voltage and current in a capacitor and an inductor, for in an inductor current lags voltage, while in a capacitor current leads the voltage.

SUMMARY

1. In a series RL circuit as X_L increases, with R constant, Z increases.

2. X_L increases or decreases as the frequency F of the voltage applied to the circuit increases or decreases, respectively.
3. Therefore as F increases in a series RL circuit, Z increases. As F decreases in a series RL circuit, Z decreases.
4. As F increases in a series RL circuit, I in the circuit decreases. As F decreases, I increases.
5. In a series RC circuit, X_C increases or decreases, as the frequency F decreases or increases, respectively. That is, F has an opposite effect on X_C from what it has on X_L.
6. In a series RC circuit Z increases, as F decreases, with R and C constant. Moreover, Z decreases as F increases.
7. In a series RC circuit I decreases as F decreases; but I increases when F increases.
8. It is apparent therefore that the effects of a capacitor and an inductor on impedance and current in a series connected circuit are opposite.
9. The reactance of an inductor is $X_L = 2\pi FL$. The impedance of a series RL circuit is $Z = \sqrt{R^2 + X_L^2}$
10. The reactance of a capacitor is $X_C = (1/2\pi FC)$. The impedance of a series RC circuit is $Z = \sqrt{R^2 + X_C^2}$.
11. The frequency response of a reactive circuit is the graph of I versus F.

SELF-TEST

Check your understanding by answering these questions:
1. The X_L of a 3.19-H choke at 100 Hz is _____ Ω.

Fig. 60-2. Z and I versus F in a series RL circuit.

2. The Z of a series RL circuit, where $R = 1000\ \Omega$ and $L = 3.19$ H, at 100 Hz is _____ Ω.
3. The Z of the same circuit as in question 2, at 200 Hz, is _____ Ω.
4. A generalization concerning a series RL circuit is that Z _____ (increases/decreases) with an increase in frequency, and vice versa.
5. In the circuit of question 2, if the applied voltage $V = 22.35$ V, the current I in the circuit is _____ A.
6. In the circuit of question 3, if the applied voltage $V = 22.35$, the current I in the circuit is _____ A.
7. A generalization concerning a series RL circuit is that I _____ (increases/decreases) with an increase in frequency, and vice versa.
8. In a series RC circuit, the X_C of the capacitor is 1000 Ω at 100 Hz. The X_C of the same capacitor at 200 Hz is _____ Ω, while at 50 Hz the X_C of the capacitor is _____ Ω.
9. The Z of a series RC circuit, with R and C constant, _____ (increases/decreases) with an increase in frequency, and vice versa.
10. If the Z of the RC circuit in question 8 at 100 Hz = 1,410 Ω, the Z of the same circuit at 50 Hz is _____ (greater/less than) 1,410 Ω.
11. The current I in the RC circuit of question 8 _____ (increases/decreases) when F increases from 50 Hz to 100 Hz.

MATERIALS REQUIRED

- Equipment: AF signal generator; electronic voltmeter
- Resistor: ½-W 10,000-Ω
- Capacitor: 0.1 μF
- Inductance: 8 H at 50-mA direct current

PROCEDURE

Impedance and Current versus Frequency in an RL Circuit

1. Connect the circuit of Fig. 60-1. L is an inductance rated 8 H at 50-mA direct current. R is a 10,000-Ω resistor. The voltage source is an audio-frequency signal generator.
2. Set the frequency of the generator at 50 Hz, and the output V at 10 V rms. With an EVM, measure, and record in Table 60-2, the output voltage V applied across the circuit. Measure also and record the voltage V_R across R. Compute and record the current I in the circuit ($I = V_R/R$, where V_R is in volts and R is the resistance R in the circuit). Compute and record the impedance Z of the circuit ($Z = V/I$, where V is the voltage applied to the circuit and I is the calculated current). Show your computations.
3. Set the frequency of the generator at 100 Hz, output V at the same level as in the preceding step. Again measure the voltages V and V_R. Compute I and Z and record in Table 60-2.

NOTE: In the remaining checks, step 4, be certain that V is kept at the same level as in steps 2 and 3.

4. Repeat step 3 for each of the frequencies shown in the table.

5. From the data in Table 60-2, draw a graph of Z versus F, using the vertical axis for Z and horizontal axis for F. Label the graph Z_L.
6. Using the same coordinate axes, but labeling the vertical axis as I, draw a graph of I versus F. Label the graph I_L.

TABLE 60-2.

F, Hz	V, volts	V_R, volts	$I = \dfrac{V_R}{R}$ amperes	$Z = \dfrac{V}{I}$
50				
100				
150				
200				
250				
300				
350				
400				
450				
500				

Impedance and Current versus Frequency in an RC Circuit

7. Connect the circuit of Fig. 60-3. Set the frequency of the generator at 500 Hz and output V at 10-V rms. With an EVM, measure the output voltage V applied across the circuit and record in Table 60-3. Measure also and record the voltage V_R across R. Compute and record the current I in the circuit, and the impedance Z of the circuit. Show your computations.

8. Repeat step 7 for the other frequencies shown in Table 60-3. Be certain that the voltage V is maintained at the same level as in step 7. You need not show your computations.

9. From the data in Table 60-3, draw a graph of Z versus F. Label the graph Z_C. Let Z be the vertical axis, F the horizontal.

10. From the data in Table 60-3, draw a graph of I versus F. Label the graph I_C. Use the same coordinate axes as in step 9, but let I be the vertical axis.

QUESTIONS

1. Refer to your graph of Z_L versus F. How does Z_L vary with F? Why?
2. Is any portion of the graph Z_L versus F linear? Identify that portion, if any, and explain its linearity.
3. In the graph I_L versus F, how does I vary with respect to F? Why?
4. How does Z_C vary with respect to F? Why?
5. Compare the graphs Z_C versus F with Z_L versus F. How are they different?
6. How does I_C vary with respect to F? Why?
7. Compare the graphs of I_L and I_C versus F. How are they different?

Extra Credit

8. At any specific frequency, what do you think would be the effect of connecting L, C, and R in series (in the same circuit) on:
   (a) Impedance
   (b) Current
   assuming the same source V is used as in this experiment? Why?
9. Suggest a formula for impedance in a series LCR circuit.

Fig. 60-3. Series RC circuit to determine effect of change of frequency F on impedance Z and current I.

10. What is the minimum impedance of a series LCR circuit? Why?
11. What would be the frequency when the impedance is minimum in a series LCR circuit? Why?

Answers to Self-Test

1. 2,000
2. 2.235
3. 4,120
4. increases
5. 0.01
6. 0.0054
7. decreases
8. 500; 2000
9. decreases
10. greater
11. increases

TABLE 60-3.

F, Hz	V, volts	V_R, volts	$I = \dfrac{V_R}{R}$ amperes	$Z = \dfrac{V}{I}$ ohms
500				
450				
400				
350				
300				
250				
200				
150				
100				
50				

61

OBJECTIVE

To verify experimentally that the impedance Z of a series RLC circuit is $Z = \sqrt{R^2 + (X_L - X_C)^2}$

INTRODUCTORY INFORMATION

Impedance of an RLC Circuit

To understand the effects on alternating current of series-connected RLC circuits, it is necessary to recall the individual effect of each of these components.

The effects of a resistor in an ac circuit are the same as in a dc circuit, since alternating current and voltage are in phase in a resistor. In a resistive divider, ac voltage drops across each of the resistors may be added arithmetically to get the value of applied voltage.

When a capacitor is connected in series with a resistor, the reactance of the capacitor together with the resistance of the resistor determines the effect on alternating current. Capacitive reactance depends on frequency. Hence, the effects of a capacitor are determined both by its size (capacitance) and by the frequency of the current in the circuit. In a totally capacitive circuit, i.e., one which contains only capacitance, alternating current leads voltage by 90°. In a series RC circuit, the alternating current leads the voltage by a phase angle less than 90°.

The effects of an inductor connected in series with a resistor in an ac circuit also depend on the frequency of the current and the size of the inductor. In a totally inductive circuit, i.e., one which contains only inductance, current lags voltage by

90°. In a series RL circuit, the current lags the voltage by an angle less than 90°.

It may be seen from the characteristics of an inductance and a capacitance that their effects on current and voltage in an ac circuit are opposite. This fact is demonstrated in the vector representation of X_L and X_C in a series-connected RLC circuit (see Fig. 61-1).

The vector sum of X_L and X_C is the same as their arithmetic difference. The resultant takes the direction of the larger reactance. Assume that X_L is larger than X_C in Fig. 61-1. The vector diagram can then be redrawn as in Fig. 61-2a.

Note that Fig. 61-2a really represents an inductive circuit whose net reactance is the difference between X_L and X_C. The effect is as though the L and C of the series-connected RLC circuit had been replaced by an inductor whose reactance is equal to $X_L - X_C$. Figure 61-2b represents a capacitive circuit whose reactance is $X_C - X_L$.

Figure 61-2a indicates that the impedance Z of the RLC circuit is given by the formula

$$Z = \sqrt{R^2 + (X_L - X_C)^2} \qquad (61\text{-}1)$$

This formula holds whether X_L is greater or smaller than X_C because the reactive term is a squared term and is therefore always positive, or is zero when $X_L = X_C$.

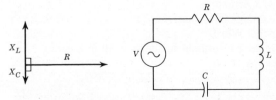

Fig. 61-1. Effect of an inductance and capacitance in a series RLC circuit.

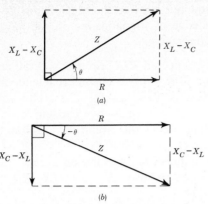

Fig. 61-2. Impedance of a series-connected *RLC* circuit.

One conclusion which can be drawn from Eq. (61-1) is that at any specific frequency the impedance of a series *RLC* circuit is always less than the impedance of a series *RL* circuit or a series *RC* circuit, assuming the same values of *R, L,* and *C*. This fact can readily be verified as follows: Connect a series *RL* circuit and apply an ac sinusoidal voltage *V* across the circuit. Measure the current I_1. Now add a capacitor *C* in series with *R* and *L*, while maintaining the same voltage *V* across the circuit. The measured current I_2 will be greater than I_1. Since $Z = V/I$, it is apparent that the *Z* of a series *RLC* circuit is smaller than that of a series *RL* circuit.

A second demonstration, starting first with a series *RC* circuit then adding an inductance *L* in series, will produce the same result.

(Optional) Impedance of an RLC Circuit Expressed Trigonometrically

Consider the case of a series *RLC* circuit which is inductive, that is, where $X_L > X_C$, as is shown in the vector diagram in Fig. 61-2a. Here the phase angle θ, between *Z* and *R*, can be used to calculate *Z*, if $(X_L - X_C)$ and *R* are known. Thus

$$\theta = \text{arc tan} \frac{X_L - X_C}{R}$$

$$(61\text{-}1)$$

and

$$Z = \frac{R}{\cos \theta} = \frac{X_L - X_C}{\sin \theta}$$

Equations (61-1) are very similar to those in a simple *RL* circuit (see Experiment 54).

If $X_L < X_C$, the circuit acts like a capacitive circuit, whose vector diagram is Fig. 61-2b. Here the phase angle θ is negative, that is, *Z* lags *R*, which is

the characteristic of an *RC* circuit. Equations (61-1) can still be used to define θ and *Z*. Note, however, that the quantity $X_L - X_C$ is negative, and therefore θ is negative.

Where $X_C > X_L$, it is possible to simplify calculations by using these equations.

$$\theta = \text{arc tan} \frac{X_C - X_L}{R}$$

$$(61\text{-}2)$$

$$Z = \frac{R}{\cos \theta} = \frac{X_C - X_L}{\sin \theta}$$

Equations (61-2) are similar to those that define θ and *Z* in an *RC* circuit (see Experiment 57).

(Optional) Impedance of an RLC Circuit Analyzed Using the j Operator

The impedance of an *RLC* circuit can be written in rectangular form. Thus

$$Z = R + jX_L - jX_C \qquad (61\text{-}3)$$

and combining the reactive terms, we get

$$Z = R + j(X_L - X_C) \qquad (61\text{-}4)$$

From Eq. (61-4) it is apparent that if $X_L > X_C$, the term $X_L - X_C$ is positive, and the circuit acts like an *RL* circuit, since *Z* has the form $Z = R + jX$ where $X = X_L - X_C$. However, if $X_L - X_C$ is negative, *Z* has the form, $Z = R - jX$, which is the characteristic of an *RC* circuit.

(Optional) Z of an RLC Circuit Expressed in Polar Form

We can transform Eq. (61-4) into the polar form. Thus

$$Z = |Z| \underline{/\theta}$$

where

$$Z = \sqrt{R^2 + (X_L - X_C)^2} = \frac{R}{\cos \theta} = \frac{X_L - X_C}{\sin \theta}$$

$$(61\text{-}5)$$

and

$$\theta = \text{arc tan} \frac{X_L - X_C}{R}$$

As we noted above, if $X_L > X_C$, θ is positive; if $X_L < X_C$, θ is negative.

SUMMARY

1. An inductor *L* and a capacitor *C* have opposite effects on current in an ac circuit. Thus in an

inductor current *lags*, while in a capacitor current *leads* the applied voltage.

2. The total reactance X of a series-connected *RLC* circuit is the algebraic sum, that is, arithmetic difference of, X_L and X_C. For $X = X_L - X_C$.

3. If $X_L > X_C$, that is, if the term $X_L - X_C$ is a positive number, the circuit acts like an *RL* circuit, whose *net* inductive reactance $X = X_L - X_C$.

4. If $X_L < X_C$, that is, if the term $X_L - X_C$ is a negative number, the circuit acts like an *RC* circuit whose net capacitive reactance $X = X_C - X_L$.

5. The impedance Z of an *RLC* circuit is expressed by the formula

$$|Z| = \sqrt{R^2 + (X_L - X_C)^2}$$

6. The phase angle θ of a series *RLC* circuit is positive (that is, Z leads R) if $X_L > X_C$; θ is negative (that is, Z lags R) if $X_L < X_C$.

7. Z may also be expressed trigonometrically by the equations

$$|Z| = \frac{R}{\cos \theta} = \frac{X_L - X_C}{\sin \theta}$$

where $\qquad \theta = \text{arc tan } \dfrac{X_L - X_C}{R}$

8. Using the j operator Z is written

$$Z = R + j(X_L - X_C)$$

9. In polar form Z of a series *RLC* circuit is written

$$Z = |Z|\ \underline{/\theta}$$

where

$$\theta = \text{arc tan } \frac{X_L - X_C}{R}$$

and

$$|Z| = \sqrt{R^2 + (X_L - X_C)^2} = \frac{R}{\cos \theta} = \frac{X_L - X_C}{\sin \theta}$$

SELF-TEST

Check your understanding by answering these questions:

NOTE: You need not answer those questions requiring a knowledge of trigonometry, if you have not studied trigonometry.

1. In a series-connected *RLC* circuit, $X_L = 50\ \Omega$, $X_C = 20\ \Omega$, and $R = 40\ \Omega$. The net reactance X of this circuit is $X = $ _____ Ω. This reactance is _____ (inductive/capacitive).

2. The impedance Z of the circuit in question 1 is _____ Ω.

3. The circuit in question 1 acts like an _____ (RC/RL) circuit, whose _____ (X_C/X_L) = _____ Ω.

4. For the same values of applied voltage V, resistance R, inductance L, and capacitance C, if $X_L > X_C$, the current in an *RL* circuit is _____ than the current in an *RLC* circuit.

5. In a series-connected *RLC* circuit, $X_L = 100\ \Omega$, $X_C = 150\ \Omega$, and $R = 120\ \Omega$. In this circuit the net _____ (inductive/capacitive) reactance is _____ Ω, and the impedance vector _____ (leads/lags) the resistance R.

6. In the circuit of question 5 the phase angle $\theta = $ _____ degrees and is _____ (negative/positive).

7. In the circuit of question 5, $|Z| = $_____ Ω.

8. The impedance Z in question 5 expressed in rectangular form is $Z = $ _____.

9. The impedance Z in question 5 expressed in polar form is $Z = $ _____.

10. The reason θ is _____ (positive/negative) is that $X_L - X_C$ is _____ (positive/negative).

MATERIALS REQUIRED

- Power supply: Line-isolated source of 6.3-V rms 60-Hz
- Equipment: EVM, oscilloscope, audio-frequency signal generator
- Resistor: ½-W 4,700-Ω
- Capacitors: 0.5 μF
- Miscellaneous: SPST switch; choke: 8 H at 50 mA direct current; fused line cord

PROCEDURE

NOTE: In this experiment the student will first determine experimentally the current and impedance in a series *RL* circuit. He will then connect a capaci-

tor C in series with L and R and will again measure the current. There should be more current in the series *RLC* circuit than there was in the *RL* circuit,

Fig. 61-3. Circuit to determine value of Z.

Fig. 61-4. Circuit to determine impedance of RLC connected in series.

because the impedance Z of the RLC circuit in this experiment is smaller than that of the RL circuit. The data will then be used to verify the impedance formula $Z = \sqrt{R^2 + (X_L - X_C)^2}$ for a series-connected RLC circuit.

1. Connect the circuit of Fig. 61-3. Use an isolated 60-Hz ac source approximately 18 V peak-to-peak. Measure this value of V with an EVM and record in Table 61-1.
2. Measure also the voltages V_R across R and V_L across L and record in Table 61-1.
3. Compute the current I_1 in the circuit and record in Table 61-1. Use the rated value of R and measured value of V_R to determine I_1. Compute and record also the impedance Z of the circuit ($Z = V_{applied}/I_1$). Show your computations.
4. Connect a 0.5-μF capacitor in series with R and L as shown in Fig. 61-4.
5. Measure the applied voltages V, V_R, V_L, and V_C and record in Table 61-2.
6. From the measured values of V_R and from the rated value of R, compute the current I_2. Also compute the values of X_L and X_C from the computed value I_2 and from the measured values V_L and V_C. Record in Table 61-2. Show your computations.

7. Compute and record the impedance Z of the RLC circuit in Fig. 61-4, using the formula

$$Z = \frac{V_{applied}}{I_2}$$

8. Compute and record Z using the formula

$$Z = \sqrt{R^2 + (X_L - X_C)^2}$$

Use the values X_L and X_C from Table 61-2.

(Optional) Extra Credit

9. Using the data in Table 61-2, record in a specially prepared Table 61-3 the impedance Z of the experimental RLC circuit in rectangular form.
10. Compute and record in Table 61-3 the phase angle θ. Show your computations.
11. Compute and record in Table 61-3 the impedance Z, in polar form, of the experimental RLC circuit. Show your computations.

QUESTIONS

1. Refer to your data in Tables 61-1 and 61-2. How was the impedance of the circuit of Fig. 61-3 affected by adding a capacitor C, as in Fig. 61-4?
2. (*a*) State the condition for which the current I will be maximum, and impedance Z minimum, in the circuit of Fig. 61-4.
   (*b*) Compute the value of maximum current.
   (*c*) What is the value of minimum impedance?
3. How does the experimental value of Z in step 7 compare with the formula value in step 8? Explain any difference.

TABLE 61-1.

V Applied, volts	V_R, volts	V_L, volts	I_1, amperes	Z, ohms

TABLE 61-2.

V Applied, volts	V_R, volts	V_L, volts	V_C, volts	I_2, amperes	X_C, ohms	X_L, ohms	Z (ohms) =	
							V/I_2	$\sqrt{R^2 + (X_L - X_C)^2}$

4. From your data in Table 61-2 is the circuit of Fig. 61-4 inductive or capacitive at line frequency? Explain why.
5. Refer to your data in Table 61-2. What is the net reactance of the circuit? Is it inductive or capacitive reactance?

(Optional) Extra Credit

6. Does the value of Z computed in procedural step 11 agree with the values of Z in Table 61-2? If not, why not?

Answers to Self-Test

1. 30; inductive
2. 50
3. RL; X_L; 30
4. less
5. capacitive; 50; lags
6. 22.6; negative
7. 130
8. $120 - j50$
9. $130 \angle -22.6°$
10. negative; negative

EXPERIMENT 62

Variation of Impedance and Current of a Series RLC Circuit with Changes in Frequency

OBJECTIVE

In a series *RLC* circuit to determine experimentally the effect on impedance and current of changes in frequency of the ac source

INTRODUCTORY INFORMATION

In Experiment 61 we verified that the impedance Z of a series-connected *RLC* circuit (Fig. 62-1) is given by the formula

$$Z = \sqrt{R^2 + (X_L - X_C)^2} \qquad (62\text{-}1)$$

In this experiment we will observe how a change in frequency of the voltage source affects circuit impedance and current.

Effect of Frequency on the Impedance of an RLC Circuit

Consider Eq. (62-1). The minimum value of impedance in an *RLC* circuit, for a given value of R, occurs when $X_L - X_C = 0$. At that point $Z = \sqrt{R^2} = R$, and the circuit acts like a pure resistance. The line current I is limited only by R. Hence I is maximum when $X_L - X_C = 0$.

Inductive reactance X_L and capacitive reactance

X_C are dependent on frequency. It is obvious that there is a frequency F_R at which $X_L = X_C$. This fact will be considered at greater length in the next experiment. At this point let us merely note the effect on impedance of frequencies greater and smaller than F_R.

In the series *RLC* circuit we have been analyzing, as the frequency F of the sinusoidal voltage source increases beyond F_R, X_L increases and X_C decreases. Hence the circuit acts like an inductance whose reactance $(X_L - X_C)$ increases as F increases. As the frequency F decreases below F_R, X_C increases while X_L decreases. The circuit now acts like a capacitance, whose reactance $(X_C - X_L)$ increases as F decreases.

A graph of impedance versus frequency of a series *RLC* circuit is shown in Fig. 62-2. This curve shows minimum Z at F_R. It shows also that Z is capacitive (Z_C) for frequencies less than F_R, and inductive (Z_L) for frequencies greater than F_R.

The variation of current versus frequency is also sketched graphically in Fig. 62-2.

Fig. 62-1. Circuit to determine the effect of frequency on impedance of *RLC* connected in series.

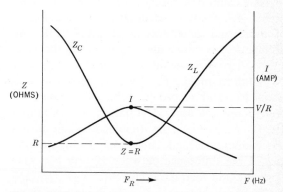

Fig. 62-2. Variation of Z and I versus F in a series-connected *RLC* circuit.

SUMMARY

1. In a series RLC circuit there is a frequency F_R of the voltage source V, at which $X_L = X_C$. At this frequency the impedance Z of the RLC circuit is minimum and $Z = R$, the resistance in the circuit.

2. For frequencies higher than F_R, the X_L of the inductor increases while X_C decreases and the circuit acts like an RL circuit, with $Z = \sqrt{R^2 + (X_L - X_C)^2}$. As F increases beyond F_R, Z increases beyond R.

3. For frequencies lower than F_R, the X_C of the *capacitor* increases while X_L decreases. The circuit acts like an RC circuit, with $Z = \sqrt{R^2 + (X_L - X_C)^2}$. As F decreases below F_R, Z increases beyond R.

4. In a series RLC circuit the current $I = V/Z$, where V is the applied voltage.

SELF-TEST

Check your understanding by answering these questions:

1. In the circuit of Fig. 62-1, for a particular frequency F_R, $X_L = X_C = 1,000 \ \Omega$. The impedance Z of the circuit is _____ Ω.

2. The circuit in question 1 acts like a pure _____.

3. For the frequency $F_R + 100$, the impedance of the circuit in Fig. 62-1 is _____ (greater/less than) 4,700 Ω and the circuit is _____ (inductive/capacitive).

4. For the frequency $F_R - 100$, the impedance of the circuit in Fig. 62-1 is _____ (greater/less than) 4,700 Ω and the circuit is _____ (capacitive/inductive).

5. As F increases above F_R, or decreases below F_R, the current I in the circuit _____ (increases/decreases/remains the same).

MATERIALS REQUIRED

- Equipment: Electronic voltmeter, oscilloscope, audio-frequency sine-wave generator
- Resistor: ½-W 4,700-Ω
- Capacitor: 0.1 μF
- Miscellaneous: SPST switch; choke, 8 H at 50-mA direct current

PROCEDURE

Effect of Frequency on Impedance

1. Connect the circuit of Fig. 62-1. The ac voltage source is an audio-frequency signal generator. Set the generator output at three-quarters of maximum voltage and the frequency at 150 Hz.

2. With the EVM connected across R and an oscilloscope calibrated in rms volts connected across points AB, increase or decrease the frequency of the signal generator, as required, until the voltage V_{AB} is minimum. At this point, the voltage V_R across R should be maximum.

3. Note the frequency F_R at which there is a minimum V_{AB} and record in Table 62-1. Measure and record the applied voltage V, the voltages V_R, V_L, V_C, and V_{AB} (the combined voltage across the inductance and capacitance). Compute and record the difference between V_L and V_C. Also compute and record the current $I = (V_R/R)$, and the impedance $Z = (V_{applied}/I)$. Show your computations.

4. Decrease the frequency of the generator by 20 Hz and record the new frequency, $F_R - 20$. Set the output V of the generator at the same level as for F_R in step 3. Repeat the measurements and computations of step 3 at the frequency $F_R - 20$ and record in Table 62-1.

5. Repeat step 4 for each of the frequencies shown in Table 62-1. Be *certain* that for each frequency the output of the generator is kept at the same voltage level as in step 3.

6. From the data in Table 62-1 draw a graph of:
   a. Z versus F
   b. I versus F

Extra Credit

7. For each value of F in Table 62-1, determine the values of X_L, X_C, and Z of the circuit using formulas $X_L = V_L/I$, $X_C = V_C/I$, and $Z = \sqrt{R^2 + (X_L - X_C)^2}$. Record your results in Table 62-2, which you will draw. For comparison,

TABLE 62-1.

Frequency, Hz	V Applied, volts	V_R, volts	V_L, volts	V_C, volts	V_{AB}, volts	$V_L - V_C$ or $V_C - V_L$, volts	I, amperes	Z, ohms
$F_R - 100$								
$F_R - 80$								
$F_R - 60$								
$F_R - 40$								
$F_R - 20$								
F_R								
$F_R + 20$								
$F_R + 40$								
$F_R + 60$								
$F_R + 80$								
$F_R + 100$								

transcribe to Table 62-2 the computed values of Z from Table 62-1.

QUESTIONS

1. Refer to the data in Table 62-1 and to the graph of Z versus F. Explain in your own words the effect on Z of a change in F.
2. Refer to the data in Table 62-1 and to the graph of I versus F. Explain in your own words the effect on I of a change in F.
3. In Fig. 62-1, what would be the effect on Z, if any, of interchanging L and C? Why?
4. In Table 62-1 comment on the relationship, if any, between the measured voltage V_{AB} and the voltage $V_L - V_C$, at any specific frequency. Explain any unexpected results.

Extra Credit

5. Discuss the results of your computations in step 7 on verification of the impedance formula for an RLC circuit [Eq. (62-1)]. Explain any unexpected results.
6. Derive a general formula for the frequency at which $X_L = X_C$ in a series RLC circuit.
7. Assume that the external resistor R in Fig. 62-1 is short-circuited. What will limit the value of current I, when $X_L = X_C$?

Answers to Self-Test

1. 4,700
2. resistance
3. greater; inductive
4. greater; capacitive
5. decreases

63

Impedance of a Parallel RL and of a Parallel RC Circuit

OBJECTIVES

1. To determine experimentally the impedance of a parallel *RL* circuit
2. To determine experimentally the impedance of a parallel *RC* circuit

INTRODUCTORY INFORMATION

Impedance of a Parallel RL Circuit

The circuit of Fig. 63-1 consists of two parallel branches, a resistive branch R and an inductive branch L. The same voltage V appears across each branch. V causes a current I_R in R, whose value is $I_R = V/R$. V also causes a current I_L in L, whose value is $I_L = V/X_L$.

What is the relationship among V, I_R, I_L, and I_T (the total line current) in the circuit? Figure 63-2 is a phasor diagram of the relationship among these quantities. Since voltage is common to each branch of the circuit, V is shown as the reference phasor. In the resistive branch, voltage and current are in phase. I_R is therefore shown in phase with V. In a pure inductance, current lags the applied voltage by 90°. I_L is therefore shown lagging V by 90°.

The currents I_R and I_L add vectorially to form the line current I_T, as shown in Fig. 63-2. Since I_L and

I_R are legs of a right triangle and I_T is the hypotenuse of this triangle,

$$I_T = \sqrt{I_R^2 + I_L^2} \qquad (63\text{-}1)$$

Figure 63-2 shows also that the line current I_T lags the applied voltage V by some angle θ. Theta may be determined from the equation

$$\theta = \arctan \frac{I_L}{I_R} \qquad (63\text{-}2)$$

The impedance of the circuit (Fig. 63-1) may be found by substituting the measured values of applied voltage and line current in the equation

$$Z = \frac{V}{I_T} \qquad (63\text{-}3)$$

Impedance of a Parallel RC Circuit

A resistive branch R and a capacitive branch C constitute the paths for current in the parallel RC circuit of Fig. 63-3. Analysis of this circuit is similar to that of the parallel RL circuit.

In Fig. 63-3 the applied voltage V is common to both branches. The current I_R in R is given by the equation $I_R = V/R$. The current I_C in C is given by the equation $I_C = V/X_C$. The relationship between V, I_C, I_R, and I_T (the line current) is shown in the phasor diagram (Fig. 63-4). Again V is used as the reference phasor. The current I_R in R is shown in phase with V. The current I_C in C is shown *leading* V by 90°. The line current I_T is the phasor sum of I_R and I_C. Since I_R and I_C are the legs of a right triangle and I_T the hypotenuse of this triangle,

$$I_T = \sqrt{I_R^2 + I_C^2} \qquad (63\text{-}4)$$

Fig. 63-4 also shows that the line current I_T *leads* the applied voltage by an angle θ, whose value may be determined from the equation

Fig. 63-1. Total current in a parallel *LR* circuit.

Fig. 63-2. Total current I_T in a parallel *LR* circuit is the phasor sum of I_R and I_L.

$$\theta = \text{arc tan } \frac{I_C}{I_R} \qquad (63\text{-}5)$$

The impedance Z of the circuit in Fig. 63-3 may be found by substituting the measured values of applied voltage and line current in the equation

$$Z = \frac{V}{I_T} \qquad (63\text{-}6)$$

It is interesting to note that the line current in a parallel *RC* circuit is a leading current (with reference to the applied voltage) whereas the line current in a parallel *RL* circuit is a lagging one.

SUMMARY

1. In a parallel *RL* or *RC* circuit the voltage across each branch of the circuit is the same and is therefore used as the reference phasor in an analysis of the relationships between the currents in the circuit.
2. The branch currents I_R, I_C, and I_L are phasors, and the total (line) current I_T is the phasor sum of I_R and I_L in an *RL* circuit; and I_R and I_C in an *RC* circuit.
3. In a parallel *RL* circuit the current in *L* lags the current in *R* by 90°. Therefore, the numerical value of I_T can be computed from the formula

$$I_T = \sqrt{I_R^2 + I_L^2}$$

4. In a parallel *RL* circuit the total current *lags* the applied voltage V by some angle θ, where

$$\theta = \text{arc tan } \frac{I_L}{I_R}$$

Fig. 63-3. Total current in a parallel *RC* circuit.

5. In a parallel *RC* circuit, the current in *C leads* the current in *R* by 90°. Therefore, the numerical value of I_T can be computed from the formula $I_T = \sqrt{I_R^2 + I_C^2}$.
6. In a parallel *RC* circuit the total current *leads* the applied voltage V by some angle θ, where

$$\theta = \text{arc tan } \frac{I_C}{I_R}$$

7. The impedance Z of an *RL* or an *RC* circuit may be computed from the equation

$$Z = \frac{V}{I_T}$$

where V is the applied voltage and I_T is the value of total or line current.

SELF-TEST

Check your understanding by answering these questions:

1. In a parallel *RL* circuit, the applied voltage $V =$ 100 V; the measured value of $I_L = 0.2$ A and of $I_R = 0.5$ A. The numerical value of $I_T =$ _____ A.
2. The angle θ by which total current I_T (in question 1) _____ (leads/lags) the applied voltage V is _____ degrees.
3. The impedance of the circuit in question 1 is _____ Ω.
4. In a parallel *RC* circuit, the applied voltage $V =$ 30 V, and the measured values of $I_C = 0.1$ A and $I_R = 0.2$ A. The numerical value of $I_T =$ _____ A.
5. The angle θ by which the total current I_T (in question 4) _____ (leads/lags) the applied voltage V is _____ degrees.
6. The impedance of the circuit in question 4 is _____ Ω.

MATERIALS REQUIRED

- Power supply: Source of line-isolated variable ac voltage (variable autotransformer plugged into line-isolation transformer)

Fig. 63-4. Total current I_T in a parallel *RC* circuit is the phasor sum of I_R and I_C.

- Equipment: 0–25 mA ac meter; electronic voltmeter
- Resistor: 1-W 5,000-Ω
- Capacitor: 0.5 μF

- Miscellaneous: Two SPST switches; iron-core choke rated 8 H at 50-mA direct current; fused line cord

PROCEDURE

Impedance of a Parallel RL Circuit

1. Connect the circuit of Fig. 63-5. Switches S_1 and S_2 are open. V is the output of a variable autotransformer plugged into a line-isolation transformer (1:1) which derives its power from the 120-V ac line. M is a 0–25 mA ac meter or the equivalent of a multirange milliammeter. V_1 is an EVM set on ac voltage. L is an iron-core choke rated at 8 H, with resistance of 250 Ω.

2. Close S_1. Slowly increase the output of the power source until the voltage V_{AB} across the 5,000-Ω resistor reads 50 V rms. M now measures the current I_R. Record I_R in Table 63-1.

3. Open S_1. Meter M should read zero. Close S_2. Monitor the voltage V_{AB} and reset at 50 V if it has changed. Measure and record the current I_L.

4. Close S_1. Both switches are now closed. L and R now constitute a parallel circuit. Reset the voltage V_{AB} at 50 V, if necessary. Measure and record the line current I_T. **Power off.** Open S_1 and S_2. Compute and record the impedance Z of the circuit by substituting the measured values V and I_T in $Z = V/I_T$.

5. Substitute the measured values of I_R and I_L in the equation

$$I_T = \sqrt{I_R{}^2 + I_L{}^2}$$

Compute this value of I_T and record in Table 63-1. Show your computations.

Impedance of a Parallel RC Circuit

6. Replace L by a 0.5-μF capacitor as in Fig. 63-6. Close S_1. Only R is in the circuit. Set the voltage V_{AB} at 60 V. M now reads the current I_R. Record I_R in Table 63-2.

7. Open S_1. Close S_2. Only C is now in the circuit. Monitor the voltage V_{AB} and reset at 60 V if it has changed. Measure and record the current I_C.

8. Close S_1. Both switches are now closed. R and C are in the circuit. Reset the voltage V_{AB} at 60 V, if necessary. Measure and record the line current I_T. **Power off.** Compute and record the impedance Z of the circuit by substituting the measured values of V and I_T in the equation $Z = V/I_T$. Substitute the measured values of I_R and I_C in the equation $I_T = \sqrt{I_R{}^2 + I_C{}^2}$. Compute this value of I_T and record in Table 63-2. Show your computations.

QUESTIONS

1. Refer to Table 63-1. Does the measured value of I_T equal the computed value $\sqrt{I_R{}^2 + I_L{}^2}$? Should it? Explain any unexpected results.

2. Refer to Table 63-2. Does the measured value of I_T equal the computed value $\sqrt{I_R{}^2 + I_C{}^2}$? Should it? Comment on any unexpected results.

3. What are the sources of error in this experiment?

4. In computing the impedance of a parallel RL or RC circuit by the equation $Z = V/I_T$, you found only the arithmetic value of Z. Compute the phase angle of Z and in each case indicate whether it is a leading or lagging angle.

5. Draw a phasor diagram of the branch currents in each circuit with both switches closed. Show total current and phase angle between applied voltage and total current. Draw to scale.

Answers to Self-Test

Answers to slide-rule accuracy.

1. 0.538
2. 21.8°
3. 186
4. 0.224
5. 26.5
6. 134

Fig. 63-5. Experimental circuit for measuring the total current of a resistor R in parallel with an inductor L.

Fig. 63-6. Experimental circuit for measuring the total current of a resistor R in parallel with a capacitor C.

TABLE 63-1.

V_{AB}, volts	I_R, amperes	I_L, amperes	I_T, amperes	$Z = \dfrac{V}{I_T}$ ohms	$\sqrt{I_R{}^2 + I_L{}^2}$ amperes
50					

TABLE 63-2.

V_{AB}, volts	I_R, amperes	I_C, amperes	I_T, amperes	$Z = \dfrac{V}{I_T}$ ohms	$\sqrt{I_R{}^2 + I_C{}^2}$ amperes
60					

64

EXPERIMENT   Impedance of a Parallel RLC Circuit

OBJECTIVE

To determine experimentally the impedance of a circuit containing a resistance R in parallel with an inductance L, in parallel with a capacitance C

INTRODUCTORY INFORMATION

The simplest type of parallel RLC circuit is illustrated in Fig. 64-1. The ac voltage V is common to each leg in the parallel network. The current I_R in R may be calculated from the equation

$$I_R = \frac{V}{R} \qquad (64\text{-}1)$$

where I is in amperes, V is the rms voltage, and R is the resistance in ohms of the resistor. Similarly the currents I_C in C and I_L in L may be calculated respectively from the equations

$$I_C = \frac{V}{X_C} \qquad (64\text{-}2)$$

$$I_L = \frac{V}{X_L} \qquad (64\text{-}3)$$

It will be noticed that the voltage across each leg is the applied voltage V.

The total line current I_{line} is equal to the phasor sum of I_R, I_C, and I_L. This fact is consistent with the analysis and results of Experiment 63, where parallel RL and RC circuits were considered.

Figure 64-2 is a phasor diagram showing the phase relationship between the applied voltage V and the currents in each leg. Thus, I_R is in phase with V, I_C leads V by 90°, and I_L lags V by 90°. Since I_C and I_L are 180° out of phase, they may be subtracted arithmetically to give the reactive current in the circuit, as in Fig. 64-3. If I_L is larger than I_C, the circuit is inductive. If I_C is larger than I_L, the circuit is capacitive, as in Fig. 64-3. The line current may be obtained by finding the phasor sum of I_R and $I_C - I_L$, as in Fig. 64-3.

The line current may also be calculated by substituting the values of I_R, I_C, and I_L in the equation

$$I_{\text{line}} = \sqrt{I_R{}^2 + (I_L - I_C)^2} \qquad (64\text{-}4)$$

This formula holds as long as there is no resistance in series with L or C in the circuit of Fig. 64-1.

The line current I_{line} is the sum of the instantaneous values of the resistive and reactive currents and is not in phase with either current. It differs from I_R, the resistive current taken as reference, by the angle θ. This angle may be computed from the formula

$$\tan \theta = \frac{I_C - I_L}{I_R}$$

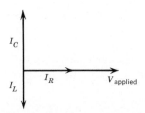

Fig. 64-2. Phase relations between applied voltage and current in each leg of a parallel RLC circuit.

Fig. 64-1. Parallel RLC circuit.

Fig. 64-3. Solving for line current in a capacitive *RLC* circuit.

$$\theta = \text{arc tan } \frac{I_C - I_L}{I_R} \qquad (64\text{-}5)$$

Having found the line current I_{line}, we can now compute the value of impedance Z in ohms of the circuit of Fig. 64-1 by substituting the values of I in amperes and V in volts in the equation

$$Z = \frac{V}{I_{\text{line}}} \qquad (64\text{-}6)$$

Since I_L and I_C are dependent on X_L and X_C, respectively, it is apparent that Z is a function of the frequency of the voltage V applied to the circuit.

NOTE: One important difference between parallel dc and ac circuits should be pointed out. In parallel dc circuits the total line current I_T increases with every parallel branch added. In parallel ac circuits the line current does not always increase as more branches are added. For example, if Fig. 64-1 initially consisted of a resistive branch in parallel with a capacitive branch, the effect of adding a third inductive branch would be to *reduce* the line current, a fact which is evident from Eq. (64-4).

A corollary to this is that the total impedance of parallel ac circuits is not always less than the smallest branch impedance, but is sometimes *greater* than the impedance of any one branch.

A more complex circuit is shown in Fig. 64-4. The ac generator applies a voltage V to two parallel branches AB and CD. Each branch is a series network. Thus the capacitive branch AB contains R_C and C connected in series. The inductive branch CD contains R_L and L connected in series. The impedance of each branch may be found by the methods outlined in the previous experiments dealing with series-connected *RL* and *RC* circuits. Thus, the impedance of branch AB is

$$Z_{AB} = \sqrt{R_C{}^2 + X_C{}^2} \qquad (64\text{-}7)$$

The current I_{AB} in this branch may be computed as follows:

$$I_{AB} = \frac{V}{Z_{AB}} \qquad (64\text{-}8)$$

Fig. 64-4. A parallel *RLC* network consisting of two series-connected branches.

where the value of Z_{AB} is taken from Eq. (64-7).

The current I_{AB} is a leading current with relation to the applied voltage V, and the phase angle θ is given by the formula

$$\theta_1 = \text{arc tan } \frac{X_C}{R_C}$$
$$(64\text{-}9)$$
or $\qquad \theta_1 = \text{arc tan } \dfrac{V_C}{V_{R_C}}$

The impedance of branch CD is

$$Z_{CD} = \sqrt{R^2 + X_L{}^2} \qquad (64\text{-}10)$$

The current I_{CD} in this branch is

$$I_{CD} = \frac{V}{Z_{CD}} \qquad (64\text{-}11)$$

where the value of Z_{CD} is taken from Eq. (64-10). The current in branch CD is a lagging current whose phase angle θ_2 may be found from the formula

$$\theta_2 = \text{arc tan } \frac{X_L}{R_L}$$
$$(64\text{-}12)$$
or $\qquad \theta_2 = \text{arc tan } \dfrac{V_L}{V_{R_L}}$

After the currents in each branch have been found, it is possible to determine line current (I_{line}) by finding the phasor sum of I_{AB} and I_{CD} as in Fig.

Fig. 64-5. Finding line current in a parallel *RLC* network.

64-5. Note that the phase angle of line current, θ_3, is different from θ_1 or θ_2.

The line current in Fig. 64-5 is shown as a lagging (inductive) current because in our illustration we assumed that Z_{CD} was smaller than Z_{AB}. Hence, in this case, the inductive current in branch CD is larger than the capacitive current in branch AB. Since the inductive current predominates, the total current may be considered inductive.

The value of total impedance Z that a generator sees "looking" into the circuit of Fig. 64-4 may be computed from the formula

$$Z = \frac{V}{I_{line}} \qquad (64\text{-}13)$$

It is evident that Eqs. (64-6) and (64-13) are identical and that the method of finding the impedance of a parallel RLC circuit, regardless of the number or type of branches involved, is the same.

SUMMARY

A method for finding the impedance of a parallel RLC circuit requires the following steps:
1. Calculate the impedance of each branch or leg of the parallel network by the methods already discussed in previous experiments. Determine both the numerical value $|Z|$ and the phase angle θ.
2. Find the current in each branch and the phase angle between the applied voltage V and the branch current. Branch current is computed using the equation

$$I = \frac{V}{|Z|}$$

Note that the phase angle (in degrees) is the same as the angle θ in summary statement 1, but is opposite in sign. Thus the phase angle of a capacitive current is positive (+), of an inductive current is negative (−).
3. Obtain the phasor sum of all branch currents. The resultant is the line current, whose amplitude and phase angle are determined from the phasor diagram which has been drawn to scale.
4. If the parallel RLC circuit has no resistance in the capacitive or inductive branch, that is, if the circuit is exactly as in Fig. 64-1, then the net reactive current is $I_C - I_L$ if the capacitive current is larger than the inductive current; or $I_L - I_C$ if the inductive current is larger than the capacitive current.
5. The total current I_{line} for the circuit in summary statement 4 is found by using the formula

$$I_{line} = \sqrt{I_R{}^2 + (I_L - I_C)^2}$$

and the phase angle θ between the applied voltage V and the total current is

$$\theta = \text{arc tan } \frac{I_L - I_C}{I_R}$$

6. Find the impedance $|Z|$ of the parallel RLC circuit using the formula $|Z| = V/I_{line}$, where V is the voltage across the parallel network and I_{line} is the line current.

SELF-TEST

Check your understanding by answering these questions:
1. In the circuit of Fig. 64-1, $R = 120 \ \Omega$, $X_C = 150 \ \Omega$, and $X_L = 180 \ \Omega$. The applied voltage $V = 18$ V.
   (a) The current I_R in $R = $ _____ A.
   (b) The phase angle between V and I_R is _____ degrees.
2. For the circuit in question 1,
   (a) The current I_C in $C = $ _____ A.
   (b) The phase angle between V and $I_C = $ _____ degrees.
3. For the circuit in question 1,
   (a) The current I_L in $L = $ _____ A.
   (b) The phase angle between V and I_L is _____ degrees.
4. The net reactive current in the circuit in question 1 is _____ A and it is _____ (capacitive/inductive).
5. The total current (I_{line}) in the circuit of question 1 = _____ A.
6. The phase angle between the applied voltage V and I_{line} in the circuit of question 1 is _____ degrees and its sign is _____ (+ or −).
7. The total impedance in the circuit of question 1 is _____ Ω.

MATERIALS REQUIRED

■ Power supply: Source of line-isolated variable ac voltage (variable autotransformer plugged into line-isolation transformer)
■ Equipment: 0–25 mA ac meter; electronic voltmeter
■ Resistor: 1-W 5,000-Ω
■ Capacitor: 0.5 μF
■ Miscellaneous: Three SPST switches; iron-core choke rated 8 H at 50-mA direct current; fused line cord

Fig. 64-6. Experimental circuit for determining the impedance of R, L, and C connected in parallel.

PROCEDURE

1. Connect the circuit Fig. 64-6. All switches are open. V is the output of a variable autotransformer plugged into a line-isolation transformer which derives its power from the 120-V ac line. V is set at 0 V. M is a 0–25 mA ac meter, or the equivalent of a multirange milliammeter. V_1 is an EVM set on ac voltage. L is an iron-core choke rated 8 H at 50-mA direct current.

2. Close S_1. Slowly increase the voltage until $V_{AB} = 60$ V. Measure the current I_R and record in Table 64-1.

3. Open S_1. Close S_2. Readjust V if necessary until $V_{AB} = 60$ V. Measure and record the current I_L.

4. Open S_2. Close S_3. Readjust V if necessary until $V_{AB} = 60$ V. Measure and record the current I_C.

5. Now close S_1. At this point S_1 and S_3 are closed. R and C are in the circuit. Readjust V if necessary until $V_{AB} = 60$ V. Measure and record the current I_{RC}.

6. Now close S_2. At this point all switches are closed. Readjust V if necessary until $V_{AB} = 60$ V. Measure and record the line current I_{RLC}. Note if I_{RLC} is larger or smaller than I_{RC}. **Power off.**

7. Find I_{line} by substituting the measured values I_R, I_L, and I_C in the formula

$$I_{line} = \sqrt{I_R^2 + (I_L - I_C)^2}$$

Record in Table 64-1.

8. Compute Z by substituting in the formula

$$Z = \frac{V_{AB}}{I_{RLC}}$$

the measured values of I_{RLC} and V_{AB}. Record in Table 64-1.

9. Compute and record the angle θ, the phase angle between V_{AB} and I_{line}, by substituting in the equation $\theta = \arctan{(I_C - I_L)/I_R}$, the measured values of I_C, I_L, and I_R.

Extra Credit

10. Explain in detail the method you would use to verify the phase angle θ between line current and applied voltage in the experimental RLC circuit. Set up the circuit. Measure and record the angle θ. _____ degrees.

QUESTIONS

1. In Table 64-1 how does the measured value I_{RLC} compare with the formula value $\sqrt{I_R^2 + (I_L - I_C)^2}$? Should they be equal? Explain any unexpected results.
2. Which current is larger, I_{RC} or I_{RLC}? Why?
3. Is the impedance of the parallel RLC circuit higher or lower than the impedance of the parallel RC circuit? RL circuit? Why?
4. Is the line current in Fig. 64-6 capacitive or inductive? Why?
5. Draw a series-equivalent circuit for Fig. 64-6. Compute the value of resistance and reactance of the series-equivalent circuit.

TABLE 64-1.

V_{AB}, volts	I_R, amperes	I_L, amperes	I_C, amperes	I_{RL}, amperes	I_{RC}, amperes	I_{RLC}, amperes	I_{line} (amperes) $= \sqrt{I_R^2 + (I_L - I_C)^2}$	Z, ohms	θ, degrees
60									

65

Resonant Frequency and Frequency Response of a Series-resonant Circuit

OBJECTIVES

1. To determine experimentally that the resonant frequency F_R of a series-resonant LC circuit is given by the formula

$$F_R = \frac{1}{2\pi\sqrt{LC}}$$

2. To measure the frequency response of a series-resonant LC circuit

INTRODUCTORY INFORMATION

Resonant Frequency of Series LC Circuit

Consider the circuit of Fig. 65-1. Assume V is an ac generator whose frequency and output voltage are manually adjustable. For a particular frequency F and output voltage V there will be a current I such that $I = V/Z$, where Z is the impedance of the circuit. The voltage drops across R, L, and C will be given by IR, IX_L, and IX_C respectively.

If the generator frequency is changed but V remains constant, the current I and the voltage drops across R, L, and C will change.

There is a frequency F_R called the *resonant* frequency at which

$$X_L = X_C \qquad (65\text{-}1)$$

Fig. 65-1. Series *RLC* Circuit.

It is a simple matter to determine, analytically, the resonant frequency F_R. Consider the following:

$$X_L = 2\pi FL$$
$$\qquad (65\text{-}2)$$
and $$X_C = \frac{1}{2\pi FC}$$

At F_R, when $X_L = X_C$,

$$2\pi F_R L = \frac{1}{2\pi F_R C} \qquad (65\text{-}3)$$

Equation (65-3) can be put into the form

$$F_R{}^2 = \frac{1}{(2\pi)^2 LC} \qquad (65\text{-}4)$$

Taking the square root of both sides of Eq. (65-4) leads to the formula for the resonant frequency of a series-resonant circuit, namely

$$F_R = \frac{1}{2\pi\sqrt{LC}} \qquad (65\text{-}5)$$

Here F_R is given in hertz, L in henrys, and C in farads. Equation (65-5) is used for finding the resonant frequency of a series circuit containing a given value of L and of C.

Example

Find the resonant frequency of a series LC circuit where $L = 8$ H and $C = 0.01$ μF.

Solution.

1. $$F_R = \frac{1}{2\pi\sqrt{LC}}$$

Substituting the given values in 1 results in

2. $$F_R = \frac{1}{2\pi\sqrt{8 \times 0.01 \times 10^{-6}}}$$

3.
$$F_R = \frac{1}{2\pi\sqrt{8 \times 1.0 \times 10^{-8}}}$$

4.
$$F_R = \frac{10^4}{2\pi\sqrt{8}}$$

5.
$$F_R = 563 \text{ Hz}$$

Characteristics of a Series-resonant Circuit

Some characteristics of a series-connected RLC circuit were investigated in a previous experiment. The facts observed were:

1. The voltage drop across a reactive component is equal to the product of the current I in the circuit and the reactance X of the component.
2. Capacitive reactance X_C tends to cancel the effects of inductive reactance X_L.
3. The impedance Z of an RLC circuit is

$$Z = \sqrt{R^2 + (X_L - X_C)^2}$$

4. The impedance Z of the circuit is minimum when $X_L = X_C$ and circuit I is maximum at this point.

Fact 4 states that *at resonance, the impedance of a series LC circuit is minimum and current I in the circuit is maximum*. This fact, which was established experimentally in a preceding experiment, can also be determined analytically. For at F_R

$$Z = \sqrt{R^2 + (0)^2} = R$$

Therefore, a *minimum impedance Z = R exists at* F_R. Also, since $I = V/Z = V/R$, and V is assumed fixed, there will be *maximum current I in the circuit* at resonance. Note that at F_R the applied voltage V appears across R, and resonant current I may be computed from

$$I = \frac{V}{R} \qquad (65\text{-}6)$$

Since at resonance I is limited only by the value of R, the circuit is said to be *resistive*. For all frequencies higher than F_R, X_L is larger than X_C and the circuit is *inductive*. For all frequencies lower than F_R, X_C is larger than X_L and the circuit is *capacitive*.

At *resonance* the voltage V_L across L and the voltage V_C across C are *maximum* and equal. Theoretically, at the resonant frequency, when R becomes zero, the current I becomes infinite and the voltages V_L and V_C become infinitely large. Practically, this condition is never realized because L

has some resistance R_L which may be considered in series with it.

Frequency Response of a Series-resonant Circuit

The frequency-response characteristic can be determined experimentally by injecting a constant-amplitude signal voltage into the circuit at the resonant frequency and at frequencies on both sides of resonance. The voltage across L or C is measured, and a graph of V_L or V_C versus F is plotted. This is one form of the frequency-response characteristic of the circuit.

The circuit I can also be measured at F_R and at frequencies on both sides of F_R. A graph of I versus F is another form of the frequency-response characteristic of the circuit.

In this experiment we will make a response check of V_C versus F.

Notes on Procedure

In this experiment the student will determine the resonant frequency F_R of a series LC circuit. There will be no resistance in the circuit other than the internal resistance of the coil. Resonance will therefore be determined by tuning the signal-generator frequency for maximum voltage V_L across L, or V_C across C. An electronic voltmeter connected across C will be used as the voltage indicator. One fact must be borne in mind. The input impedance of the voltmeter may affect the resonant frequency of the circuit when the meter is connected across L or C in the circuit.

The ground connections of the electronic voltmeter and the signal generator should always be at the same common point, to avoid extraneous effects which may arise from internal connections within the instruments. The component across which a voltage is to be measured is connected so that one leg is grounded. The order in which the components are connected in the series circuit will not affect circuit operation.

When loaded by a frequency-selective circuit, the voltage output of most signal generators will not remain constant as the frequency is varied. In order to compensate for this characteristic of the generator, the generator-output level must be monitored and maintained at the same voltage output, if necessary, at each new frequency setting of the instrument. Thus, a constant signal input to the circuit under test will be maintained. In this experiment the generator-output level will be monitored with an oscilloscope and the generator-output cables will be terminated by a 100-Ω resistor to minimize generator-signal voltage variations.

SUMMARY

1. In a series RLC circuit there is a frequency F_R called the *resonant frequency*, which may be computed from the formula $F_R = 1/2\pi\sqrt{LC}$. At resonance $X_L = X_C$.
2. At F_R, the impedance of the circuit is minimum and $Z = R$.
3. At F_R, the current I in the circuit is maximum and $I = V/R$, where V is the voltage applied across the circuit.
4. At resonance the voltages V_L across L and V_C across C are maximum and equal.
5. At resonance a series RLC circuit acts as a resistance; above resonance it is inductive and below resonance, capacitive.

SELF-TEST

Check your understanding by answering these questions:

1. In a series-resonant circuit, $X_L = 1{,}200$ Ω at resonance. The value of X_C at resonance is _____ Ω.
2. When the frequency F of the signal source applied to a series RLC circuit is higher than the resonant frequency F_R, then X_L _____ (equals/>/<) X_C and the circuit is _____.
3. In a series RLC circuit $R = 10$ Ω, $L = 100$ μH, $C = 0.001$ μF, and the applied voltage $V = 1.2$ V. The resonant frequency F_R of this circuit is _____ kHz.
4. In the circuit of question 3, the circuit current $I =$ _____ A at resonance.
5. In the circuit of question 3, the voltage across L, $V_L =$ _____ V.
6. In the circuit of question 3, the impedance Z at resonance is _____ Ω.

MATERIALS REQUIRED

- Equipment: AF sine-wave generator, oscilloscope, electronic voltmeter
- Capacitors: 0.005 μF, 0.01 μF, 0.05 μF
- Miscellaneous: 30-mH (approximately) choke
- Resistor: ½ W 100 Ω

PROCEDURE

Resonant Frequency

1. Adjust the oscilloscope for proper viewing and calibrate it for voltage measurements.
2. Set frequency of AF generator at 30,000 Hz and connect the oscilloscope across the output of the signal generator. Terminate the generator-output cables with a 100-Ω resistor.
3. Connect the circuit of Fig. 65-2. Leave the scope connected across generator output. Set the output of the generator at 2 V peak-to-peak. *Maintain generator-signal output voltage at this level for the remainder of the experiment.*

4. Now connect the electronic voltmeter across C as in Fig. 65-2 and observe the rms voltage V_C across C. Vary the generator frequency first on one side of 30,000 Hz and then on the other side of 30,000 Hz until the maximum voltage V_C appears across C. Read the generator frequency. This is the resonant frequency F_R of the circuit. Record this frequency in Table 65-1 in "Measured" column.

TABLE 65-1. Resonant Frequency of a Series LC Circuit

Step	L mH	C μF	Resonant	Frequency (Hz)
			Measured	Computed $F_R = \dfrac{1}{2\pi\sqrt{LC}}$
4/7	30	0.001		
5/7	30	0.05		
6/7	30	0.01		

Fig. 65-2. Determining resonant frequency of an LC circuit.

5. Replace the 0.001 μF capacitor in Fig. 65-2 with a 0.05 μF capacitor. Following the method in steps 3 and 4 find the new resonant frequency of the circuit and record in Table 65-1.
6. Repeat step 5, using a 0.01 μF capacitor.
7. For each value of L and C used, compute the resonant frequency of the circuit using Eq. (65-5). Record each value of F_R in the column labelled "Computed."

Frequency Response

8. Leave the circuit connected as in step 6 and again find the resonant frequency F_R. (It should be the same frequency as in step 6.) Record in Table 65-2.
9. With the oscilloscope measure the generator (still connected across the circuit) output and readjust generator level, if necessary, for 2 V peak-to-peak. *Maintain generator output at this level throughout.*
10. Now measure the voltage V_C across C and record in Table 65-2.
11. Fill in the "Frequency" column in Table 65-2, performing the required frequency additions and subtractions.
12. Make a frequency-response check of the circuit by setting the generator frequency where possible to the frequencies listed in Table 65-2, noting and recording the voltage V_C across C for each frequency setting.

QUESTIONS

1. In a series RLC circuit, under what conditions is it possible for the voltages V_L and V_C to be higher than the applied generator voltage V?
2. Under the conditions in question 1, what limits V_L and V_C?
3. What is the impedance of a series LC circuit at resonance?
4. Outline a method for computing the inductance L of the coil from the experimental data in this experiment, using Eq. (65-5) for the resonant frequency of a series circuit.
5. Find the inductance L by the method described in question 4. Use the data from Table 65-1.

TABLE 65-2. Frequency Response Data

Frequency deviation, Hz	Frequency, Hz	V_C, volts rms
$F_R - 6{,}000$		
$F_R - 5{,}000$		
$F_R - 4{,}000$		
$F_R - 3{,}000$		
$F_R - 2{,}000$		
$F_R - 1{,}000$		
$F_R - 500$		
Resonance F_R		
$F_R + 500$		
$F_R + 1{,}000$		
$F_R + 2{,}000$		
$F_R + 3{,}000$		
$F_R + 4{,}000$		
$F_R + 5{,}000$		
$F_R + 6{,}000$		
v, volts peak-to-peak		2

6. Using the data in Table 65-2, draw a frequency response curve showing V_C as a function of F. What type of curve is this?
7. In making a frequency-response check why is it necessary to maintain a fixed generator output over the range of frequencies used?
8. How do the measured values of F_R compare with the computed values in Table 65-1? Explain any discrepancies.

Answers to Self-Test

Answers to slide-rule accuracy.

1. 1,200
2. >; inductive
3. 503 kHz
4. 0.12
5. 37.9
6. 10

66

EXPERIMENT

Effect of Q on Frequency Response and Bandwidth of a Series-resonant Circuit

OBJECTIVES

1. To measure the effect of circuit Q on frequency response
2. To measure the effect of circuit Q on bandwidth at the half-power points

INTRODUCTORY INFORMATION

Circuit Q and Frequency Response

In Experiment 65 we studied the response curve of a series-resonant LC circuit. In that experiment the only resistance in the circuit was that associated with the coil. Theoretically, at resonance $X_L = X_C$ and the impedance $Z = R_L$, where R_L equals the resistance of the coil.

The amount of coil resistance R_L, therefore, determines the current flow through the circuit at resonance if there is no external resistance. The R_L and X_L of the coil (assuming no other resistance) determine the quality or Q of the circuit, which is given by the formula

$$Q = \frac{X_L}{R_L} \qquad (66\text{-}1)$$

The Q of the circuit also determines the rise in voltage across L and C at the resonant frequency F_R. The voltage developed across L is given by the equations

$$V_L = IX_L = \frac{V}{R} \times X_L \qquad (66\text{-}2)$$

If the circuit resistance R is the resistance R_L associated with the coil, then

$$V_L = V \times \frac{X_L}{R_L} = VQ \qquad (66\text{-}3)$$

Also, since $X_L = X_C$ at resonance,
$$IX_L = IX_C$$
and
$$V_L = V_C$$

Therefore $\quad V_C = VQ \qquad (66\text{-}4)$

Equations (66-3) and (66-4) become significant for values of $Q > 1$. For such values V_C and V_L are greater than the applied voltage V. Obviously, also, the higher the value of Q, the greater the voltage gain in the circuit. This is the *first* example of voltage gain.

Q is also significant when we consider the frequency response of a series-resonant circuit. The frequency-response characteristic can be determined by injecting a constant-amplitude signal voltage V into the circuit at the resonant frequency and at frequencies on either side of resonance. The voltage across L or C is measured and a graph of V_L or V_C versus F is plotted. This is one form of the frequency-response characteristic of the circuit.

The circuit current I can also be determined. A graph of I versus F is another form of the frequency-response characteristic of the circuit.

Circuit Q and Bandwidth

Figure 66-1 is a graph of the response of a series-resonant circuit. Three significant points have been

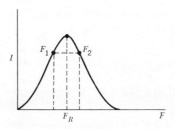

Fig. 66-1. Response curve of a series-resonant circuit.

marked on the curve. These are F_R, the resonant frequency, and F_1 and F_2. The latter points are located at 70.7 percent of maximum (maximum is at F_R) on the curve. They are called the half-power points and the frequency separation between them is $F_2 - F_1$. This frequency separation is called the bandwidth (B) of the circuit. Bandwidth, therefore, is given by the equation

$$B = F_2 - F_1 \qquad (66\text{-}5)$$

Bandwidth is related to Q. It can be shown that B is given by the equation

$$B = \frac{F_R}{Q} \qquad (66\text{-}6)$$

As was noted in the preceding experiment the resonant frequency F_R of a series RLC circuit can be computed from the formula

$$F_R = \frac{1}{2\pi\sqrt{LC}} \qquad (66\text{-}7)$$

where F_R is in hertz, L in henrys, and C in farads. Since the formula does not include R, it is apparent that the resonant frequency of the series RLC circuit is not affected by the size of the resistance R. What R does affect, however, is the bandwidth and the amplitude of the response curve. The larger the value of R, the lower is the value of Q (see Eq. 66-1) and the wider is the bandwidth. Moreover, the lower the value of Q, the lower is the value of current I in the circuit, and the lower are the voltages V_L across L and V_C across C.

Series-resonant circuits are used in radio, television, and in industrial electronics as frequency-selective circuits and as traps to eliminate unwanted signals. Normally what is required for such circuits is a highly peaked response with narrow bandwidth. To achieve such a response the Q of the circuit must be high. Hence high Q coils are employed. In such circuits the Q of the circuit is mainly determined by the Q of the coil.

However, there are some applications in electronics where wideband frequency-selective circuits are required. In such cases coil "loading" is achieved by the use of external resistors.

Notes on Experimental Procedure

In this experiment the student will determine the resonant frequency F_R of a series RLC circuit. An external resistor R will be inserted in series with L and C to simulate the coil resistance. The ef- fect of the value of R on the amplitude of current I will be investigated. The effect of the value of R on frequency response and bandwidth will also be determined.

The student will determine resonance by tuning the signal-generator frequency for maximum voltage V_R across R, or V_C across C. The oscilloscope will again be used as a voltage indicator to monitor the generator output. An electronic voltmeter will be used to measure the ac voltage across C or R. As was noted in the previous experiment the input impedance of the voltmeter may affect the resonant frequency of the circuit when the scope is placed across C or R.

Also noted in the preceding experiment, *the ground connections of the meter, the scope, and the signal generator must always be at the same common point, to avoid extraneous effects which may arise from internal connections within the instruments.*

Also, as in the preceding experiment, the signal-generator output must be maintained at a constant level. To accomplish this the output voltage of the generator will be monitored with the oscilloscope and as the generator is turned to each new frequency, manual adjustments will be made in output level, if necessary. The generator-output cables will also be terminated by a 100-Ω resistor.

SUMMARY

1. The Q or quality of a coil is defined by the formula $Q = X_L/R$.
2. In a series-resonant circuit, where the only resistance in the circuit is coil resistance R_L, the Q of the circuit is determined by the Q of the coil.
3. In a series-resonant circuit the voltages across L and C are equal and are affected by the Q of the circuit. These voltages may be computed from the formulas

$$V_L = V_C = V \times Q$$

where V is the applied voltage and V_L and V_C are the voltages across L and C, respectively.
4. If circuit Q is > 1, then the voltages V_L and V_C are larger than the applied voltage. Hence, in a series-resonant circuit we find the first example of *amplification*.
5. Bandwidth (B) of a response curve is defined as the difference between the two frequencies F_2 and F_1 at the half-power points (Fig. 66-1). The half-power points are those points which appear

at 70.7 percent of maximum on the response curve. Maximum output occurs at the resonant frequency F_R.
6. Bandwidth is related to circuit Q and is given by the formula

$$B = \frac{F_R}{Q}$$

7. The lower the Q of the circuit, the wider is the bandwidth and the "flatter" is the response curve.
8. The lower the Q of the circuit, the lower is the amplitude of the response curve, and therefore the circuit gain is lower.

SELF-TEST

Check your understanding by answering these questions.

In the circuit of Fig. 66-2, $L = 50$ mH, $C = 0.01$ μF, and $R = 20$ Ω. Assume the resistance of the coil $R_L = 0$. The applied voltage $V = 2.5$ V. The output voltage is V_C. Answer the following questions concerning this circuit.
1. The resonant frequency $F_R =$ _____ kHz.

Fig. 66-2. Effect of Q on frequency response.

2. The X_L of the coil at resonance is _____ Ω.
3. The Q of the circuit is _____ at resonance.
4. The voltage V_L across L is _____ V at resonance.
5. The bandwidth of the circuit is _____ Hz.
6. The voltage V_C across C at the half-power points is _____ V.

MATERIALS REQUIRED

■ Equipment: AF signal generator, the same as in Experiment 65, oscilloscope, electronic voltmeter
■ Resistors: ½ W 10- and 47-Ω, and two 100-Ω
■ Capacitors: 0.01 μF
■ Miscellaneous: 30 mH (approximate) choke, the same as in Experiment 65

PROCEDURE

Effects of Circuit Q on the Frequency Response of a Series-resonant Circuit

1. Connect the circuit of Fig. 66-2. $R = 10$ Ω, $L = 30$ mH (approximately), and $C = 0.01$ μF.
2. Set the AF generator output at 2 V peak-to-peak and vary the generator frequency until the maximum voltage V_C (as measured by the EVM) appears across C. Note the generator frequency F_R at this point and record in Table 66-1. Reset, if necessary, the generator output v_G at 2 V peak-to-peak at F_R.

NOTE: In the response check which follows it is necessary to maintain the output of the generator at this constant level, 2 V peak-to-peak, over the range of frequencies $F_R \pm 6$ kHz. *The voltage v_G (2 V peak-to-peak) will be used as the standard voltage for setting the signal-generator attenuator for the remainder of this experiment. With the*

scope across the output cable of the generator, (terminated by a 100-Ω resistor), which is connected across the circuit under test, the generator attenuator is adjusted for the standard voltage v_G each time the frequency of the generator is changed.

3. The generator frequency is still set at F_R output at 2 V peak-to-peak. The EVM is still connected across C. Measure and record in Table 66-1, the rms voltage V_C across C.
4. Fill in Table 66-1 with the proper information in the column labelled "Frequency," by performing the additions and subtractions noted in the column labeled "Frequency deviation, Hz."
5. Make a frequency-response check of the circuit by setting the generator frequency, where possible, to the frequencies listed in Table 66-1. Measure and record in Table 66-1 the voltage V_C across C for each frequency setting.

6. Replace R by a 47-Ω resistor and repeat steps 3 through 5. Record your data in Table 66-2.
7. Replace R by a 100-Ω resistor and repeat steps 3 through 5. Record your data in Table 66-3.

Effect of R on Resonant Frequency

8. Rearrange the circuit as in Fig. 66-3. $R = 10\ \Omega$.
9. Keeping the generator output at 2 V peak-to-peak, vary the generator frequency until circuit resonance is indicated by a maximum voltage reading across R. Note the resonant frequency and V_R at resonance and record in Table 66-4.
10. Repeat steps 8 and 9 for $R = 47\ \Omega$ and $R = 100\ \Omega$.

Graphs of Response Curves

11. Using the data in Table 66-1 through 66-3, graph the response curves of the circuits under test. Identify each curve. These graphs indicate the effects of Q on the frequency-response characteristics of the series LC circuit in the range of resonance.

TABLE 66-1.

Frequency Deviation, Hz	Frequency, Hz	V_C, volts, rms
$F_R - 6,000$		
$F_R - 5,000$		
$F_R - 4,000$		
$F_R - 3,000$		
$F_R - 2,000$		
$F_R - 1,000$		
$F_R - 500$		
Resonance F_R		
$F_R + 500$		
$F_R + 1,000$		
$F_R + 2,000$		
$F_R + 3,000$		
$F_R + 4,000$		
$F_R + 5,000$		
$F_R + 6,000$		
v_G, volts peak-to-peak		

TABLE 66-2.

Frequency Deviation, Hz	Frequency, Hz	V_C, volts, rms
$F_R - 6,000$		
$F_R - 5,000$		
$F_R - 4,000$		
$F_R - 3,000$		
$F_R - 2,000$		
$F_R - 1,000$		
$F_R - 500$		
Resonance F_R		
$F_R + 500$		
$F_R + 1,000$		
$F_R + 2,000$		
$F_R + 3,000$		
$F_R + 4,000$		
$F_R + 5,000$		
$F_R + 6,000$		
v_G, volts peak-to-peak		

12. In the graphs you have drawn, determine and mark the half-power points. Determine the bandwidth of each graph and compute circuit Q in each case. Show your computations.

Extra Credit

13. Refer to Fig. 66-3. State how you could determine experimentally the phase relationship between applied voltage v_G and current I at resonance. Perform this measurement and show your results.

QUESTIONS

1. At resonance, what is the theoretical relationship between V_R and the applied generator voltage, assuming L has zero resistance? Is this relationship confirmed by the measurements in Table 66-4? If not, explain why.
2. What is meant by the Q of a series-resonant circuit?
3. What effect does Q have on sharpness of frequency response of a series LC circuit?
4. The "half-power points" are defined by the frequency at which the output voltage is 70.7 percent of the output voltage at resonance. What is the frequency dif-

TABLE 66-3.

Frequency Deviation, Hz	Frequency, Hz	V_C, volts, rms
$F_R - 6{,}000$		
$F_R - 5{,}000$		
$F_R - 4{,}000$		
$F_R - 3{,}000$		
$F_R - 2{,}000$		
$F_R - 1{,}000$		
$F_R - 500$		
Resonance F_R		
$F_R + 500$		
$F_R + 1{,}000$		
$F_R + 2{,}000$		
$F_R + 3{,}000$		
$F_R + 4{,}000$		
$F_R + 5{,}000$		
$F_R + 6{,}000$		
v_G, volts peak-to-peak		

Fig. 66-3. Impedance of a series-resonant circuit.

TABLE 66-4.

	R, ohms		
	10	47	100
Resonant frequency, Hz			
V_R at resonance			

ference at the half-power points for $R = 10$, 47, and 100 Ω?

5. What effect, if any, did R have on the resonant frequency in steps 8 through 10?

6. At resonance what is the phase relationship between applied voltage and current?

Answers to Self-Test

1. 7.1
2. 2,230
3. 112
4. 278
5. 63.5
6. 197

67

Characteristics of Parallel-resonant Circuits

OBJECTIVES

1. To determine experimentally the resonant frequency of a parallel *RLC* circuit
2. To measure the line current and impedance at the resonant frequency
3. To measure the effect on impedance of a parallel *RLC* circuit of variations in circuit frequency, above and below resonance (frequency response)

INTRODUCTORY INFORMATION

Characteristics of a Parallel-resonant Circuit

Resonant frequency of a high-Q circuit

Consider the circuit of Fig. 67-1*a* consisting of *C* and *L* in parallel. This theoretical circuit must be modified as in Fig. 67-1*b* to reflect the resistance R_L of coil *L*. It is assumed here that the *Q* of this circuit is high (that is, R_L is small compared with X_L) and that the resistance of *C* and the wiring resistance of the circuit are negligible and may be ignored.

There is a particular frequency at which $X_L = X_C$. This frequency may be defined as the condition for parallel resonance in a high-*Q* circuit and is similar to the condition for series resonance.

There are other definitions for parallel resonance. Thus parallel resonance may be considered as the frequency at which the impedance of the parallel circuit is maximum. Also, parallel resonance may be considered as the frequency at which the parallel impedance of the circuit has unity power factor. These three definitions may lead to three different frequencies, each of which may be considered as the resonant frequency. In circuits whose *Q* is greater than 10, however, the three conditions lead to the same resonant frequency.

In a high-*Q* circuit, the formula for the resonant frequency F_R is the same as in the case of series resonance and is given by

$$F_R = \frac{1}{2\pi\sqrt{LC}} \qquad (67\text{-}1)$$

Line current

Consider again the circuit of Fig. 67-1*b*. If the resistance R_L of the inductance *L* is small, then at resonance the impedance X_C of the capacitive branch is practically equal to the impedance

$$\sqrt{X_L{}^2 + R_L{}^2}$$

of the inductive branch. The current in each branch may therefore be considered equal. These currents, however, are practically 180° out of phase. Hence their phasor sum, which is the line current, is very

(a)

(b)

Fig. 67-1. Parallel-resonant circuits.

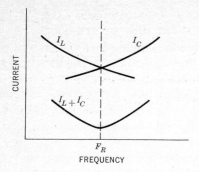

Fig. 67-2. Current versus frequency in a parallel-resonant circuit.

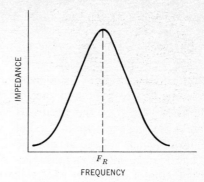

Fig. 67-3. Impedance versus frequency in a parallel-resonant circuit.

small. Moreover, since the impedance which this parallel-resonant circuit presents to the circuit is defined as $Z = V/I_{\text{line}}$, it is apparent that the impedance is very high (since line current is low). As opposed to line current, however, the circulating current in the parallel-resonant circuit is high at resonance.

Frequency response

The characteristics of a parallel-resonant circuit to frequencies on either side of the resonant frequency may now be analyzed. Thus, for a frequency F_a higher than F_R, the reactance X_C of the capacitive branch is lower than the reactance X_L of the inductive branch. There is therefore a greater capacitive than inductive current and the circuit may be considered capacitive. Similarly, it may be shown that for a frequency F_b lower than F_R the circuit is inductive. The graph of Fig. 67-2 illustrates these relationships. Individual graphs for I_L, I_C, and I_{line} are shown. It is evident that at resonance line current is minimum, that I_L equals I_C, and that I_L and I_C are each greater than line current.

Impedance

Figure 67-3 is a graph of impedance versus frequency in a parallel-resonant circuit. This graph

resembles the frequency-response characteristic of a series-resonant circuit. It shows that circuit impedance is maximum at resonance and falls off on either side of resonance.

In this experiment you will determine the resonant frequency F_R of a parallel LC circuit (see Fig. 67-4). A resistor R is inserted in series with the parallel LC circuit. You will measure the signal voltage V_R across R. You can then compute the line current from the relationship

$$I_R = \frac{V_R}{R} = I_{\text{line}} \qquad (67\text{-}2)$$

You will determine resonance by tuning for minimum voltage V_R across R, hence for minimum line current. You will then vary the signal generator on either side of resonance, monitoring and maintaining a constant generator output. By measuring V_R, computing I_{line}, and plotting the values of I_{line}, you will secure a graph of line current versus frequency.

You will measure the signal voltage V_T across the tank circuit at resonance and for frequencies on both sides of resonance. You will then calculate the value of the impedance Z of the tank circuit from the relationship

$$Z = \frac{V_T}{I_{\text{line}}} \qquad (67\text{-}3)$$

(a)

(b)

Fig. 67-4. Experimental parallel-resonant circuits.

and will plot a graph for the Z of the tank circuit from the calculated values of Z.

You will find that the signal voltage across the parallel-resonant (tank) circuit is maximum at resonance. It falls off on either side of resonance. The amplitude of the signal across the tank circuit is directly proportional to the Q of the circuit. If we "load" the circuit by placing a resistor in parallel with it, the amplitude of signal voltage across the circuit decreases. The loading effect increases as the size of the parallel resistor decreases.

You will also observe the reactive characteristic of the tank circuit as the signal-generator frequency is varied on both sides of resonance; that is, you will observe whether the circuit is inductive or capacitive for frequencies on both sides of resonance.

In this experiment an audio oscillator will be used as the signal source, since the values of L and C selected provide a circuit whose resonant frequency is in the AF range.

SUMMARY

1. In a parallel LC circuit, if the Q of the circuit is greater than 10, the resonant frequency of the circuit may be computed from the formula

$$F_R = \frac{1}{2\pi\sqrt{LC}}$$

At this frequency $X_L = X_C$.

2. In a high-Q circuit the impedance of the inductive branch equals (approximately) the impedance of the capacitive branch.

3. Since the voltage across each of the parallel branches in a high-Q circuit is the same, and the impedances of both branches are equal, the currents in both branches are equal. That is, $I_L = I_C$.

4. The currents, however, are approximately $180°$ out of phase, since one is capacitive and the other inductive. They therefore tend to cancel each other, and the line current, which is the phasor sum of I_C and I_L, is therefore very low, approximately zero in a very high-Q circuit.

5. The impedance of a high-Q parallel resonant LC circuit is therefore very high, since $Z =$

$V_{applied}/I_{line}$ and I_{line} is approximately zero.

6. For frequencies higher than F_R, the current in the capacitive branch is higher than the current in the inductive branch, and the circuit acts like a capacitive circuit.

7. For frequencies lower than F_R, current in the inductive branch is higher than current in the capacitive branch, and the circuit acts like an inductance.

8. The impedance response curve of a high-Q parallel-resonant circuit resembles the frequency-response curve of a series-resonant circuit.

SELF-TEST

Check your understanding by answering these questions:

1. The Q of a parallel-resonant circuit in Fig. 67-1a is 100. At resonance the current in the inductive branch is 50 mA and the voltage across the tank circuit is 100 V. The current in the capacitive branch is _____ mA.

2. In the parallel-resonant circuit of Fig. 67-4 the line current at resonance is 0.05 mA. The voltage across the tank circuit is 2.5 V. The impedance of the tank circuit is _____ Ω.

3. For the circuit in question 2, at frequencies higher than F_R, the impedance of the parallel-resonant circuit is _____ (higher/lower) than the impedance at F_R.

4. In the circuit of Fig. 67-4, the resistance R_L of L is 12 Ω. The resonant frequency of the circuit $F_R =$ _____ Hz.

5. The Q of the coil in the circuit of question 4 is

_____.

6. The voltage across R, in Fig. 67-4, is 1.5 V at resonance. The line current in the circuit is _____ mA.

MATERIALS REQUIRED

- Equipment: Audio generator, oscilloscope
- Resistors: Two ½-W, 33-Ω, and 10,000-Ω
- Capacitor: 0.01 μF
- Miscellaneous: 30-mH (approximately) choke

PROCEDURE

1. Resonant Frequency of a Parallel LC Circuit

a. Connect the circuit of Fig. 67-4a.
b. Adjust scope controls for proper viewing and

AF generator attenuator controls for a 4-V peak-to-peak signal output. Connect scope across the 10,000-Ω resistor, points B and C,

TABLE 67-1. Response Characteristics of a Parallel-resonant Circuit

Frequency Deviation, Hz	Frequency, Hz	v_R	v_T	I_{line} (computed) = $\dfrac{V_R\ (rms)}{R}$	Z (computed) = $\dfrac{V_T\ (rms)}{I_{line}}$
$F_R - 6{,}000$					
$F_R - 5{,}000$					
$F_R - 4{,}000$					
$F_R - 3{,}000$					
$F_R - 2{,}000$					
$F_R - 1{,}000$					
$F_R - 500$					
F_R					
$F_R + 500$					
$F_R + 1{,}000$					
$F_R + 2{,}000$					
$F_R + 3{,}000$					
$F_R + 4{,}000$					
$F_R + 5{,}000$					
$F_R + 6{,}000$					

with ground of scope going to ground of the circuit.

c. Set generator frequency at 10,000 Hz. Observe response on scope.

d. Increase or decrease generator frequency until a frequency F_R is reached where minimum signal voltage V_R appears across R. This point may be identified by the fact that V_R increases on either side of F_R. F_R is the resonant frequency of the circuit. Record in Table 67-1.

2. Line Current and Impedance of a Parallel-resonant Circuit

a. Check the generator output v and maintain it at 4 V peak-to-peak during the remainder of this exercise.

b. Measure the voltage v_R across R (BC) at resonance, and record in Table 67-1.

c. Adjust the generator frequency in turn to each of the frequencies listed in Table 67-1, maintaining a 4-V peak-to-peak output signal v. At each frequency measure and record v_R.

d. Connect the circuit of Fig. 67-4b and repeat steps **a**, **b**, and **c** for voltage measurements v_T, across the tank circuit (BC).

e. Convert each peak-to-peak value of v_R into its rms value. Compute the value of line current, $I_{line} = V_R(rms)/R$, and record in Table 67-1.

f. From the computed values of I_{line}, draw a graph of line current versus frequency of the parallel-resonant circuit.

g. Convert each peak-to-peak value of v_T into its rms value. From the rms value of V_T and the computed value of line current, compute the value of impedance Z of the parallel-resonant circuit, for every frequency in Table 67-1. Record in Table 67-1.

h. Draw a graph of Z versus frequency.

Fig. 67-5. Circuit for determining current in the inductive and capacitive legs of a parallel-resonant circuit.

TABLE 67-2. Reactance of a Parallel-resonant Circuit

Frequency Deviation, Hz	Frequency, Hz	v_{R_L}	v_{R_C}	Current in L, mA	Current in C, mA
$F_R - 6,000$					
$F_R - 5,000$					
$F_R - 4,000$					
$F_R - 3,000$					
$F_R - 2,000$					
$F_R - 1,000$					
$F_R - 500$					
F_R					
$F_R + 500$					
$F_R + 1,000$					
$F_R + 2,000$					
$F_R + 3,000$					
$F_R + 4,000$					
$F_R + 5,000$					
$F_R + 6,000$					

3. Reactance of a Parallel-resonant Circuit at Frequencies above and below Resonance

a. Connect the circuit of Fig. 67-5. Use the same values of L and C as in parts 1 and 2. Set generator output at 4 V peak-to-peak and maintain it at this value.

NOTE: The resonant frequency of this circuit may be assumed to be the same as in Fig. 67-4.

b. Measure the voltages v_{R_L} across R_L and v_{R_C} across R_C at resonance and record in Table 67-2.

c. Measure and record the voltages across R_L and R_C for every frequency shown.

d. Compute the current in the inductive branch from the relationship $I_L = V_{R_L}(\text{rms})/R_L$ for every frequency shown and record in Table 67-2.

e. Compute the current in the capacitive branch from the relationship $I_C = V_{R_C}(\text{rms})/R_C$ for every frequency shown and record in Table 67-2.

f. Draw graphs of I_L and I_C versus frequency from the computed values in Table 67-2.

QUESTIONS

1. What determines the Q of a parallel RLC circuit?
2. What is meant by a high-Q parallel RLC circuit?
3. What is the formula for resonance of a high-Q parallel RLC circuit?
4. How does line current vary with frequency in a parallel RLC circuit?
5. How does impedance vary with frequency in a parallel RLC circuit?
6. What is the nature of the impedance (that is, is it inductive, capacitive, or resistive?) of a parallel RLC circuit (a) at resonance? (b) for frequencies greater than resonance? (c) for frequencies lower than resonance?

Answers to Self-Test

Answers to slide-rule accuracy.

1. 50
2. 50,000
3. lower
4. 9,200
5. 144
6. 0.15

EXPERIMENT 68

Low-pass and High-pass Filters

OBJECTIVES

1. To determine experimentally the frequency response of a low-pass filter
2. To determine experimentally the frequency response of a high-pass filter

INTRODUCTORY INFORMATION

Frequency Filters

Frequently electronic signal currents consist of two or more frequency components. For example, the signal of a broadcast radio station consists of a radio-frequency carrier and audio-frequency intelligence which is impressed on the carrier by a process called *modulation*. The modulated signal, consisting of the high-frequency carrier and the sidebands which form a modulated wave, together carrying the intelligence contained in the low-frequency audio components, is broadcast by the radio transmitter and selected by a radio receiver. The high-frequency carrier, a unique frequency assigned to a specific station, makes it possible for the receiver to tune in or select the broadcast station. But once in the receiver, the high-frequency carrier is no longer required. Only the amplitude variations of the modulated wave at the audio frequency rate are needed. In one of the stages of the receiver the carrier frequency is eliminated; the audio variations are obtained, amplified, and left to actuate the loudspeaker. A process called *detection*, and frequency filtering are used to accomplish this result.

There are many other examples of frequency complex signals in electronics. A television-station signal is one example. An FM station signal is another. In all cases frequency-selective filters are used to separate the wanted and unwanted signal currents, and pass on the required frequency components to the circuits which process them.

There are different types of filters. Some are highly selective and will permit a single frequency or a very narrow band of frequencies through, attenuating all others. These are called "narrow-band filters." Others, called "wideband filters," will pass a wide band of frequencies while attenuating all others. Filters have other classifications, including those classified as *low-pass* and *high-pass* filters. We will examine the characteristics of the latter two types in this experiment.

High-pass Filter

Filtering is accomplished by the action of capacitors, inductors, and resistors, in combination. *LC* resonant circuits are one example of frequency filters, for these highly selective circuits favor one frequency over all others.

A capacitor, theoretically, offers an infinite reactance to a zero-frequency signal, that is, to direct current. Therefore if a capacitor is placed in series with a load resistance R_L, in Fig. 68-1b, it will pass the ac component of a complex signal but block the dc component of that signal. Fig. 68-1a shows a 6-V peak-to-peak ac signal combined with 5 V dc. The resulting combined signal varies between +8 and +2 V. Fig. 68-1b shows how a coupling capacitor C is connected in the circuit to block the dc

(a)   (b)   (c)

Fig. 68-1. (a) A 6-V ac signal imposed on a +5-V dc axis. The resulting combined signal varies between +8 and +2 V. (b) The combined dc and ac signal is connected in series with a capacitor C and a resistor R_L. (c) Capacitor C blocks the dc and permits only the ac component through. Signal variation across R_L is between +3 and −3 V, a pure ac voltage.

but permit the ac component through. Fig. 68-1c shows the pure ac sine wave across R. Note the +5 V dc has been filtered by capacitor C.

The reactance of a capacitor varies inversely with frequency. This characteristic is used to pass certain frequencies but reject others. Consider the circuit of Fig. 68-2a. The level of ac voltage v_R coupled by capacitor C to resistor R is

$$v_R = V \cos \theta = V \times \frac{R}{Z}$$

(68-1)

where

$$\theta = \text{arc tan} \frac{X_C}{R}$$

The angle θ depends on the relative values of C and R, and on the frequency F of the applied signal source. Assume there are three frequencies in the complex signal V, namely 159 Hz, 1,590 Hz, and 15,900 Hz. The reactance X_C of C at each of these frequencies is, respectively, 10,000, 1,000, and 100 Ω. The voltage V_R at 159 Hz delivered to R is less than 10 percent of V. At 1,590 Hz (the frequency at which $X_C = R$) over 70 percent of V is delivered to R, and at 15,900 Hz over 99 percent of V is delivered to R. It is evident therefore that C will permit the higher frequencies to reach R with a minimum of attenuation, but will attenuate the lower frequencies. This is one type of high-pass filter.

The circuit of Fig. 68-2a can be modified by adding an inductor L in parallel with the load resistor R to form another high-pass filter (Fig. 68-2b). L has a low reactance at the low frequencies and effectively reduces the output voltage across R at frequencies for which $X_L < R/10$ (approximately). At higher frequencies the value of X_L increases. For frequencies at which $X_L > 10R$, the parallel impedance of X_L and R approaches R, stabilizing the load at the value R. The inclusion of inductor L modifies the response characteristics of the circuit, which is still a high-pass filter.

Low-pass Filter

The reactance of an inductor varies directly with frequency. In Fig. 68-3 this characteristic of L is used to pass the low frequencies to R but to reject the high frequencies, if the input contains high and low frequencies. As in the case of the RC circuit in Fig. 68-2, the LR circuit in Fig. 68-3 constitutes a voltage divider. The amplitude of voltage delivered by V to R will depend on the inductance of L, the resistance of R, and the frequency of V. As frequency increases, X_L increases and the voltage across R decreases. As frequency decreases, X_L decreases and the voltage across R increases. This is an example of a low-pass filter.

(a)   (b)

Fig. 68-2. High-pass filters. (a) C blocks the lower and passes the higher frequency components of V to R. (b) Adding L modifies the action of the filter.

332 *Basic Electricity*

Fig. 68-3. Low-pass filter. *L* blocks the higher and passes the lower frequencies to R_L.

An example of another low-pass filter is shown in Fig. 68-4*a*. The addition of capacitor *C* in this circuit acts to increase the filtering of higher frequencies. For as the frequency *F* of the input signal increases, not only does X_L increase, reducing the output voltage (V_{out}), but X_C decreases, reducing the impedance of the output circuit, thus further reducing V_{out}.

If *L* is replaced by a resistor R_1, as in Fig. 68-4*b*, the result will still be a low-pass filter, because of the bypass (attenuating) action of capacitor *C*. Where the cut-off high frequency starts depends on the value of the two resistors and of the capacitor *C*. We will use here the definition of cut-off frequency as that frequency whose output voltage is just 70.7 percent of maximum output, that is, one whose attenuation is at least 29.3 percent of maximum.

SUMMARY

1. Complex electronic currents contain many frequency components. Electric filters are used to separate these frequency components, attenuating those that are unwanted and passing on those that are required.
2. Electric filters consist of combinations of capacitors, inductors, and resistors. The reactance of a capacitor is inversely proportional to the frequency of the signal in the circuit. The reactance of an inductor is directly proportional to frequency.

3. Figs. 68-2*a* and *b* are examples of high-pass filters. The circuits are basically voltage dividers. With increasing frequency, X_C decreases and the output voltage across *R* is accordingly higher.
4. Figs. 68-3 and 4 are low-pass filters. Here the voltage-divider action is affected by the increase in X_L with an increase in frequency. Accordingly the output voltage across *R* for low frequencies is greater than it is for high frequencies. High frequencies are attenuated and low frequencies are passed on to the output.
5. The terms low and high frequencies are relative. The frequency-response characteristic of the circuits depends on the circuit parameters, that is, on the values of *L*, *C*, and *R*, and on the circuit configuration. In all cases, the value of output voltage for each frequency component can be computed, using the methods of ac circuit analysis.

SELF-TEST

Check your understanding by answering these questions:

1. Circuits which attenuate some frequencies and pass others are called _____ _____.
2. DC can be separated from ac signals by means of a _____ _____.
3. In the high-pass circuit of Fig. 68-2*a*, the frequency at which the output voltage across *R* is 70.7 percent of maximum is the frequency at which $X_C =$ _____ Ω.
4. The cut-off frequency in the filter of Fig. 68-2*a* is _____ Hz.
5. In the low-pass filter of Fig. 68-3 the output voltage across $R_L = V \times$ _____ $= V \times$ _____.
6. In the circuit of Fig. 68-3, if $R = 500 \ \Omega$ and $L = 2$ H, all frequencies higher than _____ Hz will be attenuated more than 29.3 percent, that is, the cut-off frequency is _____ Hz.

(a)

(b)

Fig. 68-4. Another low-pass filter.

MATERIALS REQUIRED

- Equipment: AF sine-wave generator, electronic voltmeter
- Resistor: ½-W 10,000-Ω
- Capacitor: 0.1 μF
- Miscellaneous: 8-H Choke

Fig. 68-5. Experimental circuit to determine the characteristics of a high-pass filter.

PROCEDURE

High-Pass Filter

1. Connect the circuit of Fig. 68-5. An ac voltmeter is connected across the AF generator to measure its output.
2. Set generator frequency at 60 Hz and adjust generator output for 7 V rms. With the same voltmeter measure the voltage V_R across R and record it in Table 68-1.
3. Repeat step 2 for each of the frequencies in Table 68-1.

CAUTION: Be certain that *generator output* is checked at each frequency setting and adjusted for 7 V rms.

4. Draw a graph (in pencil) of V_R (vertical axis) versus F, using the data in Table 68-1. Identify the cut-off frequency.
5. Connect an 8-H choke across R and repeat steps 2 and 3.
6. Draw a graph (in ink) on the same coordinate axes as in step 4, using the data from step 5. Identify the cut-off frequency.

Low-Pass Filter

7. Connect the circuit of Fig. 68-3. L is an 8-H choke, and $R = 10,000$ Ω. An ac voltmeter is connected across the generator.

TABLE 68-1. High-pass Filter

Frequency Hz	Gen. Output Volts (rms)	V_R (across R) Volts (rms)	
		Without L	*L in parallel with R*
60	7		
80	7		
100	7		
120	7		
140	7		
160	7		
180	7		
200	7		
300	7		
500	7		
1,000	7		
2,000	7		
3,000	7		

TABLE 68-2. Low-pass Filter

Frequency Hz	Gen. Output Volts (rms)	V_R (across R) Volts (rms)	
		Without C	*C in parallel with R*
20			
30	7		
40	7		
50	7		
60	7		
80	7		
100	7		
150	7		
200	7		
300	7		
500	7		
1,000	7		
2,000	7		

8. Set generator frequency at 20 Hz, and adjust generator output for 7 V rms. With the same voltmeter, measure the voltage V_R across R and record in Table 68-2.
9. Repeat step 9 for each of the frequencies in Table 68-2.
10. Draw a graph (in pencil) of V_R (vertical axis) versus F (horizontal axis) using the data in Table 68-2. Identify the cut-off frequency.
11. Connect a 0.1-μF capacitor in parallel with R as in Fig. 68-4a and repeat steps 8 and 9.
12. Draw a graph (in ink) on the same coordinate axes as in step 10, using the data from step 11. Identify the cut-off frequency.

QUESTIONS

1. Refer to Fig. 68-5. What is the calculated frequency F at which $X_C = R$? (Show your computations.)
2. Is the computed frequency in question 1 the same as the cut-off frequency in graph 1? Should it be? If the two frequencies are not the same, explain why they are not.
3. Refer to your data in Table 68-1 and to the two graphs. What was the effect of adding an 8-H choke in parallel with R on the frequency response of the high-pass filter?
4. Refer to procedural step 7. At what frequency does $X_L = R$? Show your calculations.
5. Is the frequency calculated in step 7 the same as the cut-off frequency on the graph in step 10? If not, explain why not.
6. Refer to your data in Table 68-2 and to the two graphs drawn from this data. What was the effect of adding a 0.1-μF capacitor in parallel with R on the frequency response of the low-pass filter?

Answers to Self-Test

1. frequency filters
2. blocking capacitor
3. 1,000
4. 1,591
5. R_L/Z; $\cos \theta$
6. 39.8; 39.8

69

Phase-shifting Networks

OBJECTIVES

1. To determine experimentally the range of phase variation in a network containing a fixed C in series with a variable R
2. To determine experimentally the range of variation in a phase-shift bridge circuit

INTRODUCTORY INFORMATION

Series RC Phase-shift Circuit

Phase-shifting networks used for electronic control are an interesting application of ac circuit theory. In its simplest form, a phase-shifting network may consist of a capacitor in series with a variable resistor, as in Fig. 69-1. The input voltage V is applied across the series circuit points FH and the output voltage V_C is taken from the capacitor across points GH. As the size of R is varied, the phase of V_C varies with respect to V applied.

A phasor analysis of the circuit will clarify the phase relationships and range of phase variation as R is varied from maximum to minimum resistance. Figure 69-2a shows the phase relationships in the circuit for R greater than X_C. V applied (FH) is the reference phasor. Since this is a capacitive circuit, current I leads the applied voltage by some angle θ.

In the resistor R, voltage and current are in phase. Hence V_R, the voltage across R, is shown lying along the phasor I. The voltage V_C across C lags I by 90°. If from H a line HG is drawn perpendicular to V_R, HG_1 will represent the phasor V_C, and FG_1 will be the phasor V_R. From the right triangle FG_1H we see that

$$V = \sqrt{V_R{}^2 + V_C{}^2} \qquad (69\text{-}1)$$

This is consistent with our knowledge of the voltage relationships in an RC circuit.

Angle ϕ_1 is the angle by which the voltage V_C lags the applied voltage. From triangle FG_1H it is evident that

$$\tan \phi_1 = \frac{V_R}{V_C} = \frac{R}{X_C} \qquad (69\text{-}2)$$

Since R is greater than X_C, V_R is greater than V_C. Therefore the ratio $V_R/V_C > 1$ and ϕ_1 is an angle greater than 45°. As R increases, ϕ_1 increases. If we set a limit on the maximum value of R at $R = 10\,X_C$, then

$$\tan \phi_1 = \frac{V_R}{V_C} = \frac{10}{1}$$

and $\phi_1 = 84.3°$. The maximum value of ϕ_1 is 90° when R becomes infinitely large.

What is the value of ϕ_2 if $R = X_C$? Figure 69-2b shows this condition. Here $V_R = V_C$ and $\tan \phi_2 = V_R/V_C = 1$. Therefore, $\phi_2 = 45°$.

Figure 69-2c shows that ϕ_3 becomes smaller than 45° when R becomes less than X_C. The limiting value of ϕ_3 is zero degrees, which occurs when $R = 0$. What this means, simply, is that when $R = 0$, the applied voltage and the voltage across C are the same, and are in phase.

The maximum range of phase delay with the RC network of Fig. 69-1 is from zero up to but not including 90°.

Fig. 69-1. *RC phase-shifting circuit.*

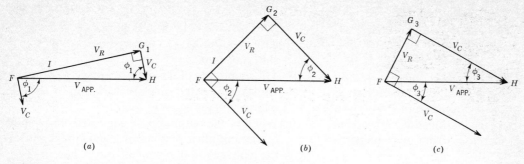

Fig. 69-2. Phasor diagrams for *RC* phase-shift circuits.

The amplitude of the voltage across *C* decreases as the phase angle between V_C and *V* increases.

Another characteristic is apparent from Fig. 69-2. Since the angle *FGH* is always a right angle, point *G* must lie along a semicircle, whose diameter is *FH*. As *R* varies continuously from infinity to the value $R = X_C$, and finally to the value $R = 0$, *G* moves in a CCW direction along the semicircle from point *H*, to G_2, and finally to point *F*.

Phase-shift Bridge Circuit

The range of variation of the angle ϕ can be increased to almost 180° by means of a bridge-type phase-shift circuit, Fig. 69-3. The applied voltage *V* appears across points *FH*, with point *P* as the center tap of the power source. That is, the voltage *FP* = voltage *PH* = *V*/2. The output voltage V_0 is taken from the points *PG*. As *R* is varied from maximum resistance to zero ohms, the phase angle ϕ of V_0 varies from some small value up to 180°. The angle ϕ is a leading angle relative to the voltage *FH*.

The phasor diagram (Fig. 69-4) illustrates these conditions. *FH* is the reference phasor (*V* applied) with *P* as center. Since current leads voltage in a capacitive circuit, the vector *I* is shown leading the

applied voltage *FH*, by some angle θ. Current *I* and voltage V_R in *R* are in phase. Hence the resistance voltage phasor *FG* lies along *I*. The voltage V_C across the capacitor lags the current *I* (and hence the voltage across *R*) by 90°. Therefore, the phasor *GH* perpendicular to *FG* represents the voltage V_C across the capacitor. *P* is the center of *FH* (since voltage *FP* = voltage *PH* = *V*/2). Hence, *P* is the center of the semicircle along which point *G* moves as the resistance of *R* is varied.

The phasor *PG* represents the output voltage V_0 from *P* to *G*, and ϕ is the angle by which V_0 leads the applied voltage *FH*. Since *P* is the center of the semicircle *FGH*, the radii *PG* and *PF* are equal and angle *PFG* = angle *PGF* = θ. In triangle *PFG* the exterior angle ϕ is equal to the sum of the two remote interior angles, or

$$\angle \phi = 2 \angle \theta \tag{69-3}$$

and

$$\angle \theta = \angle \frac{\phi}{2} \tag{69-4}$$

From right triangle *FGH*,

$$\tan \theta = \frac{V_C}{V_R} = \frac{X_C}{R} \tag{69-5}$$

Therefore

$$\tan \frac{\phi}{2} = \frac{X_C}{R} \tag{69-6}$$

Fig. 69-3. Bridge-type phase-shift *RC* network.

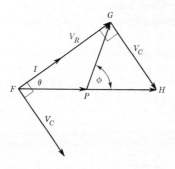

Fig. 69-4. Phasor diagram showing phase relations in bridge-type phase-shift *RC* circuit.

It is evident from Eq. (69-6) that when $R = 0$, $X_C/R = \infty$, $\phi/2 = 90°$, and $\phi = 180°$. Moreover, when $X_C = R$, $X_C/R = 1$, $\phi/2 = 45°$, and $\phi = 90°$. When R becomes very large, say $R = 10\, X_C$, $X_C/R = 1/10$, $\tan \phi/2 = 0.1000$, $\phi/2 = 5.75°$, and $\phi = 11.50°$.

Relative to the applied voltage V taken from F to H, the output voltage V_0 taken from P to G is a leading voltage and the angle of lead ϕ varies from $180°$ to $11.5°$ as R increases from zero ohms to $R = 10\, X_C$.

If the output voltage is taken from G to P, V_0 is a lagging voltage and the angle of lag equals $180° + \phi$.

SUMMARY

1. In the simple RC circuit of Fig. 69-1, as R is varied from zero ohms to its maximum value, the phase of the output voltage V_C with reference to the applied voltage V_{applied} varies from zero degrees up to some maximum value determined by the size of R and the value of X_C. The value in degrees of this phase angle θ is given by the formula

$$\tan \theta = \frac{R}{X_C}$$

2. The angle θ is the angle by which the output voltage V_C lags the applied voltage V_{applied}.
3. In the simple RC phase-shifting circuit of Fig. 69-1, $\theta = 0°$ when $R = 0\ \Omega$; when $R = 10\, X_C$, $\theta = 84.3°$. The angle θ approaches, but never equals $90°$, because R cannot be made infinitely large. We can therefore say that the range of phase variation in the simple RC circuit containing a fixed value C and a variable R is approximately $90°$.
4. The amplitude of the output voltage V_C in the simple RC phase-shifting circuit of Fig. 69-1 decreases as R increases in resistance.
5. In the phase-shift bridge circuit of Fig. 69-3 the range of variation of the phase-shift angle is increased up to but not including $180°$.
6. In the phase-shift bridge circuit the output voltage V_0 is taken from the junction of R and C (point G) and from the center point P of a low-impedance network across the applied voltage V.
7. The phase-shift angle ϕ in Fig. 69-3 is computed from the formula

$$\tan \frac{\phi}{2} = \frac{X_C}{R}$$

8. When $R = 0$, $\phi/2 = 90°$, and $\phi = 180°$. When $R = 10\, X_C$, $\phi = 11.5°$. The larger the value of R, with relation to X_C, the smaller is the phase-shift angle ϕ.
9. The amplitude of the output voltage V_0 remains constant over the range of variation of R. The output voltage $V_0 = V_{\text{applied}}/2$.
10. Whether the output phase angle in a phase-shift bridge circuit is leading or lagging depends on the voltages under comparison. Thus, in Fig. 69-3, if V_0 is taken from P to G, (V_{PG}) relative to the voltage V_{FH} taken from F to H, V_0 is a leading voltage. However, if the output voltage V_{GP} is taken from G to P, V_0 is a lagging voltage relative to the applied voltage V_{FH} taken from F to H.

SELF-TEST

Check your understanding by answering these questions.
Answer questions 1 through 4 relative to Fig. 69-1:
1. When $R = X_C$, the phase angle between V_C and V_{FG} is _____ degrees.
2. For the circuit in question 1, V_C (V_{GH}) _____ (leads/lags) V_{FH} by _____ degrees.
3. The maximum range of phase variation between V_C and V_{FH} approaches but never equals _____ degrees.
4. The amplitude of the output voltage V_C is always constant. _____ (true/false)
Answer questions 5 through 8 relative to Fig. 69-5:
5. When $R = X_C$, the phase angle between V_{GP} relative to V_{FH} is _____ degrees.
6. For the circuit in question 5, V_{PG} _____ (leads/lags) V_{FH} by _____ degrees.
7. The maximum range of phase variation between V_{GP} and V_{FH} is approximately _____ degrees.
8. The output voltage $V_{GP} =$ _____ V.

MATERIALS REQUIRED

■ Power supply: Source of 120 V 60 Hz
■ Equipment: Oscilloscope
■ Capacitor: 0.1 μF
■ Miscellaneous: 500,000-Ω 2-W potentiometer; AF output transformer (6K6 to speaker); two 33-Ω 1/2-W resistors; SPST switch; fused line cord

PROCEDURE

Series RC Circuit

1. Connect the circuit of Fig. 69-5. *T* is a voltage stepdown transformer (AF output) whose primary is connected to the 120-V 60-Hz line. (The center tap in the primary is not used.) The secondary is an untapped low impedance winding. To obtain a center tap for *P*, two 33-Ω resistors are connected as shown in Fig. 69-5. *R* is a 500,000-Ω 2-W potentiometer connected as a rheostat, and *C* is a 0.1-μF capacitor. *Do not apply power.*

2. Connect the vertical "hot" lead of the oscilloscope to point *F*, the top of the secondary, and ground lead to point *H*. Calibrate the oscilloscope for voltage measurements.

3. **Power on.** The applied-voltage sine wave should be visible on the screen. Set the oscilloscope on "line" trigger or sync. If line trigger or sync is not available, set the trigger or sync selector on "external." Ext. trigger or sync may be provided from point F on the secondary of *T*, connected to the jack marked "External Trigger or Sync." Adjust the oscilloscope's sweep and trigger or sync controls until three cycles appear stationary on the screen. Center the waveform vertically with respect to the *X* axis. Center the waveform horizontally so that the middle (approximately) sine wave is centered with respect to the *Y* axis, with the positive alternation on the left and the negative alternation on the right, as in Table 69-1. This sine wave is the reference waveform 1 and phase measurements will be made relative to this reference. Since a complete cycle is 360°, equal subdivisions of the time base represent equal angles. Thus, if each division is divided into 10 equal lengths, each length measures 18°. In the procedure

which follows, movement of the waveform to the right of this position indicates a lagging angle, movement of the waveform to the left, a leading angle. Once you have obtained the proper reference waveform *do not* readjust the sweep, trigger or sync, centering, or gain controls of the oscilloscope. Measure and record the peak-to-peak amplitude of the reference waveform.

4. Keep the ground lead of the oscilloscope at point *H*. Connect the vertical input lead to point *G*. Adjust the phase-shift control to minimum (*R* = 0). Observe and record this waveform 2 in proper time phase with the reference voltage, waveform 1. Measure and record the amplitude of waveform 2 and the number of degrees (if any) of phase shift.

5. Adjust *R* until there is a 45° phase shift. Draw this waveform 3 in proper time phase with reference waveform 1. Measure and record the amplitude of waveform 3.

6. Adjust *R* to maximum resistance. Draw waveform 4 in proper time phase with reference waveform 1. Measure and record the amplitude

TABLE 69-1.

Waveform Number	Waveform	Volts, Peak-to-Peak	Phase-shift Degrees
1			
2			
3			45°
4			

Fig. 69-5. Experimental phase-shift circuit.

of waveform 4, and the number of degrees of phase shift.

Phase-shift Bridge Circuit

7. Remove the ground lead of the oscilloscope from point *H* and connect it to point *P*. The "hot" lead is connected to point *F*. Trigger or Sync connection and all controls remain set as previously. There should be a stable presentation of three sine waves on the screen. If there is not, readjust sweep and trigger or sync controls until the three sine waves "hold." Set horizontal-centering control as in step 3 until the middle sine wave is properly centered as in Table 69-2. This is the reference waveform 1. Measure and record the peak-to-peak amplitude of the waveform.

8. Keep ground lead at point *P*. Move the hot lead of the oscilloscope to point *G*. The output appearing on the screen is the voltage V_{GP} (from *G* to *P*). Adjust the phase-shift control to minimum resistance ($R = 0$). Draw this waveform 2 in proper time phase with reference waveform 1. Measure and record the peak-to-peak amplitude of the waveform and the number of degrees (if any) of phase shift.

9. Adjust *R* until there is a 90° phase shift between V_{GP} and the reference waveform. Draw this waveform 3 in proper time phase with the reference. Measure and record the peak-to-peak amplitude of the waveform.

10. Adjust *R* to maximum resistance. Draw this waveform 4 in proper time phase with reference waveform 1. Measure and record the peak-to-peak amplitude of the waveform and the number of degrees of phase shift.

11. Reverse the vertical input leads of the oscilloscope so that the hot lead is connected to point *P* and ground lead to point *G*. The output voltage on the screen is V_{PG}. Repeat steps 8, 9, and 10. The corresponding waveforms are 5, 6, and 7. Draw 5 on the same time axis as 2; 6 on the same axis as 3; 7 on the same axis as 4. Identify the waveforms.

QUESTIONS

1. Refer to Table 69-1. Compare the experimental range of variation of the phase-shift angle with that predicted by Eq. (69-2) for the minimum and maximum values of *R*. Show your computations.

TABLE 69-2.

Waveform Number	Waveform	Volts, Peak-to-Peak	Phase-shift Degrees
1			
2, 5			
3, 6			90°
4, 7			

2. Refer to the waveforms in Table 69-1. Is the phase-shift angle leading or lagging? How do you know? Is this consistent with the theory of this circuit?

3. Refer to Table 69-2. Compare the experimental range of variation of the phase-shift angle with that predicted by Eq. (69-6) for the minimum and maximum values of *R*. Show your computations.

4. Refer to the phase-shift waveforms 2, 3, and 4 in Table 69-2. Is the phase-shift angle leading or lagging? Which should it be theoretically?

5. What is the phase relationship between waveforms (*a*) 2 and 5? (*b*) 3 and 6? (*c*) 4 and 7?

6. Explain the basic differences between the output of the simple *RC* phase-shift circuit and the *RC* bridge-type phase-shift circuit.

Answers to Self-Test

1. 45
2. lags; 45
3. 90
4. false
5. 90
6. leads; 90
7. 180
8. $V_{FH}/2$

70 EXPERIMENT

Nonlinear Resistors—Thermistors

OBJECTIVES

1. To observe the self-heating effect of current on the resistance of a thermistor
2. To determine experimentally the variation in resistance with time of current in a thermistor

INTRODUCTORY INFORMATION

We have been concerned to this point with the laws governing the flow of current in dc and ac circuits. Thus we verified Ohm's law, which shows the relationship between resistance (R), voltage (V), and current (I) in dc and ac circuits. Next we studied Kirchhoff's laws for more complex networks. Throughout our discussion and in our experimental work, we dealt with resistors whose resistance remained relatively constant despite variations in current flow and operating temperature.

There is a class of components known as "thermistors" whose resistance changes with changes in operating conditions, particularly with changes

Fig. 70-1. Thermistor types (*Western Electric Co.*).

in operating temperature. As its name implies, a thermistor is a thermally sensitive resistor. It is part of a large family of semiconductors. At this time we need merely mention that a semiconductor is a substance whose resistivity lies somewhere between that of conductors and insulators. We will study the characteristics of semiconductors in greater detail in electronics.

The thermistor is a two-terminal component which may be used in either ac or dc circuits. It is manufactured in a number of shapes, such as beads, rods, disks, washers, and flakes (see Fig. 70-1).

Temperature Characteristic

A fundamental characteristic of the thermistor is its change of resistance with changes in temperature. The manner in which the resistance varies with temperature is given by the approximate formulas

$$R = R_0 \times e^k$$

and
$$k = B\left(\frac{1}{T} - \frac{1}{T_0}\right) \qquad (70\text{-}1)$$

where R = resistance at any Kelvin temperature T
R_0 = resistance at reference temperature T_0 (°K)
B = a constant whose value depends on the thermistor material (determined from measurements at 0° and at +50°)
e = 2.7183 (base of natural logarithms)

Formula (70-1) shows that the resistance variation is nonlinear. Thermistors are frequently referred to as "nonlinear resistors."

Thermistors exhibit large negative temperature

Fig. 70-2. Temperature-resistance values of three thermistors compared with a representative metal (*Western Electric Company*).

coefficients of resistance; that is, their resistance decreases as their temperature rises and increases as their temperature falls. Figure 70-2 shows this variation for three thermistors, as compared with the resistivity of platinum under similar temperature changes. Note the tremendous variation of resistance in thermistor 1, a change of 10,000,000 to 1 over a temperature range of 500°C, while that of platinum changes by a factor of less than 10 to 1.

Figure 70-3 shows the variation in resistance of another thermistor. Here a range of 200°C causes a variation in resistance from 1,000 Ω to approximately 2 Ω.

The resistance-temperature characteristic of thermistors has been utilized in electrical and electronic applications. In particular, thermistors are used in electronics as protective devices for vacuum-tube filaments during the warmup period. They are also used for temperature measurement, control, and compensation. Thermistors are also widely used in television circuits such as the degaussing circuits for color picture tubes.

Static Volt-ampere Characteristic

There are secondary characteristics which arise from the relationship between temperature and resistance exhibited by thermistors. Among these is the static relationship between voltage and current. This secondary characteristic results from the self-heating that occurs when current flows through a thermistor. The static characteristic may be determined by use of the test circuit shown in Fig. 70-4. In this circuit a rheostat *R* is used as a limiting resistor to vary the amount of current through the thermistor. As the resistance of *R* is decreased, the current *I* increases. As the resistance of *R* is increased, the current *I* decreases. A voltmeter is used to monitor the voltage across the thermistor for every value of current. Each time *R* is varied and a change in current is effected, enough time

Fig. 70-3. Resistance versus temperature in a thermistor (*General Electric Company*).

Fig. 70-4. Test circuit for determining static *V-I* characteristic of a thermistor.

Fig. 70-5. Thermistor static volt-ampere characteristic (*General Electric Company*).

must be allowed for the voltage across the thermistor to reach a steady value. The resultant graph therefore reflects steady-state conditions of voltage versus current and is therefore called the "static volt-ampere characteristic."

The thermistor whose resistance-temperature characteristic is shown in Fig. 70-3 has the static volt-ampere characteristic shown in Fig. 70-5. A temperature of 25°C was the starting temperature for the tests from which the data for the graph were secured. The numbers on the graph, namely, 53, 73, 102, etc., indicate the thermistor temperature, measured after current had stabilized at the level shown on the graph. Thus, at 0.05 A, the voltage across the thermistor is 11 V and the temperature is 53°.

The graph for this thermistor shows that up to a temperature of 53° there is a linear relationship between voltage and current. That is, Ohm's law holds during this interval, when only small current flow is involved. The reason is that there is insufficient heat dissipated at this low level of current to affect the cold resistance of the thermistor. As current increases, however, the heat dissipated increases, thermistor temperature increases, and its resistance drops nonlinearly. The voltage across the thermistor now increases nonlinearly to a peak value V_M, designated as the "self-heat" voltage. Beyond this value, the voltage across the thermistor drops nonlinearly, with increasing current, as shown in the graph Fig. 70-5.

The static volt-ampere characteristic of thermistors is utilized in the following applications:

- Temperature alarm devices
- Pyrometer
- Flow meter
- Switching devices
- Voltage surge devices
- Gas detectors

Dynamic Characteristic

The reaction of a thermistor to a change in the external circuit requires a certain time interval. Thus a change in current through a thermistor from one value to another does not result in an instantaneous change in temperature, and hence there is no instantaneous change in resistance. These changes do occur, but they require a definite time. The thermal mass of the thermistor element will determine the time interval. A small element will change temperature more rapidly than an element with larger mass.

Figure 70-6 shows a test circuit for determining the dynamic characteristics of a thermistor. V is an ac power supply whose output is adjustable and R_L is the load resistance. An EVM set on ac volts is connected across the load R_L to monitor the voltage across R_L. When switch S is open, there is no voltage applied to the circuit. At the instant S is closed, the cold resistance of the thermistor and the value of the load resistor R_L constitute the voltage divider which determines the voltage across R_L. As

Fig. 70-6. Test circuit for determining dynamic characteristic of a thermistor.

Fig. 70-7. Dynamic characteristic of a thermistor (*General Electric Company*).

current starts to flow, the temperature of the thermistor increases, its resistance decreases, and more of the source voltage appears across the load. Within a certain time interval, the circuit is stabilized. That is, the thermistor resistance is stabilized and the voltage across the load becomes constant.

Figure 70-7 is a graph showing the dynamic characteristic of a thermistor for two different values of supply voltage and load resistance. Note that with 48 V applied and a load resistance of 17 Ω, steady state is reached in approximately 75 s. With 115 V applied and a load of 50 Ω, steady state is reached in about 15 s.

The dynamic characteristic of a thermistor is utilized in expandors and compressors (audio devices), switching devices, voltage-surge absorbers, and low-frequency negative-resistance devices.

In this experiment you will first observe the effect on resistance of a thermistor of an increase in operating temperature due to self-heating. You will then determine experimentally the dynamic characteristic of a thermistor.

SUMMARY

1. Thermistors are semiconductor devices whose resistance changes with changes in operating temperature.

2. Thermistors are called nonlinear resistors because their resistance varies exponentially and not linearly as a function of temperature.
3. The *resistance* of thermistors *decreases* as their *temperature increases*, and increases as their temperature decreases; that is, they exhibit a *negative* temperature *coefficient* of resistance.
4. Compared with this variation in resistance of a thermistor, the resistance of a carbon resistor remains relatively constant despite changes in operating temperature.
5. Thermistors exhibit unique temperature-resistance characteristics, depending on their size and construction. Thus, one particular thermistor changes in resistance from 1,000 to 2 Ω over a temperature change of 200°C. Another changes from 10,000,000 to 1 Ω over a temperature variation from −100°C to +400°C.
6. Because of the self-heating effect of current in a thermistor, this device changes resistance with changes in current.
7. A secondary characteristic of a thermistor is the change in current it exhibits in an electric circuit. A graph of current versus voltage (Fig. 70-5) of a thermistor, called a static volt-ampere characteristic, shows that current does *not* vary linearly with applied voltage.
8. A thermistor does not undergo instantaneous changes in resistance with changes in temperature. A certain time interval determined by the thermal mass of the thermistor is required to accomplish the resistance change. A thermistor with a smaller mass will change more rapidly than one with a larger mass.
9. The dynamic characteristic of a thermistor is a graph of the time it takes a thermistor to stabilize its resistance as a function of the applied voltage (Fig. 70-7).
10. Thermistors are used in temperature alarm devices, pyrometers, switching devices, voltage surge devices, etc.

SELF-TEST

Check your understanding by answering these questions:

1. A thermistor is a _____ (linear/nonlinear) resistor.
2. The resistance of a thermistor _____ (increases/decreases) with a decrease in temperature.

Fig. 70-8. Experimental control circuit.

3. A thermistor has a _____ (positive/negative/zero) temperature coefficient.
4. It takes _____ for a thermistor to undergo a change in resistance, with a change in temperature.
5. The larger the thermal mass of a thermistor the _____ (more/less) time it will take to change its resistance.

6. A(n) _____ (increase/decrease) in current in a thermistor will cause a decrease in its resistance.
7. Some applications of thermistors are: (a) _____, (b) _____, and (c) _____.
8. Current in a thermistor varies linearly with voltage across it. _____ (true/false)

MATERIALS REQUIRED

■ Power supply: Manually variable ac voltage source
■ Equipment: Two meters (EVM and VOM)
■ Resistors: 4-W 250-Ω; 10-W 300-Ω
■ Miscellaneous: Thermistor with following characteristics: 50-Ω at 150 mA, 300-Ω cold; SPST switch; fused line cord

PROCEDURE

NOTE: It will be necessary for three students to work together as a team in this experiment. One will call time and will need a watch with a seconds hand. One will make ac (rms) voltage readings across the thermistor (T). The last student will take ac (rms) readings across the limiting resistor R. The meters should be set at the range where most accurate readings are possible. It should be noted that the voltage across R will rise, with time, up to 80 percent (approximately) of the applied voltage. The

voltage across the thermistor will fall, with time. The sum of the voltages across R and across the thermistor should equal the applied voltage.

Control Experiment

1. Measure the cold resistance of the thermistor and record it in Table 70-1.
2. Connect the circuit of Fig. 70-8. Do not apply

TABLE 70-1. Thermistor Characteristic Data

Step	Thermistor resistance	t	0	15 s	30 s	45 s	1 min	1½ min	2 min	2½ min	3 min	4 min	5 min	Applied volts
1		X	X	X	X	X	X	X	X	X	X	X	X	0
3, 4	X	V_{R_1}		X	X	X	X	X	X	X	X	X		56
		V_{R_2}		X	X	X	X	X	X	X	X	X		
8	X	V_T												28
		V_{R_2}												
9		X	X	X	X	X	X	X	X	X	X	X	X	0
12	X	V_T												56
		V_{R_2}												
13		X	X	X	X	X	X	X	X	X	X	X	X	0

power as yet to the circuit. Set the adjustable power supply to 56 V rms.

3. **Power on.** Monitor power-supply voltage. It should be set at 56 V. Measure, and record in Table 70-1, the voltage across R_1 (V_{R_1}) and the voltage across R_2 (V_{R_2}).

4. Allow power to remain **on** for 5 min. Now measure the voltage across R_1 and the voltage across R_2 and record in Table 70-1.

5. **Power off.**

Dynamic Characteristics

NOTE: Before applying power to the circuit, be certain that you understand the remaining checks which must be made. The timing factor is critical. Read carefully the following steps and refer to Table 70-1.

6. Remove R_1 from the circuit and substitute the thermistor for it, as in Fig. 70-9.

7. Set output of variable ac supply at 28 V rms. Maintain it at this value during step 8.

8. **Power on.** Measure, and note in Table 70-1, the rms voltage across the thermistor (V_T) and across R_2 (V_{R_2}) at $t = 0$ (that is, as soon as power is applied, $t = 15$ s, etc.). The series of checks for this step is complete at $t = 5$ min (that is, five minutes after power is applied).

9. **Power off.** Quickly remove the thermistor from the circuit. Measure its resistance at once, while it is still hot. Record in Table 70-1. Let thermistor cool for about ten minutes.

10. Reconnect the thermistor in the circuit (Fig. 70-9).

11. Set output of ac supply at 56 V rms. Maintain it at this value during step 12.

12. Repeat step 8, with the supply set at 56 V.

13. Repeat step 9.

14. From the data in Table 70-1, draw a graph showing the relationship between voltage across load resistor, V_{R_2}, and time, for (1) $V_{\text{applied}} = 28$ V; (2) $V_{\text{applied}} = 56$ V.

QUESTIONS

1. What is meant by negative temperature coefficient? Positive temperature coefficient?

2. What is the temperature coefficient of an ordinary

Fig. 70-9. Experimental circuit for determining dynamic characteristic.

constant-value resistor? Substantiate your answer by referring to the data in Table 70-1.

3. What is the temperature coefficient of a thermistor?

4. How can you determine the temperature-versus-resistance characteristic of a thermistor? Is time involved?

5. What is meant by the dynamic characteristic of a thermistor?

6. From the data in Table 70-1 how would you compute the hot resistance of a thermistor without actually measuring it?

7. Compute the hot resistance of the thermistor after 5-min operation in the circuit of Fig. 70-9, with (a) 28 V applied; (b) 56 V applied. Why is there a difference, if any, in these two values?

8. How do the computed hot resistance values of the thermistor compare with the measured values? Explain any discrepancies.

9. What limitations are there on the amount of current which may be permitted to flow through a thermistor?

10. What are some uses of thermistors?

11. What is the relationship between the voltage across the thermistor and the voltage across the limiting resistor R_2, at any instant of time?

12. From the data in Table 70-1 compute the
(a) Maximum current through the thermistor.
(b) Minimum current. Show all computations.

Answers to Self-Test

1. nonlinear
2. increases
3. negative
4. time
5. more
6. increase
7. (a) temperature alarm devices; (b) pyrometers; (c) voltage surge devices (or switching devices)
8. false

71

EXPERIMENT Nonlinear Resistors—Varistors (VDRs)

OBJECTIVES

1. To determine experimentally the volt-ampere characteristic of a varistor
2. To determine experimentally the relationship between voltage and resistance

INTRODUCTORY INFORMATION

In the previous experiment you discovered that there is a class of resistors call thermistors, whose resistance is dependent on temperature. There is another group of nonlinear resistors whose resistance is voltage-dependent. These voltage-sensitive resistors are called varistors or VDRs. The current in a varistor varies as a power of the applied voltage and for a particular varistor may increase as much as 64 times when the applied voltage is doubled.

Varistors are made of silicon carbide mixed with a ceramic binder, then fired at a high temperature. A metallic coating is sprayed on and electrical contact is made to this coating. Typical shapes and sizes of varistors include rods, discs, and cylinders. The smaller sizes are usually supplied with tinned copper leads. This two-terminal device can be connected in a circuit like any linear resistor. Varistors have wide application in voltage regulation circuits.

Volt-ampere Characteristic

The volt-ampere characteristic of a silicon-carbide varistor is given by the approximate equation

$$I = KV^n \qquad (71\text{-}1)$$

where I is the current in amperes

K is a constant (amperes at 1 V)
n is an exponent

The value of K depends on the resistivity and the dimensions of the varistor, while n depends upon certain factors in the manufacturing process.

Values of n may range from 2 to 6. The higher the value of n, the more nonlinear is the current and resistance characteristic. For example, doubling the voltage of a varistor will increase the current by a ratio of 8 if $n = 3$, and by a ratio of 64 if $n = 6$. The value of n is not constant over the entire voltage range. Therefore it is usually specified between an upper and lower voltage limit.

Other relationships between voltage, current, and resistance which may be derived by application of Ohm's law are:

$$V = CI^b \qquad (71\text{-}2)$$

$$RI^a = C \qquad (71\text{-}3)$$

and

$$R = \frac{1}{KV^{n-1}} \qquad (71\text{-}4)$$

where $b = 1 - a$, $n = 1/b$, and $K = 1/C^n$.

The values of n and K are the same as those in Eq. (71-1).

Equations (71-1, 2, and 3) are straight lines when graphed on logarithm paper.

Electrical Specifications

Varistors are rated for the continuous power they can dissipate at a specific temperature (e.g., 1 W at 40°C) and for the maximum voltage which may be applied. In addition, other electrical characteristics may be specified in one or more of the following ways:

Voltage-current Ratio. The constant current level

at a standard voltage and tolerance is given, together with a specified voltage ratio. The ratio replaces the n in the equation $I = KV^n$.

Example

1 mA at 68 V direct current ± 15 percent,

$$\text{Ratio } \frac{V \text{ at } 1.10 \text{ mA}}{V \text{ at } 0.1 \text{ mA}} = 2.20 \text{ max}$$

Voltage-current n is specified instead of the maximum voltage ratio, in addition to the constant current level.

Example

1 mA at 10 V dc ± 15 percent,

$n = 2.88$ min between 5 and 10 V direct current

or

$n = 3.20 \pm 10$ percent between 5 and 10 V

Resistance-voltage n. The resistance plus tolerance is given for a specified constant voltage plus the n value between a specified voltage range.

Example

10,000 $\Omega \pm 20$ percent at 10 V direct current,

$n = 2.88$ min between 5 and 10 V direct current

or

$n = 3.20 \pm 10$ percent between 5 and 10 V

Resistance at Two Voltages. Resistance plus tolerance is specified at two voltages.

Fig. 71-1. Protective circuit using varistor.

Example

10,000 $\Omega \pm 20$ percent at 10 V direct current,

45,000 $\Omega \pm 30$ percent at 5 V direct current

Other facts which the technician should know about a varistor are:

1. It possesses a negative temperature coefficient.
2. It is sensitive to high humidity. A standard varnish coating reduces the humidity effect appreciably.
3. The equivalent circuit for a varistor is a resistor and capacitor in parallel.
4. Though the standard varnish coating provides good insulation, the varistor should be mounted away from all conducting surfaces.
5. The varistor is a nonpolarized device.
6. The varistor is unaffected by pressure or vibration.
7. Varistors are used as in Fig. 71-1, to protect circuits and components from voltage surges when an ac or dc magnetic or inductive circuit current is suddenly interrupted. Varistors used in lightning arrestors serve a similar protective function. They may also be used as shunts across switch contacts to prevent sparking. In combination with other resistors, varistors may be used to provide voltage and current regulation.

SUMMARY

1. Varistors are voltage-dependent resistors (VDR). They are nonlinear devices whose resistance is dependent on the voltage across them.
2. The current in a varistor is an exponential function of the voltage across it. Thus

$$I = K \times V^n$$

   where K is a constant, depending on the type of varistor, and n is the exponent whose range is from 2 to 6.
3. Varistors are rated for their continuous power dissipation at a specified temperature and for the maximum voltage they can tolerate.
4. Other electrical specifications may include one of the following:
   (*a*) Voltage-current ratio, together with the constant current level at a standard voltage level and tolerance
   (*b*) Voltage-current n and the constant current level at a standard voltage level and tolerance

(c) Resistance-voltage n, between a specified voltage range, together with the resistance plus tolerance for a specified constant voltage

(d) Resistance plus tolerance at two specified voltages

5. A varistor has a negative temperature coefficient.
6. A varistor is sensitive to high humidity.
7. A varistor is a nonpolarized device which is unaffected by pressure or vibration.
8. Varistors are used as protective devices in lightning arrestors, and in ac and dc circuits to protect components against voltage surges.

SELF-TEST

Check your understanding by answering these questions:

1. The resistance of a varistor _____ (increases/decreases) as the voltage across it increases.
2. The equation which defines the volt-ampere characteristic of a varistor is $I = K \times V^n$. _____ (true/false)
3. Referring to the equation which defines the volt-ampere characteristic of a varistor, if $n = 2$ and the voltage across the varistor is tripled, the current in the varistor will _____ (increase/decrease) _____ (how many) times.
4. If the temperature of a varistor increases, its resistance _____ .
5. In a protective circuit, varistors are most reliable in a _____ (humid/dry) climate.
6. Polarity must be observed in connecting a varistor in a dc circuit. _____ (true/false)

MATERIALS REQUIRED

- Power supply: Variable regulated dc source
- Equipment: Multirange dc micromilliammeter (the current ranges of a 20,000-Ω/V VOM will suffice); electronic voltmeter
- Resistors: $1/2$-W 5,600-Ω
- Miscellaneous: Varistor (Carborundum, Electronics Division, Globar Plant, Niagara Falls, N.Y.) Part #333 BNR-4, or the equivalent; SPST switch

NOTE: If a high-voltage variable regulated dc source is not available, a variable ac source may be used instead. If an ac supply is used, an ac multirange micromilliammeter (the ac current ranges of a solid-state VOM will suffice) will be required.

PROCEDURE

1. Connect the circuit of Fig. 71-2. M is a multirange dc micromilliammeter (the current ranges on a 20,000-Ω/V VOM may be used), and V_1 is an EVM set on dc voltage to measure V_{AB}. V is set at zero volts.
2. Close S. Increase the output of power supply V so that V_{AB} takes, in turn, each of the voltages shown in Table 71-1. Measure and record the current I at each of these voltages. Measure also and record the voltage V_{AB} for which $I = 0.1 \times 10^{-3}$ A (0.1 mA), and $I = 1.0 \times 10^{-3}$ A (1 mA).
3. Compute and record the resistance of the varistor at each voltage setting, using Ohm's law.
4. Plot a graph of V versus I, with V as the vertical and I as the horizontal axis. Use log-log graph paper.

Fig. 71-2. Experimental circuit for determining volt-ampere characteristic of a varistor.

TABLE 71-1.

V_{AB}, volts	I, amperes	$R = \dfrac{V}{I}$, ohms	V_{AB}, volts	I, amperes	$R = \dfrac{V}{I}$, ohms
0			85		
5			90		
10			95		
15			100		
20			105		
25			110		
30			115		
35			120		
40			125		
45			130		
50			135		
55			140		
60			145		
65			150		
68			155		
70			160		
72				(0.1 mA) 0.1×10^{-3}	
75				1.0×10^{-3} (1.0 mA)	
80					

On the same graph paper plot V versus R, using R as the horizontal axis.

5. Compute and record the ratio

$$\frac{V \text{ at } 1 \times 10^{-3}\text{ A}}{V \text{ at } 0.1 \times 10^{-3}\text{ A}}$$

Show your computation.

Extra Credit

6. Compute also and record the values of n and K (within the range 0.1×10^{-3} A to 1.0×10^{-3} A). Show your computations [see Eq. (71-1)].
7. Compute also and record the values b, a, and c [see Eqs. (71-1) to (71-3)].

QUESTIONS

1. What is the difference between a linear and nonlinear resistor?

2. Explain how a varistor of the proper value can protect the circuit, in Fig. 71-1, against a voltage surge.
3. Comment on and explain the significance of your graph of V versus I. Why was the graph made on log paper?
4. From your graph of V versus R, comment on the relationship between V and R.

Extra Credit

5. Accepting the empirical equation $I = KV^n$, derive Eqs. (71-2), (71-3) and (71-4).

Answers to Self-Test

1. decreases
2. true
3. increase; 9
4. decreases
5. dry
6. false

APPENDIX

Parts Requirements

Components with asterisk (*) are added new parts for this edition.

Resistors, ½ W

Resistance, ohms	Quantity	Resistance, ohms	Quantity	Resistance, ohms	Quantity
10	1	1,800	1	18 kΩ	1
33	2	2,200	1	22 kΩ	1
47	1	2,700	1	27 kΩ	1
*100	2	3,300	1		
330	1	4,700	1	56 kΩ	1
		5,100 (5%)	2	100 kΩ	2
470	1	5,600	1	120 kΩ	1
560	1	6,800	1	330 kΩ	1
820	1	8,200	1	1 MΩ	2
1,000	3	10 kΩ	1	2.2 MΩ	1
1,200	3	12 kΩ	1		
1,500	1	15 kΩ	1		

Resistors, Higher Wattage

Resistance, ohms	Watts	Quantity
*15	25	1
250	4	1
300	10	1
500	5	1
5,000	5	1

Capacitors

Capacitance, μF	Quantity	Voltage
0.005	1	400
0.01	1	400
0.05	1	400
0.1	2	400
0.5	1	400
*1.0	1	400

- *Potentiometers* (2-W): One of each unless otherwise specified.

NOTE: Potentiometers with symbol ‡ will be required only if resistance decade box is not available.

‡100-Ω, 500-Ω, 10 kΩ (two), 100 kΩ, 500 kΩ
- *Tube & Tube Socket:* Any type, for identification only
- *Transistor:* Any type, for identification only
- **Mounted Transistor Socket* (for Experiment 2)
- *Solid State Diodes:* Any type, for identification only (1N2615)
- *Thermistor:* 50-Ω/150-mA, 300-Ω cold, Philco part #33-1343-3 or equivalent globar
- *Varistor:* Carborundum #333-BNR or equivalent
- *Transformers, Chokes, & Coils:*
  Choke: Two, 8-H @ 50 mA dc
  Audio Output Transformer: 6K6 GT, push-pull to 3.2 Ω
  RF Coil: 30 mH
  *Filament Transformer: 117 V primary, 6.3 V secondary
  *NOTE: Filament Transformer is required only if 6.3-V/60-Hz source is not otherwise available.
- *Miscellaneous:*
  (1) Breadboarding device with terminals and connectors
  (2) Prewired PC circuit board (very simple, for Experiment 2)
  (3) Prewired PC circuit board with resistors and capacitors mounted on it (for Experiment 4)
  (4) Wire: (*a*) #18 and #22 stranded hookup wire
       *(*b*) 20 ft of #18 varnish insulated copper wire
       *(*c*) *Nichrome wire:* #40 Standard

(60% nickel, 16% chrome, 24% iron), 2 ft
       *(*d*) *Copper wire:* #40, 1 ft
       (*e*) RG 59U coaxial cable, 3 ft
       (*f*) Shielded wire
       (*g*) Instrument lead wire
(*5) *Solenoid:* 100 turns of #18 varnished insulated copper wire wound in three layers on a 3-in. long × 1-in. diameter hollow cardboard or plastic cylindrical form
(*6) *Coil Form & Core:* (*a*) 2-in. long × ½-in. diameter, hollow cardboard or plastic cylinder, or copper tubing
       (*b*) 2-in. soft-iron core to fit inside coil form
(7) *Switches:* SPDT (one), SPST (two)
(8) *Fuses:* Two different types, for identification only
(9) *Loudspeaker:* PM, 3.2 Ω (for identification only)
(*10) *Magnets:* two bar magnets, one horseshoe magnet
(*11) *Iron filings in salt shaker type dispenser*
(12) *Dry cells and cell holders:* four 1½-V
(13) *Microphone connector*
(14) *Alligator clips:* three
(15) *Defective resistors:* for Experiment 23 (These may be made by changing color code on good resistors.)
(*16) *Cardboard or glass plane*, 8½ × 11 in. (one)
(*17) *Insulators for Experiment 9:*
       (*a*) Rubber strip, 2 × ¼ in.
       (*b*) Plastic strip, 2 × ¼ in.
       (*c*) Wooden strip, 2 × ¼ in.
(*18) *Fused (1 A) line cord*
(19) *Meter movement:* 0–1 mA, 50-Ω, 50-mV
(20) *Resistance decade box (2 W):* Adjustable in 1-Ω steps from 1 to 99,999-Ω
(*21) *Magnetic compass*